AIRO Springer Series

Volume 8

The **AIRO Springer Series** focuses on the relevance of operations research (OR) in the scientific world and in real life applications.

The series publishes peer-reviewed only works, such as contributed volumes, lectures notes, and monographs in English language resulting from workshops, conferences, courses, schools, seminars, and research activities carried out by AIRO, Associazione Italiana di Ricerca Operativa – Optimization and Decision Sciences: http://www.airo.org/index.php/it/.

The books in the series will discuss recent results and analyze new trends focusing on the following areas: Optimization and Operation Research, including Continuous, Discrete and Network Optimization, and related industrial and territorial applications. Interdisciplinary contributions, showing a fruitful collaboration of scientists with researchers from other fields to address complex applications, are welcome.

The series is aimed at providing useful reference material to students, academic and industrial researchers at an international level.

Should an author wish to submit a manuscript, please note that this can be done by directly contacting the series Editorial Board, which is in charge of the peer-review process.

THE SERIES IS INDEXED IN SCOPUS

Lavinia Amorosi • Paolo Dell'Olmo • Isabella Lari
Editors

Optimization in Artificial Intelligence and Data Sciences

ODS, First Hybrid Conference, Rome, Italy, September 14-17, 2021

AIRO
ASSOCIAZIONE ITALIANA DI RICERCA OPERATIVA
OPTIMIZATION AND DECISION SCIENCE

Springer

Editors
Lavinia Amorosi
Department of Statistical Sciences
Sapienza University of Rome
Rome, Italy

Paolo Dell'Olmo
Department of Statistical Sciences
Sapienza University of Rome
Rome, Italy

Isabella Lari
Department of Statistical Sciences
Sapienza University of Rome
Rome, Italy

ISSN 2523-7047 ISSN 2523-7055 (electronic)
AIRO Springer Series
ISBN 978-3-030-95379-9 ISBN 978-3-030-95380-5 (eBook)
https://doi.org/10.1007/978-3-030-95380-5

This Springer imprint is published by the registered company Springer Nature Switzerland AG
The registered company address is: Gewerbestrasse 11, 6330 Cham, Switzerland

Preface

The International Conference on Optimization and Decision Science—ODS2021 (Rome, Italy, September 14–17) has been the 50th annual meeting and the first hybrid conference organized by AIRO, the Italian Operations Research Society, and the Department of Statistical Sciences, Sapienza University of Rome, and supported by Springer. ODS2021 aimed at being a unique opportunity for researchers and practitioners focused on today's problems of knowledge, growth, sustainability, and operational excellence from various sectors (quantitative management, engineering, applied mathematics, statistics, computer science, economics and finance, medicine and healthcare), private and public companies, industries, and policy makers, to present and share ideas, experiences, and knowledge, as well as to enforce or create a new cooperation network. Despite the difficult pandemic period, we had more than 200 participants from more than 10 nations across Europe, Asia, and North America, of which about half participated on-site and the rest online, and four plenary speakers, of which one participated online and three onsite. The conference contributions were related to the wide field of analytics, optimization, problem-solving, and decision-making methods. However, a special focus was on optimization in artificial intelligence and data science.

This book contains a selection of short papers submitted to ODS2021, which have been peer reviewed by experts in operations research and related fields.

We thank the authors, for their work and dedication, all members of the Program Committee, and auxiliary reviewers, who helped by offering their expertise and time. We express our gratitude to the plenary speakers, Teodor G. Crainic, Matthias Ehrgott, Pitu B. Mirchandani, and Dolores Romero Morales, for their invaluable contributions.

Rome, Italy
September 2021

Lavinia Amorosi
Paolo Dell'Olmo
Isabella Lari

Contents

Editors and Contributors

About the Editors

Lavinia Amorosi is Assistant Professor of Operations Research in the Department of Statistical Sciences at Sapienza University of Rome. She received her PhD in operations research from Sapienza University of Rome in 2018. She has been visiting PhD student at Lancaster University Management School, Lancaster UK, in 2016 and at the Institute of Mathematics of the University of Seville, Spain, in 2017. Her research area is mainly on combinatorial optimization, with particular interest in network optimization, multi-objective programming, and data science, with applications to real network problems in telecommunication and transportation areas, where she has published articles in international journals. She is a founder of the Young Researchers Chapter of the Italian Operations Research Society (AIROYoung), of which she is outgoing coordinator. She is one of the awardees of the YoungWomen4OR EURO program in 2020.

Paolo Dell'Olmo is Full Professor of Operations Research in the Department of Statistical Sciences at Sapienza University of Rome. He has been formerly department head, coordinator of the PhD program in operations research, director of the master's program in statistics for the management of information systems, coordinator of the Italian Inter-University Center for Operations Research, member of the scientific committee of Fondazione Roma Sapienza, and member of the board of directors of Fondazione Roma Sapienza. He is currently director of the master's program in data intelligence and strategic decisions. His research interests are mainly in combinatorial optimization, also applied to real-life problems. He has been scientific coordinator of a number of national research projects, and he is author of several books and about 80 papers published in international journals.

Isabella Lari obtained her PhD degree in operations research from Sapienza University of Rome in 1994. She is a researcher in the Department of Statistics at Sapienza University of Rome. Dr. Lari teaches optimization mathematical methods and, in the past, has held courses on elements of computer science, data structure and algorithms, and simulation techniques. Her research activity is mainly focused on combinatorial optimization with particular attention to location and partitioning problems on graphs. She has published in several major journals in the field.

Contributors

Lavinia Amorosi Department of Statistical Sciences, Sapienza University of Rome, Rome, Italy

Roberto Aringhieri Department of Computer Engineering, University of Torino, Torino, Italy

Annamaria Barbagallo Department of Mathematics and Applications "R. Caccioppoli", University of Naples Federico II, Naples, Italy

Andrea Bettinelli OptIt, Bologna, Italy

Daniel Bienstock IEOR, Columbia University, New York, NY, USA

Sara Bigharaz Department of Industrial Economics and Technology Management, Faculty of Economics and Management, Trondheim, Norway

Beatrice Bolsi Department of Sciences and Methods for Engineering, University of Modena and Reggio Emilia, Reggio Emilia, Italy

Paola Cappanera DINFO, University of Florence, Florence, Italy

Giulia Caselli School of Doctorate E4E (Engineering for Economics - Economics for Engineering), University of Modena and Reggio Emilia, Modena, Italy

Martina Cerulli LIX CNRS Ecole Polytechnique, Institut Polytechnique de Paris, Palaiseau, France

Marc-Antoine Coindreau GSEM, University of Geneva - Uni Mail, Geneva, Switzerland

Domenico Conforti de-Health Lab, Department of Mechanical, Energy and Management Engineering, University of Calabria, Rende, Italy

Patrizia Daniele Department of Mathematics and Computer Science, University of Catania, Catania, Italy

Thiago Alves de Queiroz Institute of Mathematics and Technology, Federal University of Catalão, Catalão, GO, Brazil

Fulvio De Santis Department of Statistical Sciences, Sapienza University of Rome, Rome, Italy

Maxence Delorme Department of Econometrics and Operations Research, Tilburg University, Tilburg, AB, The Netherlands

Lorenzo Di Rocco Department of Statistical Sciences, Sapienza University of Rome, Rome, Italy

Doncho Donchev GATE Institute, Sofia University "St. Kliment Ohridski", Sofia, Bulgaria

Davide Duma Department of Mathematics "Felice Casorati", University of Pavia, Pavia, Italy

Mauro Escobar LIX CNRS Ecole Polytechnique, Institut Polytechnique de Paris, Palaiseau, France

Georgia Fargetta Department of Mathematics and Computer Science, University of Catania, Catania, Italy

Giovanni Fasano Department of Management, University Ca' Foscari of Venice, Venice, Italy

Olivier Gallay Faculty of Business and Economics (HEC Lausanne), Department of Operations, University of Lausanne, Lausanne, Switzerland

Marco Gavanelli DE, University of Ferrara, Ferrara, Italy

Maria Carmela Groccia de-Health Lab, Department of Mechanical, Energy and Management Engineering, University of Calabria, Rende, Italy

Alberto Guastalla Computer Science Department, University of Turin, Turin, Italy

Stefania Gubbiotti Department of Statistical Sciences, Sapienza University of Rome, Rome, Italy

Rosita Guido de-Health Lab, Department of Mechanical, Energy and Management Engineering, University of Calabria, Rende, Italy

Manuel Iori DISMI, University of Modena and Reggio Emilia, Reggio Emilia, Italy

Arthur Kramer Department of Production Engineering, Federal University of Rio Grande do Norte, Natal, RN, Brazil

Lorenzo Lampariello Department of Business Economics, Roma Tre University, Rome, Italy

Giacomo Lanza Department of Computer Science, University of Pisa, Pisa, Italy

Gilbert Laporte HEC Montréal, Montréal, QC, Canada

Roel Leus Faculty of Economics and Business, Leuven, Belgium

Leo Liberti LIX CNRS Ecole Polytechnique, Institut Polytechnique de Paris, Palaiseau, France

Carlo Alberto Magni School of Doctorate E4E (Engineering for Economics - Economics for Engineering), University of Modena and Reggio Emilia, Modena, Italy

Maddalena Nonato DE, University of Ferrara, Ferrara, Italy

Mauro Passacantando Department of Computer Science, University of Pisa, Pisa, Italy

Umberto Ferraro Petrillo Department of Statistical Sciences, Sapienza University of Rome, Rome, Italy

Gianluca Priori Department of Computer, Control and Management Engineering "A. Ruberti", Sapienza University of Rome, Rome, Italy

Fabio Raciti Department of Mathematics and Computer Science, University of Catania, Catania, Italy

Luigi Rarità Department of Management and Innovation Systems, University of Salerno, Fisciano, Italy

Department of Science and Technology, University of Sannio, Benevento, Italy

Giovanni Righini Department of Computer Science, University of Milan, Milan, Italy

Marco Roma DE, University of Ferrara, Ferrara, Italy

Massimo Roma Department of Computer, Control and Management Engineering "A. Ruberti", Sapienza, University of Rome, Rome, Italy

Simone Sagratella Department of Computer, Control and Management Engineering "A. Ruberti", Sapienza University of Rome, Rome, Italy

Daniele Sciacca Department of Mathematics and Computer Science, University of Catania, Catania, Italy

Laura Scrimali Department of Mathematics and Computer Science, University of Catania, Catania, Italy

Maria Grazia Scutellà Department of Computer Science, University of Pisa, Pisa, Italy

Demir Tonchev GATE Institute, Sofia University "St. Kliment Ohridski", Sofia, Bulgaria

Vassil Vassilev Cyber Security Research Centre, London Metropolitan University, London, UK

Fabrizio Venturelli FDL Servizi, Breno, Italy

Marie-Sklaerder Vié GSEM, University of Geneva - Uni Mail, Geneva, Switzerland

Pier Giorgio Villani Department of Emergency and Critical Care, Hospital of Cremona, Cremona, Italy

Nicolas Zufferey GSEM, University of Geneva - Uni Mail, Geneva, Switzerland

A Variational Inequality Approach to a Class of Network Games with Local Complementarities and Global Congestion

Mauro Passacantando and Fabio Raciti

Abstract We investigate a class of network games with strategic complements and congestion effects, by using the variational inequality approach. Our contribution is twofold. We first express the boundary components of the Nash equilibrium by means of the Katz-Bonacich centrality measure. Then, we propose a new ranking of the network nodes based on the social welfare at equilibrium and compare this solution-based ranking with some classical topological ranking methods.

Keywords Network games · Nash equilibrium · Network centrality measures · Social welfare

1 Introduction

The topic of Network Games is relatively recent and was formulated in a general framework in the influential paper by Ballester et al. [1], where the authors proposed to model the social and economic interactions among individuals with the help of a graph where each individual (player) is identified with the node of a graph and can interact only with his/her neighbors in the graph, while congestion effects are due to all the players in the network. The solution concept considered is the Nash equilibrium of the game and is related to the so called Katz-Bonacich centrality measure, in the case of interior solution. Although the topic has grown at a high pace in the last fifteen years (see, e.g., [4]), only recently some authors have proposed to use the variational inequality approach to investigate this kind of games. In particular, in [9], the authors make an in-depth analysis of uniqueness and sensitivity of equilibrium, with particular emphasis on its connection with the spectral properties

M. Passacantando
Department of Computer Science, University of Pisa, Pisa, Italy
e-mail: mauro.passacantando@unipi.it

F. Raciti (✉)
Department of Mathematics and Computer Science, University of Catania, Catania, Italy
e-mail: fraciti@dmi.unict.it

L. Amorosi et al. (eds.), *Optimization in Artificial Intelligence and Data Sciences*, AIRO Springer Series 8, https://doi.org/10.1007/978-3-030-95380-5_1

of the adjacency matrix of the graph, while in [8] a generalized Nash equilibrium problem is proposed within the framework of variational inequalities.

In this note, we consider a game where the generic player is influenced by his/her neighbors through a local strategic complement term and experiments a global congestion effect. We first derive a new representation formula for the Nash equilibrium, in the case where some of its components reach their upper bound, which makes use of the Katz-Bonacich vector. Then, we focus on the problem of assessing the importance of the players and propose a new centrality measure based on the social welfare computed at the Nash equilibrium.

More specifically, the paper is structured as follows. In Sect. 2 we first introduce the notation and the basic definitions, as well as the essential tools from variational inequality theory needed for our investigation. We then introduce the utility functions which describe a quadratic model with local complementarities and global congestion. Moreover, we recall the classical Katz-Bonacich formula for the interior solution case, where the strategy set of each player is $\mathbb{R}_+$. In Sect. 3, we assume that the strategy sets are bounded also from above and derive a representation formula for the solution in the case where some of its components lie on the boundary, which is based on the Katz-Bonacich centrality. In Sect. 4, we propose to assess the importance of a player by measuring the variation of the social welfare at equilibrium when the player is removed from the network. Moreover, we compare the ranking thus obtained with that one obtained using some classical topological measures in the literature. A small concluding section ends the paper.

2 Network Games

2.1 Game Formulation and Variational Inequality Approach

In Network Games players are represented by the nodes of a graph (V, E), where V is the sets of nodes and E is the set of edges formed by pairs of nodes (v, w). Here, we only consider undirected simple graphs. Two nodes v and w are said to be adjacent if they are connected by an edge, i.e., if (v, w) is an edge. The information about the adjacency of nodes can be stored in the adjacency matrix G whose elements g_{ij} are equal to 1 if (v_i, v_j) is an edge, 0 otherwise. G is thus a symmetric and zero-diagonal matrix. Given a node v, the nodes connected to v with an edge are called the *neighbors* of v, and are grouped in the set $N_v(g)$. The number of elements of $N_v(g)$ is the *degree* of v and will be denote by $deg_v(g)$. A *walk* in the graph g is a finite sequence of the form $v_{i_0}, e_{j_1}, v_{i_1}, e_{j_2}, \ldots, e_{j_k}, v_{j_k}$, which consists of alternating nodes and edges of the graph, such that $v_{i_{t-1}}$ and v_{i_t} are end nodes of e_{j_t}. The *length* of a walk is the number of its edges. Let us remark that it is allowed to visit a node or go through an edge more than once. The indirect connections between any two nodes in the graph are described by means of the powers of the adjacency matrix G. Indeed, it can be proved that the element $g_{ij}^{[k]}$

of G^k gives the number of walks of length k between v_i and v_j. We now define some common topological measures used to assess the importance of a node i in a network:

- *degree centrality*: $DC_i = deg_i$;
- *closeness centrality*: $CC_i = \frac{1}{\sum_{j \neq i} d_{ij}}$, where d_{ij} is the shortest path length between i and j;
- *betweennes centrality*: $BC_i = \sum_{s,t \neq i} \frac{n_{st}(i)}{n_{st}}$, where n_{st} is the number of shortest paths between s and t and $n_{st}(i)$ is the number of such paths that pass through node i.

In the sequel, the set of players will be denoted by $\{1, 2, \ldots, n\}$ instead of $\{v_1, v_2, \ldots, v_n\}$. We denote with $A_i \subset \mathbb{R}$ the action space of player i, while $A = A_1 \times \cdots \times A_n$. For each $a = (a_1, \ldots, a_n)$, $a_{-i} = (a_1, \ldots, a_{i-1}, a_{i+1}, \ldots, a_n)$ and the notation $a = (a_i, a_{-i})$ will be used when we want to distinguish the action of player i from the action of all the other players. Each player i is endowed with a payoff function $u_i : A \to \mathbb{R}$ that he/she wishes to maximize. The notation $u_i(a, G)$ is often utilized when one wants to emphasize that the utility of player i also depends on the actions taken by her/his neighbors in the graph.

The solution concept that we consider here is the Nash equilibrium of the game, that is, we seek an element $a^* \in A$ such that for each $i \in \{1, \ldots, n\}$:

$$u_i(a_i^*, a_{-i}^*) \geq u_i(a_i, a_{-i}^*), \qquad \forall\, a_i \in A_i. \tag{1}$$

According to how variations of the actions of player's i neighbors affect his/her marginal utility, two classes of game can be defined. Specifically, the game has the property of *strategic complements* if: $\frac{\partial^2 u_i}{\partial a_j \partial a_i}(a) > 0,\ \forall (i,j) : g_{ij} = 1, \forall\, a \in A$, while it has the property of *strategic substitutes* if: $\frac{\partial^2 u_i}{\partial a_j \partial a_i}(a) < 0,\ \forall (i,j) : g_{ij} = 1, \forall\, a \in A$.

For the subsequent development it is important to recall that if the u_i are continuously differentiable functions on A, and $u_i(\cdot, a_{-i})$ are concave, the Nash equilibrium problem is equivalent to the variational inequality $VI(F, A)$: find $a^* \in A$ such that

$$F(a^*)^\top (a - a^*) \geq 0, \qquad \forall\, a \in A, \tag{2}$$

where

$$[F(a)]^\top := -\left(\frac{\partial u_1}{\partial a_1}(a), \ldots, \frac{\partial u_n}{\partial a_n}(a)\right) \tag{3}$$

is also called the *pseudo-gradient* of the game. For an account of variational inequalities the reader can refer to [6, 7]. We recall here some monotonicity properties.

Definition 1 $T : \mathbb{R}^n \to \mathbb{R}^n$ is said to be monotone on A iff:

$$[T(x) - T(y)]^\top (x - y) \geq 0, \qquad \forall\, x, y \in A.$$

If the equality holds only when $x = y$, T is said to be strictly monotone. T is said to be β-strongly monotone on A iff there exists $\beta > 0$ such that

$$[T(x) - T(y)]^\top (x - y) \geq \beta \|x - y\|^2, \qquad \forall\, x, y \in A.$$

For linear operators on $\mathbb{R}^n$ the two concepts of strict and strong monotonicity coincide and are equivalent to the positive definiteness of the corresponding matrix.

Conditions that ensure the unique solvability of a variational inequality problem are given by the following theorem (see, e.g. [7]).

Theorem 1 *If $K \subset \mathbb{R}^n$ is compact and convex, and $T : \mathbb{R}^n \to \mathbb{R}^n$ is continuous on K, then the variational inequality problem $VI(F, K)$ admits at least one solution. In the case that K is unbounded, the existence of a solution may be established if the following coercivity condition holds, for $x \in K$ and some $x_0 \in K$:*

$$\lim_{\|x\| \to +\infty} \frac{[T(x) - T(x_0)]^\top (x - x_0)}{\|x - x_0\|} = +\infty.$$

Furthermore, if T is strictly monotone on K the solution is unique.

2.2 *The Linear-Quadratic Model with Local Complementarities and Global Congestion*

Let $A_i = \mathbb{R}_+$ for any $i \in \{1, \dots, n\}$, hence $A = \mathbb{R}^n_+$. The payoff of player i is given by:

$$u_i(a, G) = \alpha a_i - \frac{1}{2} a_i^2 + \phi \sum_{j=1}^{n} g_{ij} a_i a_j - \gamma \sum_{j=1}^{n} a_i a_j, \qquad \alpha, \phi, \gamma > 0. \tag{4}$$

In this model α and ϕ take on the same value for all players, which then only differ according to their position in the network. The third term describes the interaction between neighbors and since $\phi > 0$ this interaction falls in the class of strategic complements. On the other term, since $\gamma > 0$ the last term falls in the class of strategic substitutes and models the overall congestion effects in the network. Thus, without further hypotheses, this model does not belong to any of the above mentioned class. The pseudo-gradient's components of this game are easily computed as: $F_i(a) = (1+\gamma)a_i - \alpha - \phi \sum_{j=1}^n g_{ij}a_j + \gamma \sum_{j=1}^n a_j$, $i \in \{1, \ldots, n\}$, which can be written in compact form as $F(a) = [(1+\gamma)I - \phi G + \gamma U]a - \alpha \mathbf{1}$, where $U_{ij} = 1$ for any $i, j = 1, \ldots, n$ and $\mathbf{1} = (1, \ldots, 1)^\top \in \mathbb{R}^n$. We will seek Nash equilibrium points by solving the variational inequality:

$$F(a^*)^\top (a - a^*) \geq 0, \qquad \forall\, a \in \mathbb{R}^n_+. \tag{5}$$

Since the constraint set is unbounded, to ensure solvability we require that F be strongly monotone, which also guarantees the uniqueness of the solution. In the next lemma we recall a well known result about matrices.

Lemma 1 *Let T be a symmetric matrix and $\rho(T)$ the spectral radius of T. If $\rho(T) < 1$, then the matrix $I - T$ is positive definite and $(I - T)^{-1} = \sum_{k=0}^{\infty} T^k$.*

We now give an important definition for our analysis.

Definition 2 For any weight $w \in \mathbb{R}^n_+$, and $\phi > 0$ the weighted vector of Katz-Bonacich [2] for the network G is given by:

$$b_w(G, \phi) = M(G, \phi) = (I - \phi G)^{-1} w = \sum_{p=0}^{\infty} \phi^p G^p w. \tag{6}$$

In the case where $w = \mathbf{1}$, the (non weighted) centrality measure of Katz-Bonacich of node i can be interpreted as the total number of walks in the graph, which start at node i, exponentially damped by ϕ. The connection between the Katz-Bonacich vector and the Nash equilibrium is given in the following theorem.

Theorem 2 (See Theorem 1 in [1]) *If $\phi\rho(G) < 1 + \gamma$, then the unique Nash equilibrium of the game with utility functions (4) and $A = \mathbb{R}^n_+$ is interior and given by:*

$$a^* = \frac{\alpha}{1 + \gamma + \gamma \sum_{i=1}^{n} \left(b_{\mathbf{1}}\left(G, \frac{\phi}{1+\gamma}\right)\right)_i} b_{\mathbf{1}}\left(G, \frac{\phi}{1+\gamma}\right). \tag{7}$$

For the subsequent developments we also need to define the social welfare:

$$W(a, G) = \sum_{i=1}^{n} u_i(a, G). \tag{8}$$

3 A Katz-Bonacich Representation Formula

We now assume that the strategies of each player have an upper bound and derive a Katz-Bonacich type representation of the solution, in the case where exactly k components take on their maximum value.

Theorem 3 *Let u_i be defined as in (4), $a_i \in [0, L_i]$ for any $i \in \{1, \dots, n\}$, and $\phi\rho(G) < 1+\gamma$. Then there exists a unique Nash equilibrium a^* of the game and $a_i^* > 0$ holds for any $i \in \{1, \dots, n\}$. Moreover, assume that exactly k components of a^* take on their maximum value: $a_{i_1}^* = L_{i_1}, \dots, a_{i_k}^* = L_{i_k}$, and denote with $\tilde{a}^* = (\tilde{a}_{i_{k+1}}^*, \dots, \tilde{a}_{i_n}^*)$ the subvector of the nonboundary components of a^*. We then get:*

$$\tilde{a}^* = \left(\frac{1}{1+\gamma}\right) b_{\mathbf{w}}\left(G_1, \frac{\phi}{1+\gamma}\right) - \left(\frac{\gamma}{1+\gamma}\right) \frac{\sum\limits_{m=k+1}^{n} \left(b_w\left(G_1, \frac{\phi}{1+\gamma}\right)\right)_{i_m}}{1+\gamma+\gamma \sum\limits_{m=k+1}^{n} \left(b_{\mathbf{1}_{n-k}}\left(G_1, \frac{\phi}{1+\gamma}\right)\right)_{i_m}} b_{\mathbf{1}_{n-k}}\left(G_1, \frac{\phi}{1+\gamma}\right), \tag{9}$$

where G_1, G_2, U_1, U_2 are submatrices of G and U defined as follows:

$$G = \begin{array}{c} \\ i_1 \\ \vdots \\ i_k \\ i_{k+1} \\ \vdots \\ i_n \end{array} \begin{array}{c} \begin{array}{cc} i_1 \dots i_k & i_{k+1} \dots i_n \end{array} \\ \left(\begin{array}{c|c} * & * \\ \hline G_2 & G_1 \end{array}\right) \end{array}, \qquad U = \begin{array}{c} \\ i_1 \\ \vdots \\ i_k \\ i_{k+1} \\ \vdots \\ i_n \end{array} \begin{array}{c} \begin{array}{cc} i_1 \dots i_k & i_{k+1} \dots i_n \end{array} \\ \left(\begin{array}{c|c} * & * \\ \hline U_2 & U_1 \end{array}\right) \end{array},$$

$w = [\alpha \mathbf{1}_{n-k} + (\phi G_2 - \gamma U_2)\, L]/(1+\gamma)$ *and* $L = (L_{i_1}, \dots, L_{i_k})^\top$.

Proof Let us notice that the matrix γU is positive semidefinite and $(1+\gamma)I - \phi G$ is positive definite by Lemma 1, hence the map F is strongly monotone on $\mathbb{R}^n$ and the game has a unique Nash equilibrium a^*, which solves the variational inequality

$$F(a^*)^\top (a - a^*) \geq 0, \qquad \forall\, a \in K, \tag{10}$$

where $K = [0, L_1] \times \dots \times [0, L_n]$. Let us notice that $a^* \neq 0$, otherwise we have $0 \leq -\alpha \mathbf{1}^\top a$ holds for any $a \in K$, that is impossible. Define the set $S \subseteq \{1, \dots, n\}$ such that $a_i^* > 0,\ \forall i \in S,\ a_i^* = 0,\ \forall i \notin S$. We then get that a^* solves the KKT system associated with $VI(F; K)$:

$$
\begin{aligned}
&[(1+\gamma)I - \phi G + \gamma U]a^* - \alpha\mathbf{1} - \lambda + \mu = 0, && \\
&\lambda_i a_i^* = 0, \qquad \lambda_i \geq 0, && i = 1, \ldots, n, \\
&\mu_i^*(a_i^* - L_i) = 0, \qquad \mu_i^* \geq 0, && i = 1, \ldots, n,
\end{aligned}
$$

which implies:

$$(1+\gamma)a_i^* - \phi \sum_{j \in S} g_{ij} a_i^* + \gamma \sum_{j \in S} a_i^* - \alpha + \mu_i = 0, \qquad \forall\, i \in S, \tag{11}$$

$$-\phi \sum_{j \in S} g_{ij} a_i^* - \alpha + \gamma \sum_{j \in S} a_i^* - \lambda_i = 0, \qquad \forall\, i \notin S. \tag{12}$$

We then get: $((1+\gamma)I_S - \phi G_S)a_S^* = (\alpha - \gamma \sum_{j=1}^n a_j^*)\mathbf{1}_S - \mu_S$, and because $\phi\rho(G_S) \leq \phi\rho(G) < 1+\gamma$, we also have $((1+\gamma)I_S - \phi G_S)^{-1} \geq 0$, hence:

$$
\begin{aligned}
0 < a_S^* &= \left(\alpha - \gamma \sum_{j=1}^n a_j^*\right)((1+\gamma)I_S - \phi G_S)^{-1}\mathbf{1}_S - ((1+\gamma)I_S - \phi G_S)^{-1}\mu_s \\
&\leq \left(\alpha - \gamma \sum_{j=1}^n a_j^*\right)((1+\gamma)I_S - \phi G_S)^{-1}\mathbf{1}_S,
\end{aligned}
$$

which implies $\alpha - \gamma \sum_{j=1}^n a_j^* > 0$. If there exists an index $i \notin S$, then from (12) we get the contradiction: $0 < \alpha - \gamma \sum_{j=1}^n a_j^* = -\phi \sum_{j \in S} g_{ij} a_i^* - \lambda_i \leq 0$. Therefore, $a_i^* > 0$ holds for any $i \in \{1, \ldots, n\}$.

Let $\tilde{K}$ denote the face of K obtained intersecting K with the hyperplanes: $a_{i_1} = L_{i_1}, \ldots, a_{i_k} = L_{i_k}$. Moreover, let $\tilde{a} = (a_{i_{k+1}}, \ldots, a_{i_n})$, $\tilde{a}^* = (\tilde{a}^*_{i_{k+1}}, \ldots, \tilde{a}^*_{i_n})$ and define $\tilde{F} : \mathbb{R}^{n-k} \to \mathbb{R}^{n-k}$ such that $\tilde{F}_{i_l}(\tilde{a})$ is obtained by fixing $a_{i_1} = L_{i_1}, \ldots, a_{i_k} = L_{i_k}$ in $F_{i_l}(a)$. We consider now the restriction of (10) to $\tilde{K}$, which reads:

$$\sum_{l=k+1}^n \tilde{F}_{i_l}(\tilde{a}^*)(\tilde{a}_{i_l} - \tilde{a}^*_{i_l}) \geq 0, \qquad \forall\, \tilde{a} \in \tilde{K}.$$

Since we are assuming that exactly k components of the solution a^* reach their upper bounds, it follows that $\tilde{a}^*$ lies in the interior of $\tilde{K}$, hence $\tilde{F}(\tilde{a}^*) = 0$, that is equivalent to

$$(1+\gamma)a_{i_l}^* - \phi \sum_{m=k+1}^n g_{i_l i_m} a_{i_m}^* + \gamma \sum_{m=k+1}^n a_{i_m}^* = \alpha + \phi \sum_{m=1}^k g_{i_l i_m} L_{i_m} - \gamma \sum_{m=1}^k L_{i_m},$$

for any $l = k+1, \dots, n$, which yields $[(1+\gamma)I_{n-k} - \phi G_1 + \gamma U_1]\tilde{a}^* = \alpha \mathbf{1}_{n-k} + \phi G_2 L - \gamma U_2\, L$, which can be written as

$$\left[I_{n-k} - \frac{\phi}{1+\gamma} G_1 + \frac{\gamma}{1+\gamma} U_1 \right] \tilde{a}^* = \frac{1}{1+\gamma} w,$$

with $w = \alpha \mathbf{1}_{n-k} + (\phi G_2 - \gamma U_2)L$. To derive $\tilde{a}^*$ let us first notice that $U_1 \tilde{a}^* = \left(\sum_{m=k+1}^{n} \tilde{a}^*_{i_m} \right) \mathbf{1}_{n-k}$ and the matrix $(I_{n-k} - \frac{\phi}{1+\gamma} G_1)$ is not singular because $\frac{\phi}{1+\gamma} \rho(G_1) < 1$, thus we get:

$$\tilde{a}^* = \frac{1}{1+\gamma} \left[I_{n-k} - \frac{\phi}{1+\gamma} G_1 \right]^{-1} w - \frac{\gamma}{1+\gamma} \left(\sum_{m=k+1}^{n} \tilde{a}^*_{i_m} \right) \left[I_{n-k} - \frac{\phi}{1+\gamma} G_1 \right]^{-1} \mathbf{1}_{n-k}, \tag{13}$$

which, from the definition of the Katz-Bonachich vector (6), yields:

$$(1+\gamma)\, \tilde{a}^* + \gamma \left(\sum_{m=k+1}^{n} \tilde{a}^*_{i_m} \right) b_{\mathbf{1}_{n-k}} \left(G_1, \frac{\phi}{1+\gamma} \right) = b_w \left(G_1, \frac{\phi}{1+\gamma} \right), \tag{14}$$

which can be exploited to derive $\sum_{m=k+1}^{n} \tilde{a}^*_{i_m}$. Indeed, summing up on the components of both the left and right handside of (14) we get:

$$\sum_{m=k+1}^{n} \tilde{a}^*_{i_m} = \frac{\sum_{m=k+1}^{n} (b_w(G_1, \frac{\phi}{1+\gamma}))_{i_m}}{1 + \gamma + \gamma \sum_{m=k+1}^{n} (b_{\mathbf{1}_{n-k}}(G_1, \frac{\phi}{1+\gamma}))_{i_m}}. \tag{15}$$

Inserting (15) into (13), we finally obtain (9) and the proof is completed. □

4 A Social Welfare Centrality Measure

In this section we propose a new centrality measure of nodes (players) of a network game based on the social welfare computed at the Nash equilibrium. Specifically, for any node $i \in \{1, \dots, n\}$, we measure the importance of i as the percentage variation of the social welfare computed at the Nash equilibrium after i is removed from the network, that is

$$SWC(i) = 100 \cdot \frac{W(NE(G)) - W(NE(G \setminus \{i\}))}{W(NE(G))}, \tag{16}$$

Fig. 1 Network topology of Example 1

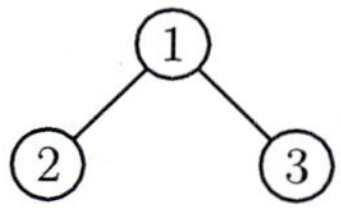

where $NE(G)$ is the Nash equilibrium in the network G, $NE(G \setminus \{i\})$ is the Nash equilibrium in the network G where node i has been removed, and W is the social welfare function (8). Note that $SWC(i)$ can be negative if the total social welfare of the network increases after removing the node i (as Example 1 below shows). This situation can be compared to the well-known Braess paradox [3], where the efficiency of a network improves due to the removal of a link.

Example 1 We consider the network game on the small network shown in Fig. 1 with three nodes and two links. We assume that the game parameters are $\alpha = 1$, $\phi = 2$, $\gamma = 3$ and $a_i \in [0, 1]$. Note that $\rho(G) = \sqrt{2}$ so that Theorem 3 guarantees the existence and uniqueness of the Nash equilibrium in the original network and the sub-networks obtained by removing one node at a time. The Nash equilibrium in the network G is $(2/17, 3/34, 3/34)$ and the corresponding social welfare is equal to $7/68$. When the node 1 is removed, the network becomes disconnected and the total social welfare at equilibrium decreases to 7/100. On the other hand, when node 2 or 3 is removed, the social welfare at equilibrium increases to 7/64, hence the social welfare centrality of nodes 2 and 3 takes on negative values. More precisely, we have $SWC = (32, -25/4, -25/4)$.

Example 2 We now compare the social welfare centrality with three well-known topological centrality measures: degree, closeness and betweenness. We consider a network game on a graph with 10 nodes, where the adjacency matrix G has been randomly generated (see Fig. 2), and the game parameters are $\alpha = 1$, $\gamma = 1$, $\phi = 0.9(1 + \gamma)/\rho(G)$ and $a_i \in [0, 1]$ for any $i = 1, \ldots, 10$. Table 1 shows the ranking of nodes according to the social welfare centrality measure and the three considered topological centrality measures. It is interesting noting that the ranking defined by the new measure is quite different from that provided by the other measures. Moreover, Fig. 2 gives a graphical representation of the values associated to the network nodes for each considered centrality measure.

5 Conclusions and Further Research Perspectives

In future work, we aim to apply our results to some specific social or economic problems. An interesting account of these applications can be found in the survey [4]. Also the theory of stochastic variational inequalities [5] could be used to cope with uncertain parameters in the model.

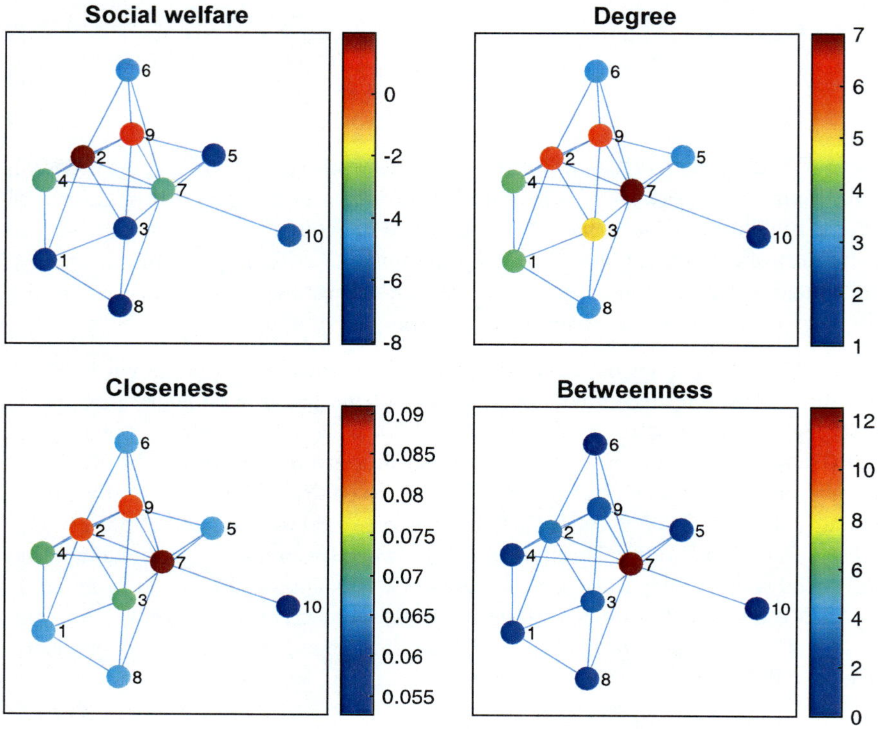

Fig. 2 Comparison between the social welfare centrality measure and degree, closeness and betweenness centrality measures

Table 1 Ranking of nodes according to the new social welfare centrality measure and the well known degree, closeness and betweenness centrality measures

	Centrality measures			
Rank	Social Welfare	Degree	Closeness	Betweenness
1	2	7	7	7
2	9	2	2	2
3	4	9	9	3
4	7	3	3	9
5	6	1	4	1
6	10	4	1	8
7	3	5	5	4
8	5	6	6	5
9	1	8	8	6
10	8	10	10	10

Acknowledgments The authors are members of the Gruppo Nazionale per l'Analisi Matematica, la Probabilità e le loro Applicazioni (GNAMPA—National Group for Mathematical Analysis, Probability and their Applications) of the Istituto Nazionale di Alta Matematica (INdAM—National Institute of Higher Mathematics). This research was partially supported by the research project "Programma ricerca di ateneo UNICT 2020-22 linea 2-OMNIA" of the University of Catania. This support is gratefully acknowledged.

References

1. Ballester, C., Calvo-Armengol, A., Zenou, Y.: Who's who in networks. Wanted: The key player. Econometrica **74**, 1403–1417 (2006)
2. Bonacich, P.: Power and centrality: a family of measures. Am. J. Sociol. **92**, 1170–1182 (1987)
3. Braess, D.: Über ein Paradoxon aus der Verkehrsplanung. Unternehmenforschung **12**, 258–268 (1968)
4. Jackson, M.O., Zenou, Y.: Games on Networks. In: Handbook of Game Theory with Economic Applications, pp. 95–163. Elsevier (2015)
5. Jadamba, B., Khan, A.A., Raciti, F.: Regularization of stochastic variational inequalities and a comparison of an L_p and a sample-path approach. Nonlinear Anal. Theory Methods Appl. **94**, 65–83 (2014)
6. Konnov, I.: Equilibrium Models and Variational Inequalities. Elsevier (2007)
7. Nagurney, A.: Network Economics A Variational Inequality Approach. Springer (1999)
8. Passacantando, M., Raciti F.: A note on generalized Nash games played on networks. In: Rassias, T.M. (ed.) Nonlinear Analysis, Differential Equations and Applications, pp. 365–380. Springer (2021)
9. Parise, F. Ozdaglar, A.: A variational inequality framework for network games: Existence, uniqueness, convergence and sensitivity analysis. Games Econ. Behav. **114**, 47–82 (2019)

A Two-Stage Variational Inequality Formulation for a Game Theory Network Model for Hospitalization in Critic Scenarios

Patrizia Daniele and Daniele Sciacca

Abstract In this paper, we introduce the theoretical structure of a stochastic Generalized Nash Equilibrium model describing the competition among hospitals with first aid departments for the hospitalization in a disaster scenario. Each hospital with a first aid department has to solve a two-stage stochastic optimization problem, one before the declaration of the disaster scenario and one after the disaster advent, to determine the equilibrium hospitalization flows to dispatch to the other hospitals with first aid and/or to hospitals without emergency rooms in the network. We define the Generalized Nash Equilibria of the model and, particularly, we consider the Variational Equilibria which is obtained as the solution to a variational inequality problem. Finally, we present a basic numerical example to validate the effectiveness of the model.

Keywords Game theory · Stochastic optimization · Hospitalization dispatching · Variational equilibrium

1 Introduction

Critic scenarios, such as earthquakes, hurricanes, fires, pandemic advents, are situations in which unexpected violent natural events or global events alter normal human activities. In such situations, it is essential that institutions, governments, humanitarian organizations have the possibility to use various tools that can help them in the management of critical situations, to mitigate the consequences.

The unpredictability of such events dictates the need to provide these entities with non-deterministic mathematical models that allow them to estimate, depending on the scenario that may occur, the best strategy to be implemented to assist the multiple phases of the disaster management.

P. Daniele · D. Sciacca (✉)
Department of Mathematics and Computer Science, University of Catania, Catania, Italy
e-mail: patrizia.daniele@unict.it; daniele.sciacca@unipa.it

L. Amorosi et al. (eds.), *Optimization in Artificial Intelligence and Data Sciences*, AIRO Springer Series 8, https://doi.org/10.1007/978-3-030-95380-5_2

In the existing literature, several network-based mathematical models, both of a deterministic and non-deterministic nature, have been developed to provide support to the disaster management phases (see, for instance, [2, 4, 5, 9–11]).

More specifically, in [2], authors provide an optimization model consisting of a dynamic supply chain network for personal protective equipment in the COVID-19 pandemic scenario in which they study the associated evolutionary variational inequality (see, for instance, [1]) in the presence of a delay function. In [4], authors propose a stochastic optimization approach for the distribution of medical supplies in emergency situations due to natural disasters, providing a two-stage stochastic programming model for which they derive a two-stage variational inequality formulation. In [5], authors present an evacuation model for which they derive a two-stage stochastic programming model. Finally, in [10], is proposed a two-stage stochastic game theory model describing the behavior of national governments in a healthcare disaster determined by COVID-19 pandemic advent and their competition for essential medical supplies in different phases of disaster preparedness.

The advent of a disaster scenario could cause an uncontrolled increase in requests for hospital care. In the moments following the disaster, the injured or registered victims are cared by the hospitals in the geographical area where the event occurred. The sudden advent of a disaster, the possible huge number of requests for assistance, the lack of medical personnel and the limited capacity of hospitalization cause inevitable overcrowding of the hospital structures and delays in the management of emergencies, as well as a large use of emergency vehicles and related costs increase. On the other hand, not all hospitals located in the geographical area of interest have emergency medical departments and this causes further inconvenience in the management of hospitalization requests. These factors, together with the total unpreparedness of hospitals, could make challenging, expensive and time-consuming the process of responding to the disastrous event.

Disaster management consists of different phases, including the preparedness and the response phases. In particular, when we consider an integrated preparation and response phase, it is of fundamental importance that decision makers make predictions on the uncertain possible disastrous scenarios that may occur and on their associated severity, so as not to be caught unprepared once the event occurs. These reasons led us to propose the two-stage stochastic optimization model described below, in which the decision makers are hospitals with emergency facilities, which seek to minimize both the total time of handling hospitalization requests from different geographical areas and the total costs due to patient transfers to other hospitals.

The paper is organized as follows. In Sect. 2, we develop the disaster stochastic game theory network model for hospitalization. We describe how hospitals with first aid departments compete to minimize their expected disutility, consisting of the total management time of an emergency and the total transfer cost to other hospitals. We describe the minimization problems that hospitals have to solve in the different phases of the management of the disaster scenario and we define the Generalized Nash Equilibrium and Variational Equilibrium of the proposed game theory model. Section 4 is dedicated to conclusion.

2 The Mathematical Model

The two-stage stochastic Supply Chain Network game theory model consists of K geographic areas, with a typical one by k, M hospitals with emergency rooms, with a generic one denoted by i and N hospitals without emergency rooms, with a generic one denoted by j. We denote by $\mathcal{M}_i$ the set of hospitals with first aid departments, except hospital i. In this model, the decision makers are hospitals with emergency rooms. Usually, in non-critical situations, emergency calls from a geographic location that are taken over by a hospital with an emergency room are handled by the receiving hospital, proceeding with the reception of the patient, anamnesis, diagnosis, possible hospitalization and subsequent discharge. In some cases (specialized centers, lack of staff, hospital overcrowding) it is possible that some emergency calls taken in charge by a hospital with first aid departments are subsequently routed to other hospitals with or without first aid departments. If, on the other hand, an emergency situation arises (outbreak of a pandemic, environmental disaster, etc.), it is highly likely that there will be a greater exchange among hospitals (for example, during the Covid-19 pandemic, due to the high number of patients in intensive care in Bolzano, Northen Italy, transfers were made to hospitals in Palermo, South Italy).

In this model, we want to provide a two-stage stochastic model, where in the pre-crisis phase, hospitals with emergency rooms consider several scenarios with different probabilities, so they are not surprised by a subsequent critical phase, trying to minimize the weighted sum between the management of emergency calls and the hospital dispatching times and the transport costs due to the transfers to other hospitals.

The three-tier network that describes the problem is represented in Figs. 1 and 2 which contain, respectively, the network topology for each hospital i and the entire

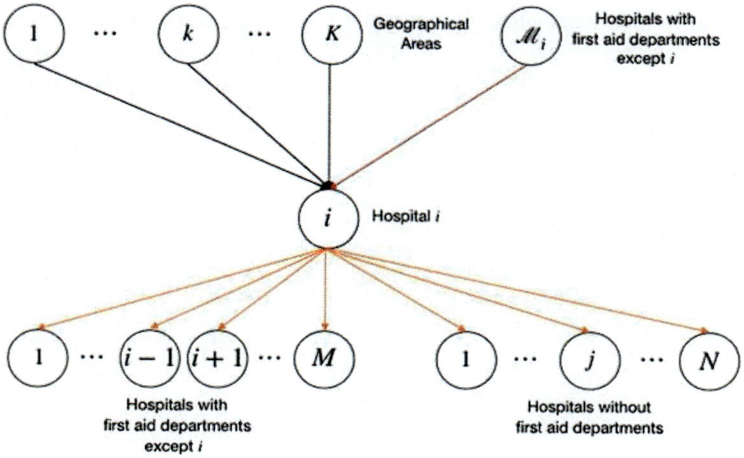

Fig. 1 Network topology for hospital i

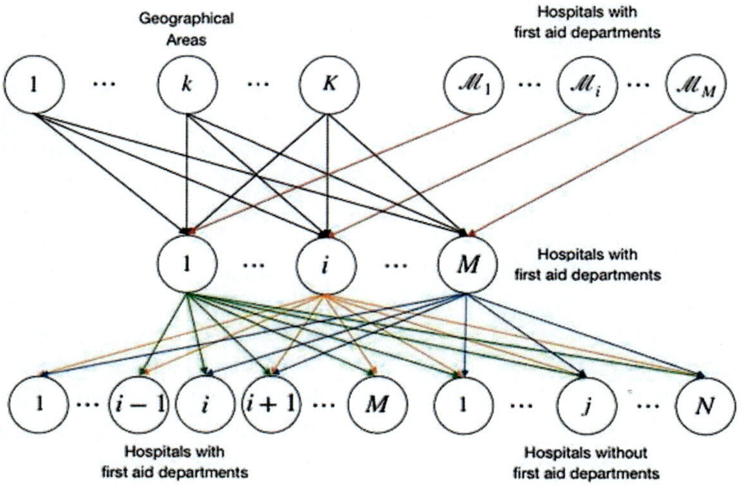

Fig. 2 Network topology

network topology. The variables and the parameters of the model are reported in Tables 1 and 2.

Facing a disaster scenario, the main goal of hospitals with first aid departments is to guarantee the treatment to patients who need medical devices and/or medical care, taking into account that requests for hospital care are all met as closely as possible. However, it is very difficult and time-consuming to ensure a complete management of the huge demand for hospital care, since the disaster scenario we are considering is global in nature.

Each hospital aims at minimizing its expected disutility which consists of the weighted sum between the expected dispatching time caused by the overcrowding of hospitals and the subsequent transportation costs of patients to other hospitals. The actions that the hospital takes in the first stage, before the disaster scenario, and the associated costs, are deterministic. However, the actions that hospitals take in phase 2, as soon as the disaster has occurred, depends on the possible scenario and the realization of probabilistic parameters.

We denote by:

- t^1_{ki} the transport time of an emergency patient from a geographic area k, $k = 1, \ldots, K$, to hospital i, $i = 1, \ldots, M$, in stage 1 and let us assume that t^1_{ki} is a function of the amount of emergency calls q^1_{ki}, namely:

$$t^1_{ki} := t^1_{ki}(q^1_{ki}), \quad \forall k = 1, \ldots, K, \ \forall i = 1, \ldots, M; \tag{1}$$

Table 1 Variables for the model

Notation	Variables
q^1_{ki}	The amount of emergency calls from the geographic area k, $k = 1, \dots, K$, handled by hospital i, $i = 1, \dots, M$, in stage 1. We group these quantities for all k into the vector $q^1_i \in \mathbb{R}^K_+$ and, in turn, we group these vectors into the vector $q^1 = (q^1_i)_{i=1,\dots,M} \in \mathbb{R}^{MK}_+$
$\hat{q}^1_i$	The amount of emergencies handled by hospital i, $i = 1, \dots, M$, dispatched by hospital with first aid departments belonging to $\mathcal{M}_i$ set in stage 1. We group these quantities into the vector $\hat{q}^1 \in \mathbb{R}^M_+$
$\tilde{q}^1_{il}$	The amount of emergencies handled by hospital i, $i = 1, \dots, M$, and subsequently dispatched to hospital $l \neq i$ in stage 1. We group these quantities for all $l \neq i$ into the vector $\hat{q}^1_i \in \mathbb{R}^{M-1}_+$ and, in turn, we group these quantities into the vector $\tilde{q}^1 \in \mathbb{R}^{M(M-1)}_+$
$\bar{q}^1_{ij}$	The amount of emergencies handled by hospital i, $i = 1, \dots, M$, dispatched to hospital j, $j = 1, \dots, N$, in stage 1. We group these quantities for all j into the vector $\bar{q}^1_i \in \mathbb{R}^N_+$ and, in turn, we group these quantities for all i into the vector $\bar{q}^1 \in \mathbb{R}^{MN}_+$
$q^{2\omega}_{ki}$	The amount of emergency calls from the geographic area k, $k = 1, \dots, K$, handled by hospital i, $i = 1, \dots, M$, in stage 2 when scenario ω occurs. We group these quantities for all k into the vector $q^{2\omega}_i \in \mathbb{R}^K_+$ and, in turn, we group these vectors into the vector $q^{2\omega} = (q^{2\omega}_i)_{i=1,\dots,M} \in \mathbb{R}^{MK}_+$. Finally, we group these vectors for all scenarios $\omega \in \Omega$ into the vector $q^2 \in \mathbb{R}^{\vert\Omega\vert MK}_+$
$\hat{q}^{2\omega}_i$	The amount of emergencies handled by hospital i, $i = 1, \dots, M$, arrived by hospital with first aid departments belonging to $\mathcal{M}_i$ set in stage 2 when scenario $\omega \in \Omega$ occurs. We group these quantities for all hospital into the vector in $\hat{q}^{2\omega} \in \mathbb{R}^M_+$ and, inturn, we group these quantities for all scenarios $\omega \in \Omega$ into the vector $\hat{q}^2 \in \mathbb{R}^{M\vert\Omega\vert}_+$
$\tilde{q}^{2\omega}_{il}$	The amount of emergencies handled by hospital i, $i = 1, \dots, N$, and subsequently dispatched to hospital $l \neq i$ in stage 2 when scenario ω occurs. We group these quantities for all $l \neq i$ into the vector $\tilde{q}^{2\omega}_i \in \mathbb{R}^{M-1}_+$ and, in turn, we group these quantities into the vector $\tilde{q}^{2\omega} \in \mathbb{R}^{M(M-1)}_+$. Finally, we group these vectors for all scenarios $\omega \in \Omega$ into the vector $\tilde{q}^2 \in \mathbb{R}^{\vert\Omega\vert M(M-1)}_+$
$\bar{q}^{2\omega}_{ij}$	The amount of emergency calls from k handled by hospital i dispatched to hospital j, $j = 1, \dots, N$, in stage 2. We group these quantities for all j into the vector $\bar{q}^{2\omega}_i \in \mathbb{R}^N_+$ and, in turn, we group these quantities for all i into the vector $\bar{q}^{2\omega} \in \mathbb{R}^{MN}_+$. Finally, we group these vectors for all scenarios $\omega \in \Omega$ into the vector $\bar{q}^2 \in \mathbb{R}^{\vert\Omega\vert MN}_+$
q	The vector $q = (Q^1, Q^2) \in \mathbb{R}^{MK+M+M(M-1)+MN+\vert\Omega\vert(MK+M+M(M-1)+MN)}_+$, where $Q^1 = (q^1, \hat{q}^1, \tilde{q}^1, \bar{q}^1)$ and $Q^2 = (q^2, \hat{q}^2, \tilde{q}^2, \bar{q}^2)$

- t^1_i the management time of an emergency patient arrived at hospital i, $i = 1, \dots, M$, in stage 1 and let us assume that t^1_i is a function of $\sum_{k=1}^{K} q^1_{ki}$ and $\hat{q}^1_i$,

Table 2 Parameters for the model

Notation	Parameters
$\omega \in \Omega$	The disaster scenario
p_ω	The probability of disaster scenario ω in stage 2, $\omega \in \Omega$
α_i	The weight in [0, 1]
C_i^1	The capacity of hospital i, $i = 1, \dots, M$, in stage 1
$C_i^{2\omega}$	The capacity of hospital i, $i = 1, \dots, M$, in stage 2 under scenario ω, $\forall \omega \in \Omega$
$\tilde{Q}_j^1$	The maximum capacity of hospital j, $j = 1, \dots, N$, in stage 1
$\tilde{Q}_j^{2\omega}$	The maximum capacity of hospital j, $j = 1, \dots, N$, in stage 2 under scenario ω, $\forall \omega \in \Omega$
β_i	The unit penalty encumbed by hospital i, $i = 1, \dots, N$, on the unmet demand
$d_i^{2\omega}$	The total demand for hospital i, $i = 1, \dots, N$, when scenario ω occurs in stage 2, $\forall \omega \in \Omega$

namely:

$$t_i^1 := t_i^1 \left(\sum_{k=1}^{K} q_{ki}^1, \hat{q}_i^1 \right) = t_i^1 \left(q_i^1, \hat{q}_i^1 \right). \tag{2}$$

This assumption suggests that the management time of an emergency call in the hospital i depends on the total flow of requests from each demand market and the total flow of requests transferred from hospitals belonging to $\mathscr{M}_i$, to hospital i;

- $\tilde{t}_{il}^1$, $l \neq i$, the transfer time of an emergency from hospital i, $i = 1, \dots, M$ to hospital $l = 1, \dots, M, l \neq i$, in stage 1 and let us assume that $\tilde{t}_{il}^1$ is a function of $\tilde{q}^1$, namely

$$\tilde{t}_{il}^1 := \tilde{t}_{il}^1(\tilde{q}); \tag{3}$$

- $\bar{t}_{ij}^1$ the transfer time of an emergency from hospital i, $i = 1, \dots, M$, to hospital j, $j = 1, \dots, N$, in stage 1 and let us assume that $\bar{t}_{ij}^1$ is a function of $\bar{q}^1$, namely

$$\bar{t}_{ij}^1 := \bar{t}_{ij}^1(\bar{q}^1) \tag{4}$$

- $\tilde{c}_{il}^1$ the transfer cost of an emergency from hospital i, $i = 1, \dots, M$, to hospital $l = 1, \dots, M, l \neq i$, in stage 1 and let us assume that $\tilde{c}_{il}^1$ is a function of $\tilde{q}_{il}^1$, namely

$$\tilde{c}^1_{il} := \tilde{c}^1_{il}(\tilde{q}^1_{il}); \tag{5}$$

- $\bar{c}^1_{ij}$ the transfer cost of an emergency from hospital i, $i = 1, \ldots, M$, to hospital j, $j = 1, \ldots, N$, in stage 1 and let us assume that $\bar{c}^1_{ij}$ is a function of $\bar{q}^1_{ij}$, namely

$$\bar{c}^1_{ij} := \bar{c}^1_{ij}(\bar{q}^1_{ij}); \tag{6}$$

Similarly, we define time and cost functions in stage 2, observing that these functions will depend on the variables of the second stage, and, therefore, will be affected by the uncertainty due to the scenario $\omega \in \Omega$.

Each hospital is faced with the following two-stage stochastic optimization model in which, as previously mentioned, it seeks to minimize the weighted sum between total expected dispatching time and the total expected costs (cf. Tables 1 and 2 for a detailed explanation of the role of each variable and parameter):

$$\begin{aligned} \text{Min} \Bigg\{ & \sum_{k=1}^{K} t^1_{ki}(q^1_{ki}) + t^1_i\left(q^1_i, \hat{q}^1_i\right) + \sum_{\substack{l=1,\ldots,M,\\ l\neq i}} \tilde{t}^1_{il}(\tilde{q}^1) + \sum_{j=1}^{N} \bar{t}^1_{ij}(\bar{q}^1) \\ & + \alpha_i \left(\sum_{\substack{l=1,\ldots,M,\\ l\neq i}} \tilde{c}^1_{il}(\tilde{q}^1_{il}) + \sum_{j=1}^{N} \bar{c}^1_{ij}(\bar{q}^1_{ij}) \right) + \mathbb{E}_\Omega[T^2_i(Q^2, \omega)] \Bigg\} \end{aligned} \tag{7}$$

subject to:

$$\sum_{\substack{l=1,\ldots,M,\\ l\neq i}} \tilde{q}^1_{il} + \sum_{j=1}^{N} \bar{q}^1_{ij} \leq \sum_{k=1}^{K} q^1_{ki} + \hat{q}^1_i, \tag{8}$$

$$\sum_{k=1}^{K} q^1_{ki} + \hat{q}^1_i \leq C^1_i, \tag{9}$$

$$\sum_{i=1}^{M} \bar{q}^1_{ij} \leq Q^1_j, \quad \forall j = 1, \ldots, N, \tag{10}$$

$$q^1_{ki},\ \hat{q}^1_i,\ \tilde{q}^1_{il},\ \bar{q}^1_{ij} \geq 0, \quad \forall k = 1, \ldots, K;\ \forall l = 1, \ldots, M,\ l \neq i;\ \forall j = 1, \ldots, N. \tag{11}$$

Constraint (8) ensures that the sum of emergencies dispatched by i to all hospitals l plus the sum of emergencies dispatched by i to all hospitals j is not grater than the number of emergencies reaching i from all geographical areas.

Constraint (9) ensures that the sum of emergency calls from all geographical areas plus the transferred emergencies from all others hospitals to hospital i is not greater than the maximum capacity of hospital i.

Constraints (10) are shared constraints and ensure that the sum of transferred emergencies from all hospitals i, $i = 1, \ldots, M$ to a hospital with no first aid department is not greater than the maximum capacity of the latter.

Constraints (11) are non-negative constraints.

The last term of objective function (7) represents the expected value of the loss to hospital i in stage 2. This loss depends also on the unmet demand from all geographical areas, that is

$$d_i^{2\omega} - \sum_{k=1}^{K} q_{ki}^1 - \hat{q}_i^1 - \sum_{k=1}^{K} q_{ki}^{2\omega} - \hat{q}_i^{2\omega}.$$

We have:

$$\mathbb{E}_\Omega[T_i^2(Q^2, \omega)] = \sum_{\omega \in \Omega} p_\omega [T_i^2(Q^2, \omega)],$$

where the loss for hospital i in stage 2 is the solution to the following second stage minimization problem:

$$\begin{aligned}
\text{Minimize } T_i^2(Q^2, \omega) = & \sum_{k=1}^{K} t_{ki}^1(q_{ki}^{2\omega}) \\
& + t_i^{2\omega}\left(q_i^{2\omega}, \hat{q}_i^{2\omega}\right) + \sum_{\substack{l=1,\ldots,M,\\ l \neq i}} \tilde{t}_{il}^1(\tilde{q}^{2\omega}) + \sum_{j=1}^{N} \bar{t}_{ij}^{2\omega}(\bar{q}^{2\omega}) \\
& + \alpha_i \left(\sum_{\substack{l=1,\ldots,M,\\ l \neq i}} \tilde{c}_{il}^1(\tilde{q}_{il}^{2\omega}) + \sum_{j=1}^{N} \bar{c}_{ij}^{2\omega}(\bar{q}_{ij}^{2\omega}) \right) \\
& + \beta_i \left[d_i^{2\omega} - \sum_{k=1}^{K} q_{ki}^1 - \hat{q}_i^1 - \sum_{k=1}^{K} q_{ki}^{2\omega} - \hat{q}_i^{2\omega} \right]
\end{aligned} \tag{12}$$

subject to the following constraints:

$$\sum_{\substack{l=1,\dots,M,\\ l\neq i}} \tilde{q}_{il}^{2\omega} + \sum_{j=1}^{N} \bar{q}_{ij}^{2\omega} \leq \sum_{k=1}^{K} q_{ki}^{2\omega} + \hat{q}_i^{2\omega}, \quad \forall \omega \in \Omega \tag{13}$$

$$\sum_{k=1}^{K} q_{ki}^{2\omega} + \hat{q}_i^{2\omega} \leq C_i^{2\omega}, \quad \forall \omega \in \Omega, \tag{14}$$

$$\sum_{i=1}^{M} \bar{q}_{ij}^{2\omega} \leq Q_j^{2\omega}, \quad \forall j = 1,\dots,N, \ \forall \omega \in \Omega, \tag{15}$$

$$\begin{aligned} q_{ki}^{2\omega}, \ \hat{q}_i^{2\omega}, \ \tilde{q}_{il}^{2\omega}, \ \bar{q}_{ij}^{2\omega} \geq 0, \forall k = 1,\dots,K; \ \forall l = 1,\dots,M, \ l \neq i; \\ \forall j = 1,\dots,N; \ \forall \omega \in \Omega. \end{aligned} \tag{16}$$

In stage 2, when the severity of a disaster is declared, each hospital i carries out restorative actions, to complete its first stage. Therefore, each hospital i seeks to minimize the total dispatching time and the total transport costs and the damage due to the unmet demand. The disaster could cause a shortage of staff, a greater demand for medical care and, therefore, a consequent decrease in the number of places available at each hospital, both with and without emergency rooms. Therefore, similarly to stage 1, in the second stage constraints (13)–(16) must be satisfied.

Following [10] and the standard stochastic optimization theory, the two optimization problems of stage 1 and stage 2 can be solved as a unique minimization problem, namely (cf. Tables 1 and 2 for a detailed explanation of the role of each variable and parameter):

$$\begin{aligned} \text{Min} \Bigg\{ & \sum_{k=1}^{K} t_{ki}^1(q_{ki}^1) + t_i^1\left(q_i^1, \hat{q}_i^1\right) + \sum_{\substack{l=1,\dots,M,\\ l\neq i}} \tilde{t}_{il}^1(\tilde{q}^1) + \sum_{j=1}^{N} \bar{t}_{ij}^1(\bar{q}^1) \\ & +\alpha_i \left(\sum_{\substack{l=1,\dots,M,\\ l\neq i}} \tilde{c}_{il}^1(\tilde{q}_{il}^1) + \sum_{j=1}^{N} \bar{c}_{ij}^1(\bar{q}_{ij}^1) \right) + \sum_{\omega \in \Omega} p_\omega \Bigg[\sum_{k=1}^{K} t_{ki}^1(q_{ki}^{2\omega}) + t_i^{2\omega}\left(q_i^{2\omega}, \hat{q}_i^{2\omega}\right) \\ & + \sum_{\substack{l=1,\dots,M,\\ l\neq i}} \tilde{t}_{il}^1(\tilde{q}^{2\omega}) + \sum_{j=1}^{N} \bar{t}_{ij}^{2\omega}(\bar{q}^{2\omega}) + \alpha_i \left(\sum_{\substack{l=1,\dots,M,\\ l\neq i}} \tilde{c}_{il}^1(\tilde{q}_{il}^{2\omega}) + \sum_{j=1}^{N} \bar{c}_{ij}^{2\omega}(\bar{q}_{ij}^{2\omega}) \right) \\ & +\beta_i \left(d_i^{2\omega} - \sum_{k=1}^{K} q_{ki}^1 - \hat{q}_i^1 - \sum_{k=1}^{K} q_{ki}^{2\omega} - \hat{q}_i^{2\omega} \right) \Bigg] \Bigg\} \end{aligned} \tag{17}$$

subject to constraints (8)–(11) and (13)–(16).

We define the feasible set of hospital i as follows:

$$\mathcal{K}_i = \Big\{ q \in \mathbb{R}_+^{MK+M+M(M-1)+MN+|\Omega|(MK+M+M(M-1)+MN)} \text{ such that: } (8), (9), (11), (13), (14), (16) \text{ hold} \Big\}, \tag{18}$$

and $\mathcal{K}^1 = \prod_{i=1}^{M} \mathcal{K}_i$. Moreover, let $\mathcal{S}$ be the set of shared constraints, namely $\mathcal{S} = \{q : (10) \text{ and } (15) \text{ hold}\}$. Finally, we define the feasible set $\mathcal{K}^2 = \mathcal{K}^1 \cap \mathcal{S}$.

The objective function (17) represents the expected disutility of hospital i.

We assume that, for each hospital i, the time and cost functions are convex and continuously differentiable. We have the following definition.

Definition 1 (Generalized Nash Equilibrium) A strategy profile $q^* \in \mathcal{K}^2$ is a Stochastic Generalized Nash Equilibrium if, for each hospital i, $i = 1, \dots, M$:

$$\begin{aligned} & E(DU_i(q_i^{1*}, \hat{q}_i^{1*}, \tilde{q}_i^{1*}, \bar{q}_i^{1*}, q_i^{2*}, \hat{q}_i^{2*}, \tilde{q}_i^{2*}, \bar{q}_i^{2*}, q_{-i}^{1*}, \hat{q}_{-i}^{1*}, \tilde{q}_{-i}^{1*}, \bar{q}_{-i}^{1*}, q_{-i}^{2*}, \hat{q}_{-i}^{2*}, \tilde{q}_{-i}^{2*}, \bar{q}_{-i}^{2*})) \\ & \leq E(DU_i(q_i^{1}, \hat{q}_i^{1}, \tilde{q}_i^{1}, \bar{q}^{1}, q_i^{2}, \hat{q}_i^{2}, \tilde{q}_i^{2}, \bar{q}_i^{2}, q_{-i}^{1*}, \hat{q}_{-i}^{1*}, \tilde{q}_{-i}^{1*}, \bar{q}_{-i}^{1*}, q_{-i}^{2*}, \hat{q}_{-i}^{2*}, \tilde{q}_{-i}^{2*}, \bar{q}_{-i}^{2*})), \\ & \qquad \forall (q_i^1, \hat{q}_i^1, \tilde{q}_i^1, \bar{q}_i^1, q_i^2, \hat{q}_i^2, \tilde{q}_i^2, \bar{q}_i^2) \in \mathcal{K}_i \cap \mathcal{S}, \end{aligned} \tag{19}$$

where

$$q_{-i}^{1*} = (q_1^{1*}, \dots, q_{i-1}^{1*}, q_{i+1}^{1*}, \dots, q_M^{1*}), \quad q_{-i}^{2*} = (q_1^{2*}, \dots, q_{i-1}^{2*}, q_{i+1}^{2*}, \dots, q_M^{2*}),$$

$$\hat{q}_{-i}^{1*} = (\hat{q}_1^{1*}, \dots, \hat{q}_{i-1}^{1*}, \hat{q}_{i+1}^{1*}, \dots, \hat{q}_M^{1*}), \quad \hat{q}_{-i}^{2*} = (\hat{q}_1^{2*}, \dots, \hat{q}_{i-1}^{2*}, \hat{q}_{i+1}^{2*}, \dots, \hat{q}_M^{2*}),$$

$$\tilde{q}_{-i}^{1*} = (\tilde{q}_1^{1*}, \dots, \tilde{q}_{i-1}^{1*}, \tilde{q}_{i+1}^{1*}, \dots, \tilde{q}_M^{1*}), \quad \tilde{q}_{-i}^{2*} = (\tilde{q}_1^{2*}, \dots, \tilde{q}_{i-1}^{2*}, \tilde{q}_{i+1}^{2*}, \dots, \tilde{q}_M^{2*}),$$

$$\bar{q}_{-i}^{1*} = (\bar{q}_1^{1*}, \dots, \bar{q}_{i-1}^{1*}, \bar{q}_{i+1}^{1*}, \dots, \bar{q}_M^{1*}), \quad \bar{q}_{-i}^{2*} = (\bar{q}_1^{2*}, \dots, \bar{q}_{i-1}^{2*}, \bar{q}_{i+1}^{2*}, \dots, \bar{q}_M^{2*}).$$

Each hospital seeks to minimize its expected disutility, that depends not only on its own decisions, but also on the strategies of the other players. According to the above definition, hospitals will be in a state of equilibrium if no player can unilaterally change his strategy without obtaining a greater disutility. Moreover, the presence of shared constraints provides an interconnection among feasible sets of players.

This formulation provides a model based on a Generalized Nash Equilibrium (see, for instance, [3]). In general, Generalized Nash Equilibrium problems can be formulated through quasi-variational inequality problems (see [6]). However, a class of Generalized Nash Equilibria, the Variational Equilibria, can be formulated as a variational inequality problem (see, for instance, [8] and [11]). As in [10], we will deal with the variational equilibrium of the model.

Definition 2 (Variational Equilibrium) A strategy vector $q^* \in \mathcal{K}^2$ is a Variational Equilibrium of the above Stochastic Generalized Nash Equilibrium problem if $q^* \in \mathcal{K}^2$ is a solution to the variational inequality:

$$
\begin{aligned}
&\sum_{k=1}^{N}\sum_{i=1}^{M}\left[\frac{\partial t_{ki}^{1}(q_{ki}^{1*})}{\partial q_{ki}^{1}}+\frac{\partial t_{i}^{1}\left(q_{i}^{1*},\hat{q}_{i}^{1*}\right)}{\partial q_{ki}^{1}}-\beta_i\right]\times(q_{ki}^{1}-q_{ki}^{1*})\\
&+\sum_{i=1}^{M}\left[\frac{\partial t_{i}^{1}\left(q_{i}^{1*},\hat{q}_{i}^{1*}\right)}{\partial \hat{q}_{i}^{1}}-\beta_i\right]\times(\hat{q}_{i}^{1}-\hat{q}_{i}^{1*})\\
&+\sum_{i=1}^{M}\sum_{\substack{l=1,\dots,M,\\ l\neq i}}\left[\frac{\partial \tilde{t}_{il}^{1}(\tilde{q}^{1*})}{\partial \tilde{q}_{il}^{1}}+\alpha_i\frac{\partial \tilde{c}_{il}^{1}(\tilde{q}_{il}^{1*})}{\partial \tilde{q}_{il}^{1}}\right]\times(\tilde{q}_{il}^{1}-\tilde{q}_{il}^{1*})\\
&+\sum_{i=1}^{M}\sum_{j=1}^{N}\left[\frac{\partial \bar{t}_{ij}^{1}(\bar{q}^{1*})}{\partial \bar{q}_{ij}^{1}}+\alpha_i\frac{\partial \bar{c}_{ij}^{1}(\bar{q}_{ij}^{1*})}{\partial \bar{q}_{ij}^{1}}\right]\times(\bar{q}_{ij}^{1}-\bar{q}_{ij}^{1*})\\
&+\sum_{\omega\in\Omega}p_\omega\sum_{k=1}^{N}\sum_{i=1}^{M}\left[\frac{\partial t_{ki}^{2\omega}(q_{ki}^{2\omega*})}{\partial q_{ki}^{2\omega}}+\frac{\partial t_{i}^{2\omega}\left(q_{i}^{2\omega*},\hat{q}_{i}^{2\omega*}\right)}{\partial q_{ki}^{2\omega}}-\beta_i\right]\times(q_{ki}^{2\omega}-q_{ki}^{2\omega*})\\
&+\sum_{\omega\in\Omega}p_\omega\sum_{i=1}^{N}\left[\frac{\partial t_{i}^{2\omega}\left(q_{i}^{2\omega*},\hat{q}_{i}^{2\omega*}\right)}{\partial \hat{q}_{i}^{2\omega}}-\beta_i\right]\times(\hat{q}_{i}^{2\omega}-\hat{q}_{i}^{2\omega*})\\
&+\sum_{\omega\in\Omega}p_\omega\sum_{i=1}^{M}\sum_{\substack{l=1,\dots,M,\\ l\neq i}}\left[\frac{\partial \tilde{t}_{il}^{2\omega}(\tilde{q}^{2*})}{\partial \tilde{q}_{il}^{2\omega}}+\alpha_i\frac{\partial \tilde{c}_{il}^{2\omega}(\tilde{q}_{il}^{2\omega*})}{\partial \tilde{q}_{il}^{2\omega}}\right]\times(\tilde{q}_{il}^{2\omega}-\tilde{q}_{il}^{2\omega*})\\
&+\sum_{\omega\in\Omega}p_\omega\sum_{i=1}^{M}\sum_{j=1}^{N}\left[\frac{\partial \bar{t}_{ij}^{2\omega}(\bar{q}^{2*})}{\partial \bar{q}_{ij}^{2\omega}}+\alpha_i\frac{\partial \bar{c}_{ij}^{2\omega}(\bar{q}_{ij}^{2\omega*})}{\partial \bar{q}_{ij}^{2\omega}}\right]\times(\bar{q}_{ij}^{2\omega}-\bar{q}_{ij}^{2\omega*})\geq 0 \quad \forall q\in\mathcal{K}^2.
\end{aligned}
\tag{20}
$$

The advantage of detecting a variational equilibrium consists in using the well-known variational inequality theory, for which theorems of existence and uniqueness of the solution are stated (see [7]).

3 An Illustrative Numerical Example

In this section, we solve an illustrative numerical example to validate the effectiveness of the model. We consider $g = 2$ geographical areas, $h = 2$ hospitals with first aid departments, $s = 3$ hospitals without first aid departments and $|\Omega| = 2$ scenarios. In the first scenario $\omega_1 = 1$, we suppose that the consequences of the advent of the disaster scenario are severe while in the second scenario $\omega_2 = 2$ we assume that the consequences are not severe. Consequently, in the first scenario the requests for hospitalization are more than in the second one.

For the computation of the optimal solution we have applied the modified projection method described in [10]. The calculations were performed using the MATLAB program. The algorithm was implemented on a laptop with 1.8 GHz Intel Core i5 dual-core and 8 GB RAM, 1600 MHz DDR3. For the convergence of the method a tolerance of $\epsilon = 10^{-4}$ was fixed. The method has been implemented with a constant step $\alpha = 0.1$.

The numerical data and the size of the problem are constructed for easy interpretation purposes. We have the following data:

$$p_{\omega_1} = 0.8,\ p_{\omega_2} = 1 - p_{\omega_1} = 0.2, \quad \beta_1 = \beta_2 = 20,$$
$$\Gamma_1^1 = 25,\ \Gamma_2^1 = 35, \quad \Gamma_1^{2\omega_1} = 45,\ \Gamma_2^{2\omega_1} = 60, \quad \Gamma_1^{2\omega_2} = 30,\ \Gamma_2^{2\omega_2} = 53,$$
$$\tilde{C}_1^1 = 10,\ \tilde{C}_2^1 = 15,\ \tilde{C}_3^1 = 25, \quad \tilde{C}_1^{2\omega_1} = 18,\ \tilde{C}_2^{2\omega_1} = 20,\ \tilde{C}_3^{2\omega_1} = 30,$$
$$\tilde{C}_1^{2\omega_2} = 15,\ \tilde{C}_2^{2\omega_2} = 18,\ \tilde{C}_3^{2\omega_2} = 25,$$
$$d_1^{2\omega_1} = d_2^{2\omega_1} = 80, \quad d_1^{2\omega_2} = d_2^{2\omega_2} = 50.$$

The equilibrium solution is shown in Table 3.

The computational time needed to calculate the equilibrium solution is 50 seconds. As shown in Table 3, in phase 1, where cost and time functions and demands are deterministic, there is not a huge transfer between hospitals with first aid departments and between hospitals with and without first aid departments.

Table 3 Equilibrium solution

Stage 1		Stage 2: scenario ω_1		Stage 2: scenario ω_2	
Solution	Value	Solution	Value	Solution	Value
q_{11}^{1*}	20.6	$q_{11}^{2\omega_1*}$	46.3	$q_{11}^{2\omega_2*}$	24.3
q_{12}^{1*}	3.8	$q_{12}^{2\omega_1*}$	46.3	$q_{12}^{2\omega_2*}$	21.8
q_{21}^{1*}	9.3	$q_{21}^{2\omega_1*}$	53.6	$q_{21}^{2\omega_2*}$	35.7
q_{22}^{1*}	26.1	$q_{22}^{2\omega_1*}$	53.6	$q_{22}^{2\omega_2*}$	33.2
$\hat{q}_1^{1*}$	0.5	$\hat{q}_1^{2\omega_1*}$	47.6	$\hat{q}_1^{2\omega_2*}$	16.1
$\hat{q}_2^{1*}$	0.5	$\hat{q}_2^{2\omega_1*}$	47.3	$\hat{q}_2^{2\omega_2*}$	15.9
$\tilde{q}_{12}^{1*}$	4.6	$\tilde{q}_{12}^{2\omega_1*}$	18.5	$\tilde{q}_{12}^{2\omega_2*}$	11.9
$\tilde{q}_{21}^{1*}$	7.1	$\tilde{q}_{21}^{2\omega_1*}$	22.7	$\tilde{q}_{21}^{2\omega_2*}$	20.2
$\bar{q}_{11}^{1*}$	3.9	$\bar{q}_{11}^{2\omega_1*}$	7.2	$\bar{q}_{11}^{2\omega_2*}$	5.2
$\bar{q}_{12}^{1*}$	3.4	$\bar{q}_{12}^{2\omega_1*}$	8.2	$\bar{q}_{12}^{2\omega_2*}$	6.2
$\bar{q}_{13}^{1*}$	3.8	$\bar{q}_{13}^{2\omega_1*}$	11.1	$\bar{q}_{13}^{2\omega_2*}$	6.6
$\bar{q}_{21}^{1*}$	3.9	$\bar{q}_{21}^{2\omega_1*}$	10.8	$\bar{q}_{21}^{2\omega_2*}$	9.7
$\bar{q}_{22}^{1*}$	3.5	$\bar{q}_{22}^{2\omega_1*}$	11.8	$\bar{q}_{22}^{2\omega_2*}$	11.8
$\bar{q}_{23}^{1*}$	3.3	$\bar{q}_{23}^{2\omega_1*}$	14.6	$\bar{q}_{23}^{2\omega_2*}$	11.1

When a disaster scenario occurs in stage 2, there is an increase in requests for hospitalization and a consequent increase in transfers between hospital structures. Particularly, under scenario ω_1, the severity of which is higher, hospital 1 fails to satisfy the total demand, having an unmet demand equal to 174.

4 Conclusion

In this paper, we presented a stochastic Generalized Nash Equilibrium model to describe the competition among hospitals with first aid departments for hospitalization in response to the advent of a disaster scenario. We obtained a two-stage stochastic optimization problem and the presence of shared constraints for all hospitals with first aid departments led us to consider a Generalized Nash Equilibrium problem for which we derived the Variational Equilibrium and the associated variational inequality problem. The results in this paper add to the growing literature of game theory and and two-stage stochastic models in disaster management. This theoretical model can be applied to any disastrous event that involves a sudden and nondeterministic increase in hospitalization, such as the recent and still current COVID-19 pandemic.

Acknowledgments The research was partially supported by the research project "Programma ricerca di ateneo UNICT 2020-22 linea 2 (OMNIA)" University of Catania. This support is gratefully acknowledged.

References

1. Daniele, P.: Dynamic Networks and Evolutionary Variational Inequalities. Edward Elgar Publishing, Cheltenham, UK (2006)
2. Daniele, P., Sciacca, D.: A dynamic supply chain network for PPE during the Covid-19 pandemic. J. Appl. Numer. Optim. **3**, 403–424 (2021)
3. Debreu, G.: A social equilibrium existence theorem. Proc. Natl. Acad. Sci. **38**(10), 886–893 (1952)
4. Fargetta, G., Scrimali, L.: A Two-Stage Variational Inequality for Medical Supply in Emergency Management. In: Cerulli, R., Dell'Amico, M., Guerriero, F., Pacciarelli, D., Sforza, A. (eds) Optimization and Decision Science. AIRO Springer Series, **7**, Springer, Cham (2021). https://doi.org/10.1007/978-3-030-86841-3_8
5. Fargetta, G., Scrimali, L.: Optimal Emergency Evacuation with Uncertainty. In: Parasidis, I.N., Providas, E., Rassias, T.M. (eds) Mathematical Analysis in Interdisciplinary Research. Springer Optimization and Its Applications. **179**, Springer, Cham (2021). https://doi.org/10.1007/978-3-030-84721-0_14
6. Fischer, A., Herrich, M., Schonefeld, K.: Generalized Nash equilibrium problems. Recent advances and challenges. Pesquisa Operacional **34**(3), 521–558 (2014)
7. Nagurney, A.: Network Economics: A Variational Inequality Approach. Kluwer Academic Publishers, Boston, MA (1993)

8. Nagurney, A., Yu, M., Besik, D.: Supply chain network capacity competition with outsourcing: A variational equilibrium framework. J. Global Optim. **69**(1), 231–254 (2017)
9. Nagurney, A, Nagurney, L.S.: A mean-variance disaster relief supply chain network model for risk reduction with stochastic link costs, time targets, and demand uncertainty. In: Kotsireas, I.S., Nagurney, A., Pardalos, P.M. (eds.) Dynamics of Disasters: Key Concepts, Models, Algorithms, and Insights. Springer International Publishing, Switzerland (2016)
10. Nagurney, A., Salarpour, M.: A multicountry, Multicommodity Stochastic Game Theory Network Model of Competiotion for Medical Supplies Inspired by the Covid-19 Pandemic. Int. J. Prod. Econ. **235**, 108074 (2021)
11. Nagurney, A., Salarpour, M., Daniele, P.: An integrated financial and logistical game theory model for humanitarian organizations with purchasing costs, multiple freight service providers, and budget, capacity, and demand constraints. Int. J. Prod. Econ. **212**, 212–226 (2019)

On Nested Affine Variational Inequalities: The Case of Multi-Portfolio Selection

Lorenzo Lampariello, Gianluca Priori, and Simone Sagratella

Abstract We deal with nested affine variational inequalities, i.e., hierarchical problems involving an affine (upper-level) variational inequality whose feasible set is the solution set of another affine (lower-level) variational inequality. We apply this modeling tool to the multi-portfolio selection problem, where the lower-level variational inequality models the Nash equilibrium problem made up by the different accounts, while the upper-level variational inequality is instrumental to perform a selection over this equilibrium set. We propose a projected averaging Tikhonov-like algorithm for the solution of this problem, which only requires the monotonicity of the variational inequalities for both the upper- and the lower-level in order to converge. Finally, we provide complexity properties.

Keywords Multi-portfolio selection · Nested variational inequality · Purely hierarchical problem · Tikhonov method · Complexity analysis

1 Introduction: Context and Motivation for the Nested Affine Variational Inequalities Model

Nested affine variational inequalities represent a flexible modeling tool for many real-world applications like, e.g., the renowned multi-portfolio selection (see, e.g. [5]). To introduce the general formulation of the model, we first briefly describe the specific instance of the multi-portfolio optimization problem.

Consider N accounts, with $\nu = 1, \ldots, N$. Each account ν's budget $b^\nu \in \mathbb{R}_+$ is invested in K assets of a market. The decision variables $y^\nu \in Y_\nu \subseteq R^K$ stand for the

L. Lampariello
Department of Business Economics, Roma Tre University, Rome, Italy
e-mail: lorenzo.lampariello@uniroma3.it

G. Priori (✉) · S. Sagratella
Department of Computer, Control and Management Engineering "A. Ruberti", Sapienza University of Rome, Rome, Italy
e-mail: priori@diag.uniroma1.it; sagratella@diag.uniroma1.it

L. Amorosi et al. (eds.), *Optimization in Artificial Intelligence and Data Sciences*, AIRO Springer Series 8, https://doi.org/10.1007/978-3-030-95380-5_3

fractions of b^ν invested in each asset, where Y_ν is a nonempty compact polyhedron containing the feasible portfolios, e.g., the standard simplex. Let $r \in \mathbb{R}^K$ indicate random variables, where r_k is the return on asset $k \in \{1, \ldots, K\}$ over a single-period investment. We define $\mu^\nu = \mathbb{E}^\nu(r) \in \mathbb{R}^K$ as expectations of the assets' returns for ν, as well as the positive semidefinite covariance matrix $\Sigma^\nu = \mathbb{E}^\nu((r-\mu^\nu)(r-\mu^\nu)^\top)$. We consider the following measures for portfolio income I_ν and risk R_ν, where we use the portfolio variance as the risk measure: $I_\nu(y^\nu) \triangleq b^\nu(\mu^\nu)^\top y^\nu$, $R_\nu(y^\nu) \triangleq \frac{1}{2}(b^\nu)^2(y^\nu)^\top \Sigma^\nu y^\nu$.

When trades from multiple accounts are pooled for common execution, individual accounts can suffer the market impact that stems from a lack of liquidity. To take account of this transaction cost effect, we introduce a positive semidefinite market impact matrix $\Omega^\nu \in \mathbb{R}^{K\times K}$ whose entry at position (i, j) is the impact of the liquidity of asset i on the liquidity of asset j. For each account ν we consider a linear market impact unitary cost function. The total transaction costs term for ν is:

$$TC_\nu(y^1, \ldots, y^N) \triangleq \underbrace{b^\nu(y^\nu)^\top}_{\text{Invested capital}} \quad \underbrace{\Omega^\nu \sum_{\lambda=1}^{N} b^\lambda y^\lambda}_{\text{Unitary transaction costs}} .$$

The multi-portfolio problem can be formulated as the following Affine Variational Inequality AVI$(M^{\text{low}}, d^{\text{low}}, Y)$: find $y \in Y \triangleq Y_1 \times \cdots \times Y_N$ such that

$$\left(M^{\text{low}} y + d^{\text{low}}\right)^\top (w - y) \geq 0 \quad \forall w \in Y, \tag{1}$$

where $d^{\text{low}} \triangleq -b^\nu \mu^\nu$ and

$$M^{\text{low}} \triangleq \begin{pmatrix} (b^1)^2[\rho^1\Sigma^1 + \Omega^1 + \Omega^{1^\top}] & b^1 b^2 \Omega^1 & \cdots & b^1 b^N \Omega^1 \\ b^2 b^1 \Omega^2 & (b^2)^2[\rho^2\Sigma^2 + \Omega^2 + \Omega^{2^\top}] & & b^2 b^N \Omega^2 \\ \vdots & & \ddots & \\ b^N b^1 \Omega^N & b^N b^2 \Omega^N & & (b^N)^2[\rho^N\Sigma^N + \Omega^N + \Omega^{N^\top}] \end{pmatrix}.$$

We assume the matrix M^{low} to be positive semidefinite and, in turn, AVI$(M^{\text{low}}, d^{\text{low}}, Y)$ to be monotone: these properties can be guaranteed under mild assumptions, see [5, Theorem 3.3]. We denote by SOL$(M^{\text{low}}, d^{\text{low}}, Y)$ the solution set of AVI$(M^{\text{low}}, d^{\text{low}}, Y)$, which is a polyhedron (see [5, Theorem 2.4.13]). Note that AVI$(M^{\text{low}}, d^{\text{low}}, Y)$ corresponds to an equivalent Nash Equilibrium Problem (NEP), where the players' objective functions are convex and quadratic. Since the set SOL$(M^{\text{low}}, d^{\text{low}}, Y)$ is not necessarily a singleton in the framework we consider, one has to discriminate among the solutions of AVI$(M^{\text{low}}, d^{\text{low}}, Y)$ according to some further upper level criterion. Thus, to model the resulting selection problem, we introduce the monotone nested affine variational inequality AVI$\big(M^{\text{up}}, d^{\text{up}}, \text{SOL}(M^{\text{low}}, d^{\text{low}}, Y)\big)$, that is the problem of calculating $y \in$

$\mathrm{SOL}(M^{\mathrm{low}}, d^{\mathrm{low}}, Y)$ that solves

$$\left(M^{\mathrm{up}} y + d^{\mathrm{up}}\right)^\top (w - y) \geq 0, \quad \forall\, w \in \mathrm{SOL}(M^{\mathrm{low}}, d^{\mathrm{low}}, Y), \tag{2}$$

where $\mathbb{R}^{NK \times NK} \ni M^{\mathrm{up}} \succeq 0$ and $d^{\mathrm{up}} \in \mathbb{R}^{NK}$. Problem (2), which has a hierarchical structure, includes as a special instance the minimization of the convex quadratic objective function $\frac{1}{2} y^\top M^{\mathrm{up}} y + {d^{\mathrm{up}}}^\top y$, where M^{up} is symmetric, over $\mathrm{SOL}(M^{\mathrm{low}}, d^{\mathrm{low}}, Y)$. It is also worth mentioning the special instance where the N accounts form an upper-level (jointly convex) NEP to select over $\mathrm{SOL}(M^{\mathrm{low}}, d^{\mathrm{low}}, Y)$; in this case, M^{up} turns out to be nonsymmetric. We refer the reader to [1] for further information about NEPs.

Remark Convergent solution procedures have been devised in the literature (see, e.g., [3, 4]) to address monotone nested AVIs when M^{up} is positive semidefinte *plus*, i.e. M^{up} is positive semidefinite and $y^\top M^{\mathrm{up}} y = 0 \Rightarrow M^{\mathrm{up}} y = 0$ (see, [2, Ex. 2.9.24]). Requiring M^{up} to be positive semidefinite plus is restrictive: for example, taking $NK = 2$, any matrix

$$M^{\mathrm{up}} = \begin{pmatrix} m_1 & 2\sqrt{m_1 m_2} + m_3 \\ -m_3 & m_2 \end{pmatrix}$$

with m_1, m_2 nonnegative scalars and $m_3 \neq -\sqrt{m_1 m_2}$, is positive semidefinite but not positive semidefinite plus. Actually, the class of semidefinite plus matrices is "slightly" larger than the ones of symmetric positive semidefinite and positive definite matrices.

Recently, a projected averaging Tikhonov-like algorithm has been proposed in [6] to cope with monotone nested VIs allowing for matrices M^{up} that are not required to be positive semidefinite plus.

We present a solution method for problem (2). We apply the results presented in [6] to the specific instance of monotone nested affine variational inequalities, taking full advantage of some strong properties AVIs enjoy, such as error bound results. This allows us to put forward an algorithm to address problems like the multi-portfolio selection in a more general framework with respect to the literature, where the upper level operator is invariably assumed to be monotone plus (see, e.g., [5]).

2 The Tikhonov Approach

We require the following mild conditions to hold:

(A1) M^{up} is positive semidefinite;
(A2) M^{low} is positive semidefinite;
(A3) Y is nonempty and compact.

The set $\text{SOL}(M^{\text{low}}, d^{\text{low}}, Y)$ is nonempty, convex, compact and not necessarily single-valued, due to (A2) and (A3), see e.g. [2, Section 2.3]. It follows that the feasible set of the nested affine variational inequality (2) is not a singleton. Moreover, thanks to (A1), the solution set of (2) can include multiple points.

Let us introduce the Tikhonov operator:

$$\Phi_\tau(y) \triangleq \left(M^{\text{low}} y + d^{\text{low}}\right) + \frac{1}{\tau}\left(M^{\text{up}} y + d^{\text{up}}\right).$$

For any $\tau > 0$, by assumptions (A1) and (A2), Φ_τ is monotone and affine.

The following finite quantities will be useful in the forthcoming analysis:

$$H \triangleq \max_{y \in Y} \|M^{\text{up}} y + d^{\text{up}}\|_2, \quad R \triangleq \max_{y \in Y} \|M^{\text{low}} y + d^{\text{low}}\|_2, \quad D \triangleq \max_{v, y \in Y} \|v - y\|_2.$$

We propose a Linear version of the Projected Averaging Tikhonov Algorithm (L-PATA) to compute solutions of (2).

Algorithm 1: Linear version of the Projected Averaging Tikhonov Algorithm (L-PATA)

Data: $w^1 = z^1 = y^1 \in Y, i \leftarrow 1, l \leftarrow 0$;

for $k = 1, 2, \ldots$ **do**

(S.1) $\varepsilon^k = i^{-2}, \tau^k = i$;

(S.2) $y^{k+1} = P_Y\left(y^k - \frac{1}{2(k-l)^{0.5}} \Phi_{\tau^k}(y^k)\right)$;

(S.3) $z^{k+1} = \dfrac{\sum_{j=l+1}^{k+1} \frac{1}{2(j-l)^{0.5}} y^j}{\sum_{j=l+1}^{k+1} \frac{1}{2(j-l)^{0.5}}}$;

(S.4) **if** $\min_{y \in Y} \Phi_{\tau^k}(z^{k+1})^\top (y - z^{k+1}) \geq -\varepsilon^k$ **then**

  $w^{i+1} = z^{k+1}, i = i + 1, l = k$;

  end

end

Index i refers to the outer iterations occurring as the condition in step (S.4) is verified, which correspond to the (approximate) solutions w^{i+1} of the AVI subproblems

$$\Phi_\tau(y)^\top(w-y) \geq -\varepsilon_{\text{sub}}, \quad \forall\, w \in Y, \tag{3}$$

with $\varepsilon_{\text{sub}} = i^{-2}$ and $\tau = i$. The sequence $\{y^k\}$ includes all the points obtained by making classical projection steps with the given diminishing stepsize rule, see step (S.2). The sequence $\{z^k\}$ consists of the inner iterations needed to compute (approximate) solutions of the AVI subproblem (3), and it is obtained by performing a weighted average on the points y^j, see step (S.3). Index l lets the sequence of the stepsizes restart at every outer iteration, while considering solely the points y^j belonging to the current subproblem for the computation of z^{k+1}. We remark that the condition in step (S.4) only requires the solution of a linear problem.

We now deal with the convergence properties of L-PATA. With the following result we relate (approximate) solutions of the AVI subproblem (3) where $\varepsilon_{\text{sub}} \geq 0$ to approximate solutions of problem (2).

Proposition 1 *Assume conditions (A1)–(A3) to hold, and let $y \in Y$ satisfy (3) with $\tau > 0$ and $\varepsilon_{sub} \geq 0$. It holds that*

$$\left(M^{up} y + d^{up}\right)^\top (w-y) \geq -\varepsilon_{up}, \quad \forall w \in SOL(M^{low}, d^{low}, Y), \tag{4}$$

with $\varepsilon_{up} = \varepsilon_{sub}\tau$, and

$$\left(M^{low} y + d^{low}\right)^\top (w-y) \geq -\varepsilon_{low}, \quad \forall w \in Y, \tag{5}$$

with $\varepsilon_{low} = \varepsilon_{sub} + \frac{1}{\tau} H D$.

Proof We have for all $w \in \text{SOL}(M^{\text{low}}, d^{\text{low}}, Y)$:

$$\begin{aligned}
-\varepsilon_{\text{sub}}\tau &\leq \left[\tau\left(M^{\text{low}} y + d^{\text{low}}\right) + \left(M^{\text{up}} y + d^{\text{up}}\right)\right]^\top (w-y) \\
&\leq \left[\tau\left(M^{\text{low}} w + d^{\text{low}}\right) + \left(M^{\text{up}} y + d^{\text{up}}\right)\right]^\top (w-y) \\
&\leq \left(M^{\text{up}} y + d^{\text{up}}\right)^\top (w-y),
\end{aligned}$$

where the first inequality is due to (3), the second one comes from (A2), and the last one is true because $y \in Y$ and then $\left(M^{\text{low}} w + d^{\text{low}}\right)^\top (y-w) \geq 0$. Hence, (4) is true.

Moreover, we have for all $w \in Y$:

$$\left(M^{\text{low}} y + d^{\text{low}}\right)^\top (w - y) = \Phi_\tau(y)^\top (w - y) - \frac{1}{\tau}\left(M^{\text{up}} y + d^{\text{up}}\right)^\top (w - y)$$
$$\geq -\varepsilon_{\text{sub}} - \frac{1}{\tau} H D,$$

where the inequality is due to (3). Therefore, we get (5). □

Here follows the convergence result for L-PATA.

Theorem 1 *Assume conditions (A1)–(A3) to hold. Every limit point of the sequence $\{w^i\}$ generated by L-PATA is a solution of problem* (2).

Proof First of all, we show that $i \to \infty$. Assume by contradiction that this is false, hence an index $\bar{k}$ exists such that either $\bar{k} = 0$ or the condition in step (S.4) is satisfied at the iteration $\bar{k} - 1$, and the condition in step (S.4) is violated for every $k \geq \bar{k}$. In this case, it is true that $i \to \bar{\imath}$, and then $\tau^k = \bar{\tau} \triangleq \bar{\imath}$ for every $k \geq \bar{k}$.

For every $j \in [\bar{k}, k]$, and for any $v \in Y$, we have

$$\begin{aligned}
\|y^{j+1} - v\|_2^2 &= \|P_Y(y^j - \tfrac{1}{2(j-\bar{k}+1)^{0.5}} \Phi_{\bar{\tau}}(y^j)) - v\|_2^2 \\
&\leq \|y^j - \tfrac{1}{2(j-\bar{k}+1)^{0.5}} \Phi_{\bar{\tau}}(y^j) - v\|_2^2 \\
&= \|y^j - v\|_2^2 + \tfrac{1}{4(j-\bar{k}+1)} \|\Phi_{\bar{\tau}}(y^j)\|_2^2 - \tfrac{1}{(j-\bar{k}+1)^{0.5}} \Phi_{\bar{\tau}}(y^j)^\top (y^j - v),
\end{aligned}$$

and, in turn,

$$\Phi_{\bar{\tau}}(y^j)^\top (v - y^j) \geq \frac{\|y^{j+1} - v\|_2^2 - \|y^j - v\|_2^2}{(j - \bar{k} + 1)^{-0.5}} - \frac{1}{4(j - \bar{k} + 1)^{0.5}} \|\Phi_{\bar{\tau}}(y^j)\|_2^2.$$

Summing, we get

$$\begin{aligned}
\frac{\displaystyle\sum_{j=\bar{k}}^{k} \frac{1}{2(j - \bar{k} + 1)^{0.5}} \Phi_{\bar{\tau}}(y^j)^\top (v - y^j)}{\displaystyle\sum_{j=\bar{k}}^{k} \frac{1}{2(j - \bar{k} + 1)^{0.5}}} &\geq \frac{\displaystyle\sum_{j=\bar{k}}^{k} \left(\|y^{j+1} - v\|_2^2 - \|y^j - v\|_2^2 - \frac{1}{4(j - \bar{k} + 1)} \|\Phi_{\bar{\tau}}(y^j)\|_2^2\right)}{\displaystyle 2\sum_{j=\bar{k}}^{k} \frac{1}{2(j - \bar{k} + 1)^{0.5}}} \\
&= \frac{\left(\|y^{k+1} - v\|_2^2 - \|y^{\bar{k}} - v\|_2^2 - \displaystyle\sum_{j=\bar{k}}^{k} \frac{1}{4(j - \bar{k} + 1)} \|\Phi_{\bar{\tau}}(y^j)\|_2^2\right)}{\displaystyle 2\sum_{j=\bar{k}}^{k} \frac{1}{2(j - \bar{k} + 1)^{0.5}}} \\
&\geq -\frac{\left(\|y^{\bar{k}} - v\|_2^2 + \displaystyle\sum_{j=\bar{k}}^{k} \frac{1}{4(j - \bar{k} + 1)} \|\Phi_{\bar{\tau}}(y^j)\|_2^2\right)}{\displaystyle 2\sum_{j=\bar{k}}^{k} \frac{1}{2(j - \bar{k} + 1)^{0.5}}},
\end{aligned} \tag{6}$$

which implies

$$\begin{aligned}
\Phi_{\bar{\tau}}(v)^\top(v - z^k) &= \frac{\displaystyle\sum_{j=\bar{k}}^{k} \frac{1}{2(j-\bar{k}+1)^{0.5}} \Phi_{\bar{\tau}}(v)^\top(v - y^j)}{\displaystyle\sum_{j=\bar{k}}^{k} \frac{1}{2(j-\bar{k}+1)^{0.5}}} \\
&\geq -\frac{\left(\|y^{\bar{k}} - v\|_2^2 + \displaystyle\sum_{j=\bar{k}}^{k} \frac{1}{4(j-\bar{k}+1)} \|\Phi_{\bar{\tau}}(y^j)\|_2^2\right)}{2\displaystyle\sum_{j=\bar{k}}^{k} \frac{1}{2(j-\bar{k}+1)^{0.5}}} \\
&\quad + \frac{\displaystyle\sum_{j=\bar{k}}^{k} \frac{1}{2(j-\bar{k}+1)^{0.5}} (\Phi_{\bar{\tau}}(v) - \Phi_{\bar{\tau}}(y^j))^\top(v - y^j)}{\displaystyle\sum_{j=\bar{k}}^{k} \frac{1}{2(j-\bar{k}+1)^{0.5}}} \\
&\geq -\frac{\left(\|y^{\bar{k}} - v\|_2^2 + \displaystyle\sum_{j=\bar{k}}^{k} \frac{1}{4(j-\bar{k}+1)} \|\Phi_{\bar{\tau}}(y^j)\|_2^2\right)}{2\displaystyle\sum_{j=\bar{k}}^{k} \frac{1}{2(j-\bar{k}+1)^{0.5}}},
\end{aligned} \tag{7}$$

where the last inequality holds thanks to the monotonicity of $\Phi_{\bar{\tau}}$. Indicating by $z \in Y$ any limit point of the sequence $\{z^k\}$, taking the limit $k \to \infty$ in the latter relation and subsequencing, the following inequality holds true:

$$\Phi_{\bar{\tau}}(v)^\top(v - z) \geq -\frac{\left(\|y^{\bar{k}} - v\|_2^2 + \displaystyle\sum_{j=\bar{k}}^{\infty} \frac{1}{4(j-\bar{k})} \|\Phi_{\bar{\tau}}(y^j)\|_2^2\right)}{2\displaystyle\sum_{j=\bar{k}}^{\infty} \frac{1}{2(j-\bar{k})^{0.5}}} = 0,$$

because $\sum_{j=\bar{k}}^{\infty} \frac{1}{2(j-\bar{k})^{0.5}} = +\infty$ and $\left(\sum_{j=\bar{k}}^{\infty} \frac{1}{4(j-\bar{k})}\right) / \left(\sum_{j=\bar{k}}^{\infty} \frac{1}{2(j-\bar{k})^{0.5}}\right) = 0$, due to [6, Proposition 4], and then z is a solution of the dual problem

$$\Phi_{\bar{\tau}}(v)^\top(v - z) \geq 0, \quad \forall v \in Y.$$

Hence, the sequence $\{z^k\}$ converges to a solution of problem (3) with $\varepsilon_{\text{sub}} = 0$ and $\tau = \bar{\tau}$, see e.g. [2, Theorem 2.3.5], in contradiction to $\min_{y \in Y} \Phi_{\bar{\tau}}(z^{k+1})^\top (y - z^{k+1}) < -\varepsilon^k = -\bar{i}^{-2}$ for every $k \geq \bar{k}$. Therefore we can say that $i \to \infty$.

Consequently, the algorithm produces an infinite sequence $\{w^i\}$ such that $w^{i+1} \in Y$ and

$$\Phi_i(w^{i+1})^\top (y - w^{i+1}) \geq -i^{-2}, \quad \forall\, y \in Y,$$

that is (3) holds at w^{i+1} with $\varepsilon_{\text{sub}} = i^{-2}$ and $\tau = i$. By Proposition 1, specifically from (4) and (5), we obtain

$$\left(M^{\text{up}} w^{i+1} + d^{\text{up}}\right)^\top (y - w^{i+1}) \geq -i^{-1}, \quad \forall y \in \text{SOL}(M^{\text{low}}, d^{\text{low}}, Y),$$

and

$$\left(M^{\text{low}} w^{i+1} + d^{\text{low}}\right)^\top (y - w^{i+1}) \geq -i^{-1}(1 + HD), \quad \forall y \in Y.$$

Taking the limit $i \to \infty$ we get the desired convergence property for every limit point of $\{w^i\}$. □

We consider the natural residual map for the lower-level AVI$(M^{\text{low}}, d^{\text{low}}, Y)$

$$V(y) \triangleq \|P_Y(y - (M^{\text{low}} y + d^{\text{low}})) - y\|_2. \tag{8}$$

Function V is continuous and nonnegative, as reminded in [4]. Also, $V(y) = 0$ if and only if $y \in \text{SOL}(M^{\text{low}}, d^{\text{low}}, Y)$. Condition

$$V(y) \leq \widehat{\varepsilon}_{\text{low}}, \tag{9}$$

with $\widehat{\varepsilon}_{\text{low}} \geq 0$, is alternative to (5) to take care of the feasibility of problem (2).

Remark Since both the variational inequalities (1) and (2) are affine, then ε_{up} and either ε_{low} or $\widehat{\varepsilon}_{\text{low}}$ give actual upper-bounds to the distances between y and $\text{SOL}\left(M^{\text{up}}, d^{\text{up}}, \text{SOL}(M^{\text{low}}, d^{\text{low}}, Y)\right)$ and $\text{SOL}(M^{\text{low}}, d^{\text{low}}, Y)$, respectively.

Theorem 2 *If $y \in SOL(M^{low}, d^{low}, Y)$ satisfies (4), then there exists $c_{up} > 0$ such that*

$$dist_{SOL\left(M^{up}, d^{up}, SOL(M^{low}, d^{low}, Y)\right)}(y) \leq c_{up} \varepsilon_{up}.$$

(continued)

Theorem 2 (continued)
If $y \in Y$ satisfies (5), then there exists $c_{low} > 0$ such that

$$dist_{SOL(M^{low}, d^{low}, Y)}(y) \le c_{low} \varepsilon_{low}.$$

If $y \in Y$ satisfies (9), then there exists $\widehat{c}_{low} > 0$ such that

$$dist_{SOL(M^{low}, d^{low}, Y)}(y) \le \widehat{c}_{low} \widehat{\varepsilon}_{low}.$$

Proof The claim follows from [2, Proposition 6.3.3] and [6, Proposition 3]. □

In view of Theorem 2, conditions (4) and either (5) or (9) define points that are approximate solutions for problem (2), also under a geometrical perspective. In particular, the lower the values of ε_{up} and either ε_{low} or $\widehat{\varepsilon}_{\text{low}}$, the closer the point gets to the solution set of the nested affine variational inequality (2).

We give an upper bound to the number of iterations needed to drive both the upper-level error ε_{up}, given in (4), and the lower-level error $\widehat{\varepsilon}_{\text{low}}$, given in (9), under some prescribed tolerance δ.

Theorem 3 *Assume conditions (A1)–(A3) to hold and, without loss of generality, $L_\Phi \triangleq \|M^{up} + M^{low}\|_2 < 1$. Consider L-PATA. Given a precision $\delta \in (0, 1)$, let us define the quantity*

$$I_{\max} \triangleq \left\lceil \frac{H+1}{\delta} \right\rceil.$$

Then, the upper-level approximate problem (4) is solved for $y = z^{k+1}$ with $\varepsilon_{up} = \delta$ and the lower-level approximate problem (9) is solved for $y = z^{k+1}$ with $\widehat{\varepsilon}_{low} = \delta$ and the condition in step (S.4) is satisfied in at most

$$\sigma \triangleq I_{\max} \left\lceil \max \left\{ I_{\max}^{8} \frac{(D+R)^4}{(1-L_\Phi)^2} C_1, \, I_{\max}^{\frac{8}{1-2\eta}} \frac{(D+R)^{\frac{4}{1-2\eta}}}{(1-L_\Phi)^{\frac{2}{1-2\eta}}} C_{2,\eta} \right\} \right\rceil,$$

iterations k, where $\eta > 0$ is a small number, and

$$C_1 \triangleq \left(D^2 + \frac{5}{4}(R+H)^2 \right)^2, \quad C_{2,\eta} \triangleq \left(\frac{(R+H)^2}{(4\eta)} \right)^{\frac{2}{1-2\eta}}. \tag{10}$$

Proof See the proof of [6, Theorem 2]. □

References

1. Facchinei, F., Kanzow, C.: Generalized Nash equilibrium problems. Ann. Oper. Res. **175**(1), 177–211 (2010)
2. Facchinei, F., Pang, J.S.: Finite-Dimensional Variational Inequalities and Complementarity Problems. Springer, New York (2003)
3. Facchinei, F., Pang, J.S., Scutari, G., Lampariello, L.: VI-constrained hemivariational inequalities: distributed algorithms and power control in ad-hoc networks. Math. Program. **145**(1–2), 59–96 (2014)
4. Lampariello, L., Neumann, C., Ricci, J.M., Sagratella, S., Stein, O.: An explicit Tikhonov algorithm for nested variational inequalities. Comput. Optim. Appl. **77**(2), 335–350 (2020)
5. Lampariello, L., Neumann, C., Ricci, J.M., Sagratella, S., Stein, O.: Equilibrium selection for multi-portfolio optimization. Eur. J. Oper. Res. **295**(1), 363–373 (2021)
6. Lampariello, L., Priori, G., Sagratella, S.: On the solution of monotone nested variational inequalities. Technical report, Preprint (2021)

Hyper-Parameter Optimization in Support Vector Machine on Unbalanced Datasets Using Genetic Algorithms

Rosita Guido, Maria Carmela Groccia, and Domenico Conforti

Abstract Hyper-parameter optimization and class imbalance are two challenging problems for machine learning in many real-world applications. A hyper-parameter is a parameter whose value is used to control the learning process and it has to be tuned in order to reach good performance. The class imbalance occurs when one class contains significantly fewer instances than the other class. Common approaches for dealing with the class imbalance problems involve modifying the data distribution or modifying the classifier. This paper presents an optimization framework that considers two evaluation measures, i.e., accuracy and G-mean, by optimizing a cost-sensitive Support Vector Machine and its hyper-parameter by a genetic algorithm. Experimental results on two benchmark datasets show that the proposed method is effective and efficient in comparison with the commonly used grid search method.

Keywords Machine learning · Support vector machine · Cost-sensitive approach · Hyper-parameter optimization · Genetic algorithms · NSGA II

1 Introduction

In machine learning (ML), hyper-parameter is a parameter whose value is used to control the learning process. Hyper-parameter should be carefully tuned in each machine learning method because its performance depends on the hyper-parameter [1, 2].

Parameters are different from hyper-parameters: the first ones are learned automatically whereas hyper-parameters are set manually to help guide the learning process. In other words, hyper-parameters strongly affect the final result. Inappropriate hyper-parameter settings may lead to poor classification results, for instance.

R. Guido (✉) · M. C. Groccia · D. Conforti
de-Health Lab, Department of Mechanical, Energy and Management Engineering, University of Calabria, Rende, Italy
e-mail: rosita.guido@unical.it; MariaCarmela.groccia@unical.it; domenico.conforti@unical.it

L. Amorosi et al. (eds.), *Optimization in Artificial Intelligence and Data Sciences*, AIRO Springer Series 8, https://doi.org/10.1007/978-3-030-95380-5_4

A manual tuning consists on changing the hyper-parameter until the result is satisfactory. It is a tedious task because of a lot of possible combinations that can be tested.

A grid search is a hyper-parameter optimization approach based on a defined subset of the hyper-parameter space. For searching a good hyper-parameter, a lower bound, an upper bound, and an incremental step can be specified in order to find the best combination among a finite number of combinations in the search space. However, this exhaustive search can be very time-consuming. Bergstra and Bengio [1] showed that random experiments are more efficient than grid experiments for hyper-parameter optimization on several data sets.

Another important and challenging problem is related to machine learning methods built on imbalanced datasets [3]. They are often unable to generalize on new data because they are biased towards negative predictions, as the majority class has no-event cases. In order to solve this problem, suitable methods for handling imbalanced classes should be used. The main methods for sampling-based imbalance correction are based on over-sampling and under-sampling approaches. Over-sampling methods add more data to the smaller class making it the same size as the larger class; under-sampling methods sample the larger class in order to have the same size as the smaller class.

How to set hyper-parameters and combinations of interacting hyper-parameters for a given dataset is thus a very challenging task. To address this problem, in this paper we use the Non-Dominated Sorting Genetic Algorithm II (NSGA-II) [4], which is an effective and efficient multi-objective search technique based on Genetic Algorithms (GAs). Our research objective is to propose a GA-based approach for optimizing Support Vector Machine (SVM) hyper-parameters and evaluate its potential by comparing the results with those of an automated grid search.

This paper is organized as follows: a brief introduction to the main features of an SVM and issues related to hyper-parameter optimization and imbalanced datasets is given in Sect. 2. Section 3 describes the proposed GA-based approach for SVM hyper-parameter optimization. Section 4 presents experimental results on two well-known datasets using the proposed method. The results are compared to those found by a grid search approach.

To the best of our knowledge, this is the first paper that addresses a cost-sensitive SVM and hyper-parameter optimization at the same time via genetic algorithms.

2 SVM, Hyper-Parameters Optimization and Issues of Imbalanced Datasets

The goal of a binary classifier is to map feature vectors $x \in X$ to class labels $y \in \{-1, 1\}$. A vector is an example, which can be either positive, denoted by a label $y = 1$, or negative, denoted by $y = -1$. In terms of functions, a classifier can be written as $h(x) = sign[p(x)]$, where the function $p : X \rightarrow R$ is denoted as the classifier predictor.

The SVM is based on statistical learning theory [5–7]. It is a class of algorithms for classification, regression and other applications, and is the most widely used ML technique available nowadays. It searches for an optimal hyperplane that separates patterns of two classes by maximizing the margin w. Let X be a dataset with N instances $X = (x_1, \ldots, x_N)$, where $x_i, i = 1, \ldots, N$, denotes an instance with m features, and $y_i \in \{\pm 1\}$ its label. Finding the optimal hyperplane means solving the quadratic programming model (1)–(3)

$$min \frac{1}{2}||w||^2 + C \sum_{1}^{N} \xi_i \tag{1}$$

$$y_i(w^T \phi(x_i) + b) - 1 + \xi_i \geq 0 \quad i = 1, \ldots, N \tag{2}$$

$$\xi_i \geq 0 \quad i = 1, \ldots, N \tag{3}$$

where C, named penalty parameter, is a trade-off between the size of the margin w and the slack variable penalty $\xi \in R_+^N$. In nonlinearly separable dataset, the SVM basically maps inputs into high-dimensional feature spaces by the so called kernel functions [8]. The performance of an SVM model depends on a kernel function, which maps original data to higher dimensional spaces to deal with non-linearly separable data. A kernel function denotes an inner product in a feature space where it measures similarity between any pair of inputs x_i and x_j. Usually, a kernel function is denoted as $K(x_i, x_j) = \langle \phi(x_i), \phi(x_j) \rangle$. The kernel function can take many different forms. It is well known that the performance of most machine learning algorithms on a given dataset depends on well-tuned hyper-parameter. In setting up an SVM model, for instance, two problems are encountered: (1) how to select the kernel function, and (2) how to select its hyper-parameter. An SVM with polynomial kernel has three parameters to optimize: the regularization parameter C, a, and the degree d. The optimization of these three parameters if 30 steps should be performed, requires an amount of time to test the total $30^3 = 27{,}000$ combinations. The greater the number of parameters to be set, the greater is the number of combinations.

The choice of hyper-parameters of an ML model can significantly affect the resulting model's performance. Generally, the hyper-parameters are adjusted for each model in order to find a hyper-parameters setting that maximizes the model performances and so that the classifier can predict unknown data accurately. The goal of hyper-parameter optimization is to find a set of values that minimizes a predefined loss function.

The Grid Search is an approach for hyperparameter optimization based on considering all hyper-parameter combinations specified in a multi-dimensional grid. The performance of a combination is evaluated using a performance metric. The configuration with the best performance is selected and used to train the ML model on the whole dataset.

2.1 Imbalanced Dataset and Cost-Sensitive Approach

Another challenging problem in applied machine learning concerns dealing with imbalanced data as event cases—usually a given disease—are underrepresented. The instances that represent event cases belong to the minority class whereas the other instances to the majority class. Indeed, in general, classifiers built on an imbalanced dataset are often unable to generalize on new data because they are biased towards predictions similar to the majority class, that is usually given by instances representative of no-events. Furthermore, misclassification errors are treated in the same way, but this is not correct when a dataset is imbalanced. Suffice it to say that in many important applications of machine learning, such as medical diagnosis, certain types of errors are much more costly than others. Misclassifying a patient as healthy implies more loss compared to the opposite loss. In the literature, several methods are used to deal with this problem. Sampling-based imbalance corrections are based on over-sampling and under-sampling approaches. Over-sampling methods add more data to the smaller class making it the same size as the larger class [9]; under-sampling methods sample the larger class in order to have the same size as the smaller class. Other approaches are cost-sensitive based [10, 11]. Among the extensions of SVM, a cost-sensitive scheme refers to a model with multiple costs which considers different error rates for misclassification. The cost-sensitive scheme is useful when misclassification cannot be considered equal, like for medical diagnosis. Therefore, a cost-sensitive scheme poses as a modified model and hereby aims at minimizing loss function instead of generalization error.

A cost matrix $Cost$ denotes the cost of each class misclassification [12]. A general cost matrix for a binary classification problem is $Cost = \begin{bmatrix} c_{++} & c_{+-} \\ c_{-+} & c_{--} \end{bmatrix}$, where c_{+-} is the cost of predicting an instance belonging to class C_+ as belonging to class C_-, i.e., the cost of a false negative misclassification, whereas c_{-+} is the cost of a false positive misclassification. Usually, as we also assume here, there is no cost for correct classifications, that is $c_{++} = c_{--} = 0$.

The goal of this type of learning is to minimize the total misclassification cost of a model on the training set. Formally, given a cost matrix $Cost$ and an instance x, the cost $R(C_+|x)$ of classifying x into class C_+ is $R(C_+|x) = \sum_j p(C_-|x)c_{+-}$, where $p(C_-|x)$ is the probability estimation of classifying an instance into class C_-.

As detailed in the next section, we coded the two misclassification costs in a chromosome.

3 A Genetic Algorithm-Based Approach for SVM Hyper-Parameters Optimization

The genetic algorithms are optimization algorithms inspired by the Darwin evolutionary process [13]. An initial population of a given number of chromosomes, hereafter even called individuals, evolves during time through generations. One chromosome has a given number of genes, which are the variables of the problem, and it is one point in the solution space. At each generation, the individuals are evaluated through a fitness function, which quantifies the quality of each individual chromosome. According to their evaluation, some individuals in the population are selected for reproduction. More specifically, like natural selection principles, the fittest individuals are selected. To produce new individuals from the selected ones, some operators are applied to genes. *Crossover* and *mutation* operators are applied during the generation of a new population to generate new chromosomes, that is, new solutions. The crossover mechanism aims to combine two chromosomes into one chromosome. This operator is applied to two individuals named parents; the two newly generated individuals are named offspring. A mutation generates a new chromosome by changing one or more genes of a chromosome. Another type of selection is *elitism*, which consists on sent a percentage of the best individuals from the current population to the next population. This selection prevents the loss of the best solutions for a population. A new population is thus given by their descendants. In the evolutionary process, all those solutions that did not fit some defined criteria will be dropped. The solution kept will continue the selection process (i.e., crossover, mutation, elitism, and evaluation) until at least one of the defined stop conditions is met. The stop conditions generally consist of a number of generations and a certain time limit.

3.1 Hyper-Parameter Optimization via NSGA-II

Genetic Algorithms are an alternative method for determining optimum SVM values in an evolutionary way. A GA can search for a better solution without trying all possible solution. As a genetic algorithm, we chose the NSGA-II [4] because it uses an explicit diversity preservation mechanism (the so-called crowding distance) and emphasizes non-dominated solutions by a fast sorting procedure based on Pareto front-ranking to promote convergence. NSGA-II is one of the first multi-objective algorithms that introduced elitism, i.e., a partner of a non-dominated individual is chosen from among the individuals of the population that it dominates. NSGA-II has been successfully applied to several domains [14, 15]. The difference between the NSGA-II and single objective GAs is in how fitness is assigned. The NSGA-II employs Pareto dominance and the crowding distance to assign fitness values. It implements a binary tournament selection to establish the mating pool (i.e., parents) and uses the Simulated Binary Crossover and the Polynomial Mutation with defined probabilities to create children from parents. This algorithm has various parameters that may impact its computational effectiveness. The most important parameters are the population size (PS), the probability of crossover (P_c), the probability of mutation (P_m) that prevents the population from being trapped in local optima. The parameter PS plays an important role because if it is small may result in a crowded population, i.e., with a number of similar solutions and not a diversified number of chromosomes. Usually, the adopted strategy is to keep a high value of P_c and a low value of P_m.

The cost-sensitive SVM (CS-SVM) assigns different values for the penalty factor to two classes

$$\underset{w,b,\xi}{\text{argmin}} \quad \frac{1}{2}||w||^2 + C[C_1 \sum_{i|y_i=1} \xi_i + C_{-1} \sum_{i|y_i=-1} \xi_i] \tag{4}$$

$$y_i(w^T x + b) \geq 1 - \xi_i \tag{5}$$

$$\xi_i \geq 0 \quad i = 1, \ldots, N \tag{6}$$

where C_1 is the cost of a false negative and C_{-1} that one of a false positive.

This paper concentrates on optimising hyper-parameter of CS-SVM. The range of possibilities for CS-SVM hyper-parameter can be huge. We perform a hyper-parameter tuning by exploring various model architectures by NSGA-II. The best population is given as a set of chromosomes with the best model performance.

4 Experimental Results and Discussion

To evaluate our proposed method, we used two benchmark datasets, the Hepatitis dataset and the Wisconsin Prognostic Breast Cancer (WPBC), from the UCI Repository of Machine Learning Databases. Both datasets are related to medical diagnosis represented as binary classification problems. The Hepatitis dataset is used to classify patients with hepatitis in the two classes, live or die. The WPBC has 198 instances that represent follow-up data for one breast cancer case, only those cases exhibiting invasive breast cancer and no evidence of distant metastases at the time of diagnosis. The WPBC is used here to classify patients as recurrences before 24 months (i.e., positive class) or nonrecurrence beyond 24 months (i.e., negative class). To this aim we remove the feature named "Time" from the dataset because it is the recurrence time for instances in the positive class and the disease-free time for the instances of the negative class. Table 1 reports the number of positive instances (diseased examples), the number of negative instances (non-diseased examples), the ratio as minority class vs the majority class, and the number of features (excluding the class), per each dataset.

Table 1 Description of the used datasets

Dataset	Positive instances	Negative instances	Ratio	Number of features
Hepatitis	32	123	1:4	19
WPBC	47	151	1:3	32

4.1 NSGA-II Parameter Setting

In setting the parameters for the NSGA-II algorithm, we carried out several preliminary tests to find suitable parameter values. We set $P_c = 0.7$ and $P_m = 0.03$. Another parameter that we set is the distribution index of mutation DI_m, which is a control parameter inversely proportional to the amount of perturbation in the design variables. The smaller the value of DI_m, the larger the perturbation and vice versa. It was bounded by 1 and 20, which is the default value. Depending on the initial population, a GA may produce different solutions. In this research, we tested NSGA-II with three initial populations: (1) randomly generated; (2) partially fixed and the rest randomly generated (as proposed the first time in [15]); (3) all chromosomes fixed.

4.2 CS-SVM Hyper-Parameter Optimization

We tested SVM with two kernel functions, i.e., polynomial kernel $K(x_i, x_j) = (x_i^T x_j + a)^d$, and Radial Basis Function (RBF) kernel $K(x_i, x_j) = exp(-\gamma||x_i - x_j||^2)$.

The hyper-parameter C and those related to the polynomial kernel and the RBF kernel were optimized by searching the best value in a defined range of values, as reported in Table 3. Figure 1 shows how hyper-parameter of CS-SVM with RBF kernel was coded as a chromosome.

To perform model assessment, we carried out a tenfold cross-validation (tenfold CV) process embedded in an evolutionary process aimed at improving a given fitness function. The tenfold CV assures that $k = 10$ independent sets are used to test the model, simulating unseen data. It consists basically in partitioning randomly the dataset into k equal sized folds. k rounds are performed: at k-th round, all the folds are used as training set except to the k-th fold, which is used as test set. The test set is never seen during the training of the model, to avoid overfitting. Each fold

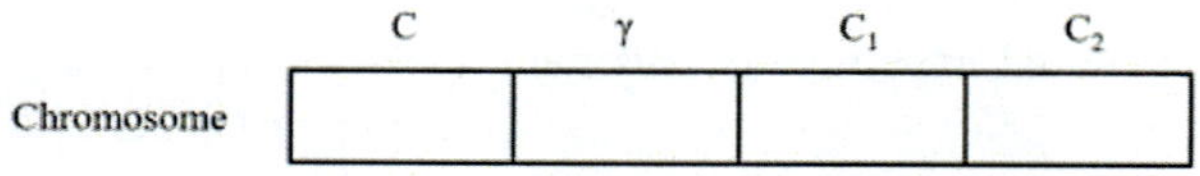

Fig. 1 Coding of CS-SVM with RBF kernel in a chromosome

is used exactly once as test set and, as consequence, each instance is used for testing exactly once. The performance metrics are averaged across the k estimates of each test fold.

Let P, be the number of the positive instances, and N the number of the negative instances of a given dataset. We evaluated the performances of the SVM classifiers with the optimized hyper-parameter by *accuracy* and *geometric mean* (*G-Mean*). Let TP and TN be, respectively, the number of instances of class C_+ and class C_- correctly classified; FP and FN be the number of instances of class C_- and class C_+ incorrectly classified, respectively. The accuracy metric, defined as $(TP + TN)/(TP + TN + FP + FN)$, is the most widely used evaluation metric for classification algorithms. However, it can be a misleading for imbalanced datasets, as we show empirically in the next section. The sensitivity, even known as true positive rate, measures how well a classifier can identify true positives. It is defined as *Sensitivity*$=TP/(TP + FN) = TP/P$. The Specificity, even known as true negative rate, measures how well a classifier can identify true negatives. It is defined as *Specificity*$=TN/(TN + FP) = TN/N$.

The *G-Mean* balances both Sensitivity and Specificity by combining them, as

$$\text{G-Mean} = sqrt(Sensitivity * Specificity)$$

NSGA-II here implemented has two objective functions, F_1 and F_2. We tested as F_1 both accuracy and G-Mean, whereas we set $F_2 = 1$. NSGA II stops after a given number of generations. We carried out the computational experiments on 43 configurations per dataset by considering $PS = \{25, 50, 100\}$ and $Gen \in \{10, 100, 200\}$. The experiments were performed by Weka 3.8.2-API [16] integrated in the Java version of NSGA-II. They were performed on a PC with 3.50 GHz Intel Xeon CPU E5 1620 and 32 GB RAM, using Windows 10 Pro operating system.

Table 2 reports the setting of NSGA-II and CS-SVM of only 12 configurations. The first column reports the computational experiment identification, which is given as *ExpID-dataset-*F_1, where $dataset \in \{H, WBCP\}$, $F_1 \in \{A, GM\}$. For instance, *Exp1-H-A* identifies the first set of the computational experiments carried out on the Hepatitis dataset; accuracy is the used fitness function. The second column of Table 2 specifies the fitness function. The third and the fourth column show the population size (the letter r means that the initial population is random) and the number of generations, respectively. The last two columns report the value of DI_m and the average computational time, in minutes, to execute all the generations.

Table 3 shows the range of values of the CS-SVM hyper-parameter investigated by NSGA-II. $c_1 = c_{+-}$ and $c_2 = c_{-+}$ denote the cost of each class misclassification.

The best results for CS-SVM with RBF kernel are summarised in Table 4 per each computational experiment. This table even shows the values related to Sensitivity, Specificity, G-mean, Accuracy, and the number of TP, FN, FP, and TN. The optimised hyper-parameter are reported in the four last columns.

In Table 4, as expected, the best accuracy does not mean the best G-mean. For instance, the optimised hyper-parameters found for $Exp1\text{-}WBCP\text{-}A$ allow to

Table 2 Hepatitis and WPBC datasets: parameter setting

Experiment	F_1	PS	Gen	DI_m	Time
Exp1-H-A	Acc	100 (r)	100	1	62.5
Exp2-H-A	Acc	50 (r)	10	20	3.1
Exp3-H-A	Acc	100 (r)	10	20	5.1
Exp1-H-GM	G-mean	100 (r)	100	1	57.2
Exp2-H-GM	G-mean	100 (r)	100	10	58.3
Exp3-H-GM	G-mean	100 (r)	10	20	5.6
Exp1-WPBC-A	Acc	100 (r)	10	20	15.1
Exp2-WBCP-A	Acc	50 (r)	10	20	7.32
Exp3-WBCP-A	Acc	50 (r)	100	20	110.3
Exp1-WBCP-GM	G-mean	100 (r)	10	20	15.6
Exp2-WBCP-GM	G-mean	50 (r)	100	20	112.5

Table 3 Hyper-parameters of CS-SVM with polynomial kernel and RBF kernel

Cost $C \in \{1 - 500\}$	Kernel
$C \in \{1 - 20\}$	$d \in \{1 - 3\}$
$c_1 \in \{1 - 20\}$	$\gamma \in \{0.001 - 5\}$
$c_2 \in \{1 - 20\}$	

Table 4 The best results, related metrics, and the optimised hyper-parameter per each computational experiment

	Best								Opt hyper-parameter			
Experiment	F_1	Sens	Spec	G-mean	TP	FN	FP	TN	C	γ	c_1	c_2
Exp1-H-A	87.742	0.656	0.935	78.317	21	11	8	115	186	0.124	13	3
Exp2-H-A	86.452	0.656	0.919	77.644	21	11	10	113	1	0.092	12	8
Exp3-H-A	87.097	0.656	0.927	77.982	21	11	9	114	454	0.144	15	3
Exp1-WBCP-A	78.283	0.106	0.993	32.443	5	42	1	150	1	4.062	5	4
Exp2-WBCP-A	77.778	0.085	0.993	29.053	4	43	1	150	1	4.459	10	9
Exp3-WBCP-A	76.768	0.255	0.927	48.619	12	35	11	140	1	1.232	17	10
Accuracy												
Exp1-H-GM	79.231	0.75	0.837	81.936	24	8	20	103	1	0.299	9	2
Exp2-H-GM	82.43	0.781	0.87	85.161	25	7	16	107	345	0.005	14	5
Exp3-H-GM	80.405	0.75	0.862	83.871	24	8	17	106	1	0.258	14	4
Exp1-WBCP-GM	60.997	0.468	0.795	71.717	22	25	31	120	199	0.449	9	7
Exp2-WBCP-GM	62.297	0.617	0.629	62.626	29	18	56	95	259	0.078	13	2
Exp3-WBCP-GM	60.997	0.468	0.795	71.717	22	25	31	120	483	0.4356	1	1

achieve the best value of accuracy with respect to the other experiments. However, the corresponding G-mean is low. If the G-mean is used as fitness function, different optimised hyper-parameters allow to improve not only the value of G-mean but even the Sensitivity. In addition, a larger number of generations allows in some cases to find better results. These results show that accuracy cannot be used as the sole criterion to evaluate the performance of a classifier when a dataset is imbalanced and that, mainly in the medical domain, other performance metrics, like G-Mean, should be considered.

Finally, we compared our best results with those found by Auto-WEKA [17] that implements a fully automated search and identifies the most appropriate machine learning algorithm and hyperparameter settings to a given dataset. The best accuracy found for the Hepatitis dataset, for instance, is 89.67 by a Simple Logistic model and a feature selection that reduced the number of features. The related G-Mean is 78.95. The best accuracy found for the WBCP is 75.75 by a Decision Tree and the related G-Mean is only 14.48 with a number of $TP = 1$.

5 Conclusion

This paper presents a CS-SVM to address imbalanced datasets and a framework for hyper-parameter tuning based on GAs. In many important applications of machine learning, such as medical diagnosis, certain types of errors are much more costly than others. The hyper-parameters of a CS-SVM are optimised by using NSGA-II. Experimental results on two benchmark datasets with different ratios of imbalance showed that the proposed method is effective. In addition, we used and compared the accuracy and G-Mean metrics to evaluate the model performance. We empirically showed that accuracy cannot be used as the sole criterion to evaluate the performance of a classifier but other performance metrics, like G-Mean, should be considered mainly in medical domain like disease diagnosis.

For future work we plan to extend the approach to other kernel functions and use NSGA-II as a multi-objective algorithm.

Acknowledgments The research has been partially supported by the research project "SI.F.I.PA.CRO.DE.—Sviluppo e industrializzazione farmaci innovativi per terapia molecolare personalizzata PA.CRO.DE. (PON ARS01_00568, CUP: B29C20000360005, CONCESSIONE RNA-COR: 4646672), Italian Ministry of University and Research, 2021.

References

1. Bergstra, J., Bengio, Y.: Random search for hyper-parameter optimization. J. Mach. Learn. Res. **13**, 281–305 (2012)
2. Yang, L., Shami, A.: On hyperparameter optimization of machine learning algorithms: theory and practice. Neurocomputing **415**, 295–316 (2020)
3. Sun, Y., Wong, A.K.C., Kamel, M. S.: Classification of imbalanced data: a review. Int. J. Pattern Recognit. Artif. Intell. **23**, 687–719 (2009)
4. Deb, K., Pratap, A., Agarwal, S., Meyarivan, T.: A fast and elitist multiobjective genetic algorithm: NSGA-II. IEEE Trans. Evol. Comput. **6**, 182–197 (2002)
5. Cortes, C., Vapnik, V.: Support-vector network. Mach. Learn. **20**, 273–297 (1995)
6. Vapnik, V.: Statistical Learning Theory. Wiley, New York (1998)
7. Cristianini, N., Shawe-Taylor, J.: An Introduction to Support Vector Machines and Other Kernel-Based Learning Methods. Cambridge University Press, Cambridge (2000)
8. Hofmann, T., Scholkopf, B., Smola, A.J.: Kernel methods in machine learning. Ann. Stat. **2008**, 1171–1220 (2008)

9. Gong, Z., Chen, H.: Model-Based oversampling for imbalanced sequence classification. In: Proceedings of the 25th ACM International on Conference on Information and Knowledge Management, pp. 1009–1018 (2016)
10. Akbani, R., Kwek, S., Japkowicz, N.: Applying support vector machines to imbalanced datasets. In: European Conference on Machine Learning, pp. 39–50 (2004)
11. Cheng, F., Zhang, J., Wen, C.: Cost-sensitive large margin distribution machine for classification of imbalanced data. Pattern Recognit. Lett. **80**, 107–112 (2016)
12. Elkan, C.: The foundations of cost-sensitive learning. In: Proceedings of the IJCAI (2001)
13. Goldberg, D.E., Holland, J.H.: Genetic algorithms and machine learning. Mach. Learn. **3**(2), 95–99 (1988)
14. Li, H., Zhang, Q.: Multiobjective optimization problems with complicated Pareto sets, MOEA/D and NSGA-II. IEEE Trans. Evol. Comput. **13**, 284–302 (2009)
15. Guido, R., Conforti, D.: A hybrid genetic approach for solving an integrated multi-objective operating room planning and scheduling problem. Comput. Oper. Res. **87**, 270–282 (2017)
16. Frank, E., Hall, M.A., Witten, I.H.: The WEKA Workbench. Online Appendix for "Data Mining: Practical Machine Learning Tools and Techniques", 4th edn. Morgan Kaufmann, Burlington (2016)
17. Kotthoff, L., Thornton, C., Hoos, H., Hutter, F., Leyton-Brown, K.: Auto-WEKA 2.0: automatic model selection and hyperparameter optimization in WEKA. J. Mach. Learn. Res. **18**(25), 1–5 (2017)

An Improvement of the Pivoting Strategy in the Bunch and Kaufman Decomposition, Within Truncated Newton Methods

Giovanni Fasano and Massimo Roma

Abstract In this work we consider the solution of large scale (possibly nonconvex) unconstrained optimization problems. We focus on Truncated Newton methods which represent one of the commonest methods to tackle such problems. In particular, we follow the approach detailed in Caliciotti et al. (Comput Optim Appl 77:627–651, 2020), where a modified version of the Bunch and Kaufman decomposition (Bunch and Kaufman, Math Comput 31:163–179, 1977) is proposed for solving the Newton equation. Such decomposition is used within SYMMBK routine as proposed by Chandra (Conjugate gradient methods for partial differential equations, Ph.D. thesis, Yale University, New Haven, 1978; see also Conn et al., Trust–Region Methods, MPS–SIAM Series on Optimization, Philadelphia, PA, 2000; HSL: A collection of Fortran codes for large scale scientific computation, https://www.hsl.rl.ac.uk/; Marcia, Appl Numer Math 58(4):449–458, 2008) for iteratively solving symmetric possibly indefinite linear systems. The proposal in Caliciotti et al. (Comput Optim Appl 77:627–651, 2020) enabled to overcome a relevant drawback of nonconvex problems, namely the computed search direction might not be gradient-related. Here we propose further extensions of such approach, aiming at improving the pivoting strategy of the Bunch and Kaufman decomposition and enhancing its flexibility.

Keywords Large scale optimization · Truncated Newton method · Bunch and Kaufman decomposition · Gradient related directions

G. Fasano (✉)
Department of Management, University Ca' Foscari of Venice, Venice, Italy
e-mail: fasano@unive.it

M. Roma
Department of Computer, Control and Management Engineering "A. Ruberti", Sapienza, University of Rome, Rome, Italy
e-mail: roma@diag.uniroma1.it

L. Amorosi et al. (eds.), *Optimization in Artificial Intelligence and Data Sciences*, AIRO Springer Series 8, https://doi.org/10.1007/978-3-030-95380-5_5

1 Introduction

Given a real valued function $f : \mathbb{R}^n \longrightarrow \mathbb{R}$, an unconstrained optimization problem consists of determining a local minimizer of f by solving

$$\begin{aligned} &\min f(x) \\ &x \in \mathbb{R}^n. \end{aligned} \tag{1}$$

In particular, we consider problems where *n is large* and the function *f is possibly nonconvex*. Moreover, we assume that both the gradient $\nabla f(x)$ and the Hessian matrix $\nabla^2 f(x)$ exist and are continuous. We do not assume any sparsity pattern on $\nabla^2 f(x)$. The iterative solution of large scale unconstrained optimization problems plays a fundamental role in many and different contexts of applied mathematics. Therefore, it is very important to have at one's disposal an efficient and robust method able to tackle also large scale difficult problems (see also [4, 14]).

As well known, in case the method of choice were a Truncated Newton method, at each iteration h, a search direction p_h and a steplength α_h are determined, so that the current point is updated according to the iterative scheme

$$x_{h+1} = x_h + \alpha_h p_h, \tag{2}$$

being $x_0 \in \mathbb{R}^n$ a given starting point. In the Truncated Newton method, the search direction p_h is often obtained by approximately solving the Newton equation

$$\nabla^2 f(x_h)p = -\nabla f(x_h), \tag{3}$$

by means of a Krylov subspace method. The iterations of the solver (called *inner iterations*) are stopped according to a suited termination criterion, still ensuring a good convergence rate of the method. This is obtained by using a particular trade–off rule between the computational burden required to solve the system (3) and the accuracy with which it is solved. The reader is referred to the seminal paper by S. Nash [18] for a survey on Truncated Newton methods.

Among the Krylov subspace methods, the Conjugate Gradient (CG) algorithm is usually the method of choice, even if it may break down when solving (3) and the matrix $\nabla^2 f(x_h)$ is indefinite. In this case, some alternative strategies have been proposed in literature (see, e.g., [8–13, 15]).

In [5] the use of the SYMMBK algorithm was proposed as an alternative to the CG method. The SYMMBK algorithm was introduced in [6, 16, 17] and it is based on the Lanczos process, which does not break down in the indefinite case. More precisely, the matrix of the Lanczos vectors is built one column at a time and (after k iterations) the resulting $n \times k$ matrix Q_k has the property that $Q_k^T \nabla^2 f(x_h) Q_k = T_k$, where T_k is tridiagonal. Then, the Bunch and Kaufman decomposition [1] of the tridiagonal matrix T_k is performed, namely $T_k = S_k B_k S_k^T$, where B_k is a block diagonal matrix with 1×1 or 2×2 diagonal blocks, and S_k is a unit lower triangular matrix. At each step k, a suited strategy is adopted for deciding whether a 1×1 or 2×2 diagonal block must be formed, in order to guarantee numerical stability.

On the other hand, the test on a pivotal element inside SYMMBK algorithm is uniquely chosen to pursue numerical efficiency and stability, inasmuch as the Bunch and Kaufman decomposition performed by SYMMBK focuses on the growth factor of the matrices resulting from decomposition (see [6]). Thus, some concerns may arise when embedding the SYMMBK algorithm within a Truncated Newton method, where a search direction must be *gradient related*, i.e., eventually *bounded* and of *sufficient descent* (see Definition 1.1 in [5] for a formal statement). The last issue was already addressed in [5], though the modification proposed therein possibly left room to further generalizations. Here we aim at filling the last gap, by proposing an enhancement with respect to [5]. In particular, we are going to propose here an update for the parameters ω and ϕ used in [5], so that they can possibly depend on the gradient vector computed at the current Truncated Newton iteration. More specifically, in the next sections we analyze and discuss the following issues:

- at step k of the Bunch and Kaufman routine, the test $|\delta_k| > \omega\eta\gamma_{k+1}^2$ discussed in [5] represents indeed a test on the curvature along the vector q_k;
- we can replace the quantity ω introduced in [5] with the sequence $\{\omega_k\}$, so that the test $|\delta_k| > \omega\eta\gamma_{k+1}^2$ turns into the test $|\delta_k| > \omega_k\eta\gamma_{k+1}^2$;
- we define a specific expression for the constant ϕ introduced in [5], so that it explicitly depends on the gradient vector currently available from the optimization framework;
- the choice of the sequence $\{\omega_k\}$ and the constant ϕ can partially be steered by the optimization framework, in case any additional knowledge is available which suggests that a better quality of the overall gradient-related direction can be sought.

The paper is organized as follows. In Sect. 2 some preliminaries on Truncated Newton methods and the Lanczos process are reported; then, the Bunch and Kaufman decomposition, as well as some basics on SYMMBK (see also [7]), are given. In Sect. 3 we show how to compute a gradient-related direction by using the Bunch and Kaufman decomposition. Finally, Sect. 4 reports some concluding remarks.

We indicate by $\|\cdot\|$ the Euclidean norm of real vectors and matrices. Moreover, $\lambda_\ell(C)$ and $\kappa(C)$ represent the ℓ-th eigenvalue and the condition number of the real symmetric matrix C, respectively. Finally, e_i is the i-th real unit vector.

2 Preliminary Results

Here we report some basic results, including introductory material on the Lanczos process and the Bunch and Kaufman factorization. Some insights on the Lancos process are mandatory, to show how SYMMBK performs an iterative decomposition of the tridiagonal matrix using the Lanczos process. For the sake of brevity, we assume that the reader is familiar with a standard Truncated Newton method which iteratively generates the sequence $\{x_h\}$ in (2). We recall the importance of an efficient truncation criterion for the inner iterations within Truncated Newton methods, as also pointed out in [8, 9, 19], and more recently in [2, 3].

Assumption 1 Let be given the function $f : \mathbb{R}^n \to \mathbb{R}$ in (1), with f twice continuously differentiable. Then, we assume that the sequence $\{x_h\}$ in (2) satisfies $\{x_h\} \subset \Omega$, being $\Omega \subset \mathbb{R}^n$ compact. □

As a very general convergence result for Truncated Newton methods, when convergence to first order stationary points is sought, we give the next proposition.

Proposition 1 *Consider the sequences $\{x_h\}$ and $\{p_h\}$ in (2), where $\{x_h\}$ satisfies Assumption 1 and the search directions are gradient-related. If an Armijo-type linesearch procedure is chosen to select the steplength α_h in (2), then*

- *$\{f(x_k)\}$ converges regardless of the choice of the initial iterate x_0;*
- *any subsequence of $\{x_k\}$ converges to a stationary point of $f(x)$.*

Definition 1 Let be given the function $f : \mathbb{R}^n \to \mathbb{R}$ in (1), with f twice continuously differentiable. Consider a vector $d \in \mathbb{R}^n \setminus \{0\}$. Then, the quantity $d^T \nabla^2 f(\bar{x}) d$ is the *normalized curvature* of f at $\bar{x}$, along the direction d.

2.1 Matrix Tridiagonalization Using the Lanczos Process

The Lanczos process [7] is a Krylov-subspace method for tridiagonalizing a symmetric indefinite matrix. Dropping the dependency of the subscript h and setting $A = \nabla^2 f(x_h)$, $b = -\nabla f(x_h)$ in (3), the application of the Lanczos process to the linear system

$$Ad = b \tag{4}$$

yields the orthogonal Lanczos vectors q_i, $i \geq 1$, according with Table 1.

After $k \leq n$ iterations the Lanczos process has generated the unit vectors $q_1, \ldots, q_k$ (the *Lanczos vectors*), along with the values $\delta_1, \ldots, \delta_k$ and $\gamma_2, \ldots, \gamma_k$, so that setting $Q_k = (q_1 \;\cdots\; q_k)$ and defining the nonsingular tridiagonal matrix

$$T_k = \begin{pmatrix} \delta_1 & \gamma_2 & & & \\ \gamma_2 & \delta_2 & \cdot & & \\ & \cdot & \cdot & \cdot & \\ & & \cdot & \delta_{k-1} & \gamma_k \\ & & & \gamma_k & \delta_k \end{pmatrix}, \tag{5}$$

Table 1 The Lanczos process for the indefinite linear system (4)

Data: $\varepsilon = \bar{\varepsilon}\|b\|$, with $\bar{\varepsilon} \in (0, 1)$. Set $k = 1$, $u_1 = b$, $q_1 = \frac{u_1}{\|u_1\|}$, $\delta_1 = q_1^T A q_1$ and compute the vector $u_2 = Aq_1 - \delta_1 q_1$.

Do

$k = k + 1$;

$\gamma_k = \|u_k\|$;

If $\gamma_k < \varepsilon$ STOP.

Else set

$q_k = u_k / \gamma_k$;

$\delta_k = q_k^T A q_k$;

$u_{k+1} = Aq_k - \delta_k q_k - \gamma_k q_{k-1}$.

End If

End Do

we have

$$A Q_k = Q_k T_k + \gamma_{k+1} q_{k+1} e_k^T; \tag{6}$$

$$Q_k^T A Q_k = T_k; \tag{7}$$

$$Q_k^T Q_k = I; \tag{8}$$

$$Q_k^T q_{k+1} = 0; \tag{9}$$

$$\text{span}\,\{q_1, q_2, \ldots, q_k\} = \text{span}\left\{u_1, Au_1, \ldots, A^{k-1}u_1\right\}. \tag{10}$$

In the case $\gamma_{k+1} = 0$ in (6), then we have from (6)–(10) (see [6])

$$\begin{cases} T_k y_k = \|b\| e_1 \\ d_k = Q_k y_k \end{cases} \tag{11}$$

which easily allow to first compute the vector y_k (from $T_k y_k = \|b\|e_1$) and then $d_k = Q_k y_k$, being d_k an *approximate* solution of (4). Finally, note that according to Definition 1 the scalar δ_k represents the normalized curvature of the function $\phi(d) = 1/2 d^T Ad + b^T d$, along the direction q_k.

3 Our Proposal of Possible Generalizations

As also reported in Sect. 2, in a Truncated Newton framework the SYMMBK algorithm can be applied for the solution of the indefinite Newton's equation (4), by exploiting the Bunch and Kaufman factorization. Unfortunately, a mere application of the last decomposition possibly provides in (11) a search direction (namely d_k)

which might not be gradient-related. Here, we suitably generalize the proposal in [5] and better exploit the SYMMBK algorithm, so that starting from an approximate solution of (3) we can compute a gradient-related direction.

We intend to slightly modify, at step k, the pivoting strategy adopted by the Bunch and Kaufman iterative decomposition $T_k = S_k B_k S_k^T$, when choosing at step k between 1×1 or 2×2 pivot. Note that the matrix B_k is a block diagonal matrix containing 1×1 and/or 2×2 blocks, while S_k is a unit lower triangular matrix. The standard pivoting strategy in SYMMBK (see [6]) consists of performing at step k a 1×1 pivot if $|\delta_k| > \eta \gamma_{k+1}^2$, otherwise a 2×2 pivot is considered (where η is a suited scalar). This strategy has also an interesting geometric interpretation suggested by Definition 1 and summarized in the next result.

Proposition 2 *Let us consider the Bunch and Kaufman decomposition $T_k = S_k B_k S_k^T$ of the tridiagonal matrix T_k in (11). Assume that at step k of the Bunch and Kaufman decomposition the generalized test $|\delta_k| > \omega_k \eta (\gamma_{k+1})^2$ is adopted, with $\omega_k > 0$. Then, the normalized curvature δ_k of the function $\phi(d) = 1/2 d^T A d + b^T d$, along the direction q_k, either satisfies $\delta_k \le (\delta_k)_-$ or $\delta_k \ge (\delta_k)_+$, where*

- *for $k = 1$, for any value of $\omega_1 > 0$ the quantities $(\delta_k)_-$ and $(\delta_k)_+$ are bounded away from zero;*
- *for $k \ge 2$, there are values of $\omega_k > 0$ such that $(\delta_k)_-$ and $(\delta_k)_+$ are bounded away from zero.*

Proof We separately analyze the cases $k = 1$ and $k \ge 2$. When $k = 1$, recalling that $\|q_i\| = 1$ for any $i \ge 1$, the test $|\delta_1| > \omega_1 \eta (\gamma_2)^2$ is equivalent to the inequalities

$$\begin{aligned} \delta_1 &< -\omega_1 \eta \left[\|Aq_1\|^2 - \delta_1^2 \right] \\ \delta_1 &> +\omega_1 \eta \left[\|Aq_1\|^2 - \delta_1^2 \right]. \end{aligned} \tag{12}$$

Recalling that $\eta = (\sqrt{5} - 1)/(2 \max_i |\lambda_i(A)|)$ (see [6]), after an easy computation the first inequality yields

$$(\delta_1)_- = \frac{1 - \left[1 + 4\omega_1^2 \eta^2 \|Aq_1\|^2\right]^{1/2}}{2\omega_1 \eta} \le \frac{1 - \left[1 + \omega_1^2 (\sqrt{5} - 1)^2 / \kappa^2(A)\right]^{1/2}}{2\omega_1 \eta},$$

$$(\delta_1)_+ = \frac{1 + \left[1 + 4\omega_1^2 \eta^2 \|Aq_1\|^2\right]^{1/2}}{2\omega_1 \eta} \ge \frac{1 + \left[1 + \omega_1^2 (\sqrt{5} - 1)^2 / \kappa^2(A)\right]^{1/2}}{2\omega_1 \eta}.$$

Similarly, by the second inequality we get

$$(\delta_1)_- = \frac{-1 - \left[1 + 4\omega_1^2\eta^2\|Aq_1\|^2\right]^{1/2}}{2\omega_1\eta} \leq \frac{-1 - \left[1 + \omega_1^2(\sqrt{5}-1)^2/\kappa^2(A)\right]^{1/2}}{2\omega\eta},$$

$$(\delta_1)_+ = \frac{-1 + \left[1 + 4\omega_1^2\eta^2\|Aq_1\|^2\right]^{1/2}}{2\omega_1\eta} \geq \frac{-1 + \left[1 + \omega_1^2(\sqrt{5}-1)^2/\kappa^2(A)\right]^{1/2}}{2\omega\eta}.$$

When $k \geq 2$, after some arrangements, the test $|\delta_k| > \omega_k\eta(\gamma_{k+1})^2$ is equivalent to the inequality

$$|\delta_k| > \omega_k\eta\|Aq_k - \delta_k q_k - \gamma_k q_{k-1}\|^2 = \omega_k\eta\left[\|Aq_k\|^2 - \delta_k^2 - \gamma_k^2\right].$$

Furthermore, an analysis similar to the case $k = 1$ holds, after distinguishing between the subcases $\|Aq_k\| > \gamma_k$ and $\|Aq_k\| \leq \gamma_k$. □

In the next section we show that, for any k, the generalized test $|\delta_k| > \omega_k\eta(\gamma_{k+1})^2$ reported in Proposition 2 allows to use an adapted SYMMBK algorithm for constructing a *gradient-related* direction.

3.1 Gradient-Related Directions Using SYMMBK

Here we show that, using the results in Proposition 2, within the Bunch and Kaufman decomposition, it is possible to guarantee that a gradient-related direction p_h at x_h for the optimization problem (1) can be computed, provided that suitable values $\{\omega_k\}$ are used. In this regard, we first recall that to iteratively compute the vector d_k in (11), the Bunch and Kaufman algorithm generates the *intermediate* vectors $\{z_i\}$ (see also [5]), $i \leq k$, such that

- if at step i a 1×1 pivot is performed by the Bunch and Kaufman decomposition, then the vector $z_i = z_{i-1} + \zeta_i w_i$ is generated,
- if at step i a 2×2 pivot is performed by the Bunch and Kaufman decomposition, then the vector $z_i = z_{i-2} + \zeta_{i-1} w_{i-1} + \zeta_i w_i$ is generated,
- when $i = k$ we have $d_k = z_k$,

being the real values $\{\zeta_i\}$ and the vectors $\{w_i\}$ computed using the entries of matrices S_k and B_k. Furthermore, as regards the vectors $\{z_i\}$ we have the next result, which represents a generalization of Proposition 3.1 in [5].

Proposition 3 *Let the matrix A in* (4) *be nonsingular, and let z_i, $i \leq k$, be the directions generated by the SYMMBK algorithm when solving the tridiagonal linear system in* (11). *Then, for any* $0 < \omega_i < 1$, $i \geq 1$, *we have that*

- *any direction in the finite sequences $\zeta_1 w_1, \ldots, \zeta_k w_k$ and $z_1, \ldots, z_k$ is bounded;*
- *the vector d_k in* (11) *coincides with z_k (and is bounded).*

Proof The proof follows guidelines similar to Proposition 3.1 in [5], so that it is omitted. □

From Proposition 3 the real vectors $\zeta_i w_i = z_i - z_{i-1}$ (respectively the vectors $\zeta_{i-1} w_{i-1} + \zeta_i w_i = z_i - z_{i-2}$) are bounded, and according with the next scheme in Table 2, they can be fruitfully used to compute the search direction p_h, at the outer iteration h of the Truncated Newton method. We remark that the scheme in Table 2 includes a relevant piece of news for the choice of the value ϕ, with respect to the *Reverse Scheme* in [5], as detailed in the proof of Proposition 4.

Table 2 Computing a gradient-related search direction with SYMMBK algorithm

Data: Set the initial vector $p_h = z_0 = 0$, along with the parameter $\phi = \bar{\phi}\|b\|$, with $\bar{\phi} > 0$;

Do $i \geq 1$

If at step i of the Bunch and Kaufman decomposition a 1×1 pivot is performed, **then**

If $\nabla f(x_h)^T(\zeta_i w_i) > 0$ **then** set $uu = -\zeta_i w_i$ **else** $uu = \zeta_i w_i$.

Set $p_h = p_h + uu$.

If at step i of the Bunch and Kaufman decomposition a 2×2 pivot is performed, **then** set

$$\tilde{\zeta}_{i-1} = \begin{cases} sgn(\zeta_{i-1}) \max\{|\zeta_{i-1}|, \phi\} & i = 2 \\ \zeta_{i-1} & i > 2. \end{cases}$$

If $\nabla f(x_h)^T(\tilde{\zeta}_{i-1} w_{i-1}) > 0$ **then** set $uu = -\tilde{\zeta}_{i-1} w_{i-1}$ **else** $uu = \tilde{\zeta}_{i-1} w_{i-1}$.

If $\nabla f(x_h)^T(\zeta_i w_i) > 0$ **then** set $vv = -\zeta_i w_i$ **else** $vv = \zeta_i w_i$.

Set $p_h = p_h + uu + vv$.

End Do

In the next proposition we show that the direction p_h, obtained by using the scheme in Table 2 at any iterate of the Truncated Newton method, is gradient-related. The next result aims at rephrasing and generalizing the results in Proposition 3.2 of [5], after introducing the sequence $\{\omega_k\}$ in place of the parameter ω and the novel definition for the parameter ϕ in Table 2. Furthermore, the forthcoming result shows that any vector in the sequence $\{p_h\}$ is of *sufficient descent* and eventually is *(uniformly) bounded* by a positive finite constant value.

Proposition 4 *Let Assumption 1 hold. Let us consider Proposition 3 where we set $A = \nabla^2 f(x_h)$ and $b = -\nabla f(x_h)$. Assume the search direction p_h in (2) is computed as in Table 2. Then, the direction d_k in (11) satisfies $\|d_k\| < \mu$, for any $k \geq 1$, with $\mu > 0$, and p_h is a gradient-related direction.*

Proof The result surely holds if in Proposition 3 the Lanczos process performs just one iteration, inasmuch as $\gamma_2 < \varepsilon$. On the other hand, in case the Lanczos process has performed at least 2 iterations, the proof follows guidelines similar to those of Proposition 3.2 in [5], so that we only report the differences. For the case in which the Lanczos process performs a 2×2 pivot, by Table 2 we obtain

$$\begin{aligned}\nabla f(x_h)^T p_h &= -sgn[\nabla f(x_h)^T(\tilde{\zeta}_1 q_1)]\nabla f(x_h)^T(\tilde{\zeta}_1 q_1)\\ &\quad + \sum_{i>2} -sgn\left[\nabla f(x_h)^T(\zeta_i w_i)\right]\nabla f(x_h)^T(\zeta_i w_i)\\ &\leq -|\tilde{\zeta}_1|\|\nabla f(x_h)\| \leq -\phi\|\nabla f(x_h)\|^2 = \bar{\phi}\|\nabla f(x_h)\|^3,\end{aligned}$$

showing that p_h is gradient related.

As regards the property of boundedness for the vectors d_k in (11) and p_h, for any $h \geq 1$, we first have $d_k = Q_k y_k$, so that $\|d_k\| = \|y_k\| \leq \|T_k^{-1}\| \cdot \|\nabla f(x_h)\| \leq \|S_k^{-T} B_k^{-1} S_k^{-1}\|\|\nabla f(x_h)\| \leq \|S_k^{-1}\|^2 \cdot \|B_k^{-1}\| \cdot \|\nabla f(x_h)\|$. Now, following the analysis of Proposition 3.2 in [5] we can similarly prove that $\|S_k^{-1}\| \leq \beta$, where

$$\begin{aligned}\beta = \left(\frac{m_1}{\bar{\omega}\eta\varepsilon}\right) + \Bigg[\frac{k-m_1}{2}\max\Bigg(&4\max_{\ell}\{|\lambda_\ell(\nabla^2 f(x_h))|\}\left(\frac{1}{\varepsilon} + \tilde{\omega}\eta\right),\\ &\frac{16}{\varepsilon^2\xi}\max_{\ell}\{|\lambda_\ell(\nabla^2 f(x_h))|\}^2\Bigg)\Bigg] + k,\end{aligned}$$

being

$$\bar{\omega} = \min_{i \text{ is } 1\times 1 \text{ pivot step}} \{\omega_i\}$$

and

$$\tilde{\omega} = \max_{i \text{ is } 2\times 2 \text{ pivot step}} \{\omega_i\}.$$

In addition, we also need to provide a suitable bound for the diagonal blocks of B_k^{-1}. From the proof of Proposition 3 and the compactness of Ω we can easily obtain the next results:

- for 1×1 pivot: δ_i^{-1} is a diagonal block of B_k^{-1} and $|\delta_i|^{-1} \leq 1/\bar{\omega}\eta\varepsilon^2$,
- for 2×2 pivot: the reader can refer to the proof of Proposition 3.2 in [5]. □

4 Conclusions

In this paper we have considered efficient Truncated Newton methods for large scale unconstrained optimization problems, where the effective use of a modified Bunch and Kaufman decomposition within the SYMMBK algorithm is considered. We slightly modified the test performed at each iteration of the Bunch and Kaufman decomposition, using the guidelines in [5], so that a more general framework with respect to the last paper is obtained. In particular, we were able to prove that the numerical efficiency of the SYMMBK routine can be suitably coupled with some mild arrangements on the Bunch and Kaufman decomposition, so that the computed search direction for the optimization framework is gradient-related.

We are persuaded that further extensions can be studied, in the case the Truncated Newton method in hand also claims for the global convergence to limit points which satisfy both first and second order necessary optimality conditions. As well known, the accomplishment of the last result needs an accurate analysis of the normalized curvature of the Hessian matrix at any iterate, along any nonzero vector.

References

1. Bunch, J., Kaufman, L.: Some stable methods for calculating inertia and solving symmetric linear equations. Math. Comput. **31**, 163–179 (1977)
2. Caliciotti, A., Fasano, G., Nash, S.G., Roma, M.: An adaptive truncation criterion, for linesearch-based truncated Newton methods in large scale nonconvex optimization. Oper. Res. Lett. **46**, 7–12 (2018)
3. Caliciotti, A., Fasano, G., Nash, S.G., Roma, M.: Data and performance profiles applying an adaptive truncation criterion, within linesearch-based truncated Newton methods, in large scale nonconvex optimization. Data in Brief **17**, 246–255 (2018)
4. Caliciotti, A., Fasano, G., Roma, M.: Preconditioned nonlinear conjugate gradient methods based on a modified secant equation. Appl. Math. Comput. **318**, 196–214 (2018)
5. Caliciotti, A., Fasano, G., Potra, F., Roma, M.: Issues on the use of a modified Bunch and Kaufman decomposition for large scale Newton's equation. Comput. Optim. Appl. **77**, 627–651 (2020)
6. Chandra, R.: Conjugate gradient methods for partial differential equations. Ph.D. thesis, Yale University, New Haven (1978). Research Report 129
7. Conn, A.R., Gould, N.I.M., Toint, P.L.: Trust–Region Methods. MPS–SIAM Series on Optimization, Philadelphia, PA (2000)

8. Dembo, R., Steihaug, T.: Truncated-Newton algorithms for large-scale unconstrained optimization. Math. Program. **26**, 190–212 (1983)
9. Dembo, R., Eisenstat, S., Steihaug, T.: Inexact Newton methods. SIAM J. Numer. Anal. **19**, 400–408 (1982)
10. Fasano, G.: Planar–conjugate gradient algorithm for large–scale unconstrained optimization, Part 1: theory. J. Optim. Theory Appl. **125**, 523–541 (2005)
11. Fasano, G.: Planar–conjugate gradient algorithm for large–scale unconstrained optimization, Part 2: application. J. Optim. Theory Appl. **125**, 543–558 (2005)
12. Fasano, G.: Lanczos-conjugate gradient method and pseudoinverse computation, in unconstrained optimization. J. Optim. Theory Appl. **132**, 267–285 (2006)
13. Fasano, G., Roma, M.: Iterative computation of negative curvature directions in large scale optimization. Comput. Optim. Appl. **38**, 81–104 (2007)
14. Fasano, G., Roma, M.: Preconditioning Newton–Krylov methods in nonconvex large scale optimization. Comput. Optim. Appl. **56**, 253–290 (2013)
15. Grippo, L., Lampariello, F., Lucidi, S.: A truncated Newton method with nonmonotone linesearch for unconstrained optimization. J. Optim. Theory Appl. **60**, 401–419 (1989)
16. HSL: A collection of Fortran codes for large scale scientific computation. https://www.hsl.rl.ac.uk/
17. Marcia, R.: On solving sparse symmetric linear systems whose definiteness is unknown. Appl. Numer. Math. **58**(4), 449–458 (2008)
18. Nash, S.: A survey of truncated-Newton methods. J. Comput. Appl. Math. **124**, 45–59 (2000)
19. Nash, S., Sofer, A.: Assessing a search direction within a truncated-Newton method. Oper. Res. Lett. **9**, 219–221 (1990)

Mixed Integer Linear Programming for a Real-World Parallel Machine Scheduling Problem with Workforce and Precedence Constraints

Giulia Caselli, Maxence Delorme, Manuel Iori, and Carlo Alberto Magni

Abstract In this work, we consider a real-world scheduling problem occurring in the engineering test laboratory of a multinational company producing hydraulic components for motion systems. Similar problems have been solved in the literature under the framework of resource constrained parallel machine scheduling problems. In our work, the tests on the hydraulic components are the jobs to be scheduled. Each job must be processed on a machine and requires an additional human resource to prepare the machine and supervise the job. Machine and workforce eligibility constraints are also included. Release and due dates are given for jobs. The aim is to minimize the total weighted tardiness. Each job has a processing time expressed in working days that depends on the machine and requires a fixed number of hours per day for its assigned worker. Moreover, precedence and contiguity relations between jobs must be respected. We propose a Mixed Integer Linear Programming formulation to model the problem and demonstrate its effectiveness on both real-world and randomly generated instances.

Keywords Parallel machine scheduling problem · Workforce · Precedence constraints · Hydraulic components · Mixed integer linear programming

---

G. Caselli (✉) · C. A. Magni
School of Doctorate E4E (Engineering for Economics - Economics for Engineering), University of Modena and Reggio Emilia, Modena, Italy
e-mail: giulia.caselli@unimore.it; magni@unimore.it

M. Delorme
Department of Econometrics and Operations Research, Tilburg University, Tilburg, AB, The Netherlands
e-mail: m.delorme@tilburguniversity.edu

M. Iori
Department of Sciences and Methods for Engineering, University of Modena and Reggio Emilia, Reggio Emilia, Italy
e-mail: manuel.iori@unimore.it

L. Amorosi et al. (eds.), *Optimization in Artificial Intelligence and Data Sciences*, AIRO Springer Series 8, https://doi.org/10.1007/978-3-030-95380-5_6

1 Introduction

In the context of automation and other industries, engineering test laboratories are crucial to deliver high-quality and customized products on time to the market. Multiple resources as machines and workforce with eligibility restrictions, job requirements and additional constraints make test laboratory scheduling a complex and time-consuming activity if done manually. In the peculiar case of hydraulic systems test laboratories, the energy consumption levels are significantly high due to huge machines working without interruption with extreme levels of pressure on hydraulic circuits. Therefore, scheduling the tests in an effective manner is critical for both economic and environmental issues and automated decision support systems for activity scheduling is a common and urgent need.

In this paper, we focus on a real-world scheduling problem for an engineering test laboratory located in Italy and serving the global market of a multinational company producing hydraulic components for motion systems. Our scheduling problem requires to schedule a set of jobs on a set of machines (also called *workbench*) with an assigned worker (also called *technician*) and respecting resource constraints and release dates. Each job has a processing time that depends on the assigned machined and a weight representing the relative priority of the job. The goal is to minimize the total weighted tardiness considering jobs for a full scheduling over a long-term period. Further constraints will be detailed later in this work.

Our main reference will be the literature on the Resource Constrained Parallel Machine Scheduling Problem (RCPMSP), where a set of jobs must be processed on a set of machines with the requirement of an additional resource. We will also refer to a specific case of the well-known Resource Constrained Project Scheduling Problem (RCPSP) having some elements in common with our problem.

In our work, we consider a RCPMSP with many additional constraints representing our real environment. First, we introduce resource eligibility constraints as machines are unrelated parallel (i.e., different features that limit their usability to any kind of job). Also, we consider workers' skills and availability during the considered period and assign a specific worker to each job. We define two types of relations between jobs: In precedence relation, two jobs must be processed in a given order but without restriction on the machine (i.e., different types of test on the same product). In contiguity relation, two jobs must be processed in a given order and on the same workbench without any other job in between (i.e., similar repeated tests on the same product) but allowing idle times.

Our goal is to solve the real-world scheduling problem proposed by our industrial partner by means of a Mixed Integer Linear Programming (MILP) model able to provide an optimal solution for the full annual scheduling of activities. The model can be easily generalized to include further constraints, thus making it a powerful tool for addressing several real-world situations. In this direction, our final goal is to further test our solution method for more general applications.

The rest of the paper is organized as follows. A concise description of the related literature is given in Sect. 2. In Sect. 3, we provide a detailed description of the problem and present our mathematical model. Section 4 reports the results obtained by the proposed solution method on our real-world instance and on randomly created instances with the aim of evaluating the model performance and scalability. Concluding remarks are provided in Sect. 5.

2 Concise Literature Review

The RCPMSP is a scheduling problem extensively studied in the literature and representing many real-world applications, where additional resources are considered in the scheduling of jobs on machines. A survey on this class of problems is provided by Edis et al. [1]. Our focus is on RCPMSPs with unrelated parallel machine environment (i.e., processing times depend on independent machines) that are mentioned to be rarely studied in the literature. Pfund et al. [2] provides a review for this class of problems. Processing, discrete and renewable additional resources (e.g., workers) are involved in the models studied by Fanjul-Peyro et al. [3], where jobs are assigned directly to machines by mean of binary integer variables and additional constraints guarantee that sufficient additional resources are available in every time slot.

Machine eligibility constraints are introduced by further studies to model more realistic scheduling environments, where each job is allowed to be processed only on a subset of machines. Many real-world scheduling problems including machine eligibility are modelled and solved in the literature, e.g., see [4].

A common objective function in RCPMSPs is the minimization of the makespan, which is easier to handle with respect to due-date-based criteria as mentioned by Edis et al. [1]. However, other objective functions are proposed in the literature. In their Parallel Machine Scheduling (PMS) problem, [5] minimize the total absolute deviation of job completion times from due dates penalizing both earliness and tardiness. Other studies only consider tardiness, such as [6] who studied a PMS problem with the goal of weighted tardiness minimization subject to machine eligibility constraints, and [7] where the total weighted squared tardiness is minimized.

Many exact approaches and heuristic procedures are proposed in the literature to solve PMS problems, most of which are proven to be NP-hard. Among many mixed integer programming formulations, time-indexed models have been used in a variety of machine scheduling problems typically in a single machine environment, as stated in the survey by Unlu and Mason [8]. More recent studies include time-indexed variables in multiple machine environments. Some works, including [9], use binary time-indexed variables to indicate whether a job starts at a specific time slot. Fanjul-Peyro et al. [3] use similar variables to indicate the job completion time. A general model reported by Edis et al. [1] introduces three-index binary variables to state whether a job completes its processing time on a specific machine at a specific time. Heuristic procedures are commonly applied to PMS problems however, they are not of interest in this part of the study.

Precedence constraints are rarely included in machine scheduling, as stated by Edis et al. [1]. Indeed, they commonly appear in project scheduling problems, where some activities have to be grouped and scheduled and scarce resources have to be allocated, subject to various constraints. A reference for the traditional RCPSP is [10].

The Test Laboratory Scheduling Problem (TLSP) has been very recently studied by Mischek and Musliu [11] as an extension of the RCPSP and inspired by a real-world industrial context similar to the one of our interest. In the TLSP, tasks have to be grouped into jobs and afterward jobs have to be scheduled assigning a mode, resources and time slots, subject to various constraints. In the restricted scheduling sub-problem, also defined as TLSP-S, precedences and links are required between some tasks, that is, a given set of *predecessors* must be completed before starting the task and a set of tasks must be assigned to the same human resource. Two similar conditions occur between jobs in our problem.

We also mention some original machine scheduling problem formulations related to real-world scenarios. Edis and Ozkarahan [12] study a RCPMSP with machine eligibility in a molding department, providing an Integer Programming formulation and alternative CP-based approaches proven to be effective in some computational experiments. A work shift scheduling problem in a potash mine with setup times and many additional constraints is investigated by Seifi et al. [13]. The authors provide a MILP model with four-index binary variables and show how their formulation outperforms some heuristic procedures from the literature.

3 Problem and Method

In this section, a detailed description of the problem is provided, followed by the proposed mathematical model.

3.1 Problem Description

In the considered problem, there is a set J of jobs to be scheduled. Each job must be processed on one machine among the set M of available machines and requires one worker among the set K of available workers to prepare and supervise the processing of the job. As each job has different machine requirements and requires a set of human skills, compatibility constraints are also included. We define a set of machines M_j that are compatible with job j and a set of workers K_{ij} that are compatible with job j and machine i. A set T of working days is given, representing the time horizon of the scheduling. A processing time p_{ij} expressed in working days is required to process job j on machine i (i.e., unrelated parallel machines). Each job has a weight w_j representing the priority of job j and requires t_j hours of human work per day. Each job has a release date r_j defined by the arrival of material (i.e., product) for testing and a desired due date d_j. The number of working hours availability u_{kt} is given for each technician k and for each working day t of the considered period. Finally, precedence and contiguity relations between jobs are defined: A *precedence* relation (j, l) exists if job l can only start once job j is completed. A *contiguity* relation (j, l) occurs if jobs j and l must be processed on the same bench and in the specified order. Idle time is allowed between two contiguous jobs, but not the processing of any other job. Each job can have at most one contiguous job (i.e., $0 \leq |Q_j| \leq 1$, where Q_j is the set of jobs contiguous to job j), while more than one precedence relation per job is allowed (i.e., $|P_j| \geq 0$, where P_j is the set of jobs that follow job j).

3.2 Mathematical Model

We introduce a set of four-index binary variables x_{ijkt} that take value 1 if job j starts on day t on machine i with worker k, and 0 otherwise ($j \in J, i \in M_j, k \in K_{ij}, t \in T, t \geq r_j$), a set of continuous variables C_j representing completion time of job j and a set of continuous variables T_j defining the tardiness of job j (i.e., equal to $C_j - d_j$ if $C_j > d_j$, and 0 otherwise). We also set

$$S_{ijt} = \begin{cases} \{0, \ldots, t\} & \text{if } t \leq p_{ij} \\ \{t - p_{ij}, \ldots, t\} & \text{if } t > p_{ij} \end{cases}$$

where $t = 0$ is the first time slot in the considered time horizon. Every set S_{ijt} is used in the model to define the set of starting times that would cause machine i to be occupied if job j starts in a time slot included in the set.

The unrelated parallel machine scheduling problem with workforce and precedence constraints is defined as follows:

$$\min \sum_{j \in J} w_j T_j \tag{1}$$

subject to

$$\sum_{i \in M_j} \sum_{k \in K_{ij}} \sum_{t \in T} x_{ijkt} = 1 \qquad j \in J \tag{2}$$

$$\sum_{j \in J} \sum_{k \in K_{ij}} \sum_{\tau \in S_{ijt}} x_{ijk\tau} \leq 1 \qquad i \in M, t \in T \tag{3}$$

$$\sum_{j \in J} \sum_{i \in M_j} \sum_{\tau \in S_{ijt}} t_j x_{ijk\tau} \leq u_{kt} \qquad k \in K, t \in T \tag{4}$$

$$C_j = \sum_{i \in M_j} \sum_{k \in K_{ij}} \sum_{t \in T} x_{ijkt}(t + p_{ij}) \qquad j \in J \tag{5}$$

$$C_j \leq C_l - \sum_{k \in K_{il}} \sum_{t \in T} \sum_{i \in M_l} p_{il} x_{ilkt} \qquad j \in J, l \in P_j \cup Q_j \tag{6}$$

$$\sum_{k \in K_{ij}} \sum_{t \in T} x_{ijkt} = \sum_{k \in K_{il}} \sum_{t \in T} x_{ilkt} \qquad i \in M, j \in J, l \in Q_j \tag{7}$$

$$\sum_{k \in K_{ij'}} \sum_{j' \in J/\{j,l\}} x_{ij'kt} \leq 1 - \sum_{k \in K_{ij}} \sum_{t'=0}^{t} x_{ijkt'} + \sum_{k \in K_{il}} \sum_{t''=0}^{t} x_{ilkt''}$$

$$i \in M, t \in T, j \in J, l \in Q_j \tag{8}$$

$$T_j \geq C_j - d_j \qquad j \in J \tag{9}$$

$$x_{ijkt} \in \{0, 1\} \qquad j \in J, i \in M_j, k \in K_{ij}, t \in T \tag{10}$$

$$C_j, T_j \geq 0 \qquad j \in J \tag{11}$$

The objective function (1) minimizes the total weighted tardiness. Constraints (2) ensure that each job is processed exactly once and is assigned exactly to one machine among the set of compatible machines and one worker among the set of compatible workers. Constraints (3) guarantee that each machine processes at most one job at the same time (i.e., one job per day). Constraints (4) guarantee that the upper limit of available working hours per day is not exceeded for each technician. Constraints (5) define the completion time of each job (i.e., starting time plus processing time on the assigned machine). Constraints (6) guarantee that the order between jobs defined by precedence and contiguity relations is respected. Constraints (7) and (8) refer to contiguity relations between jobs (i.e., processing on the same machine and without any other job in between). Constraints (9) define jobs' tardiness T_j. Finally, constraints (10) state that variables x_{ijkt} are binary and constraints (11) state that variables C_j and T_j are continuous and positive.

4 Computational Experiments

In this section, we study the performances of the model (1)–(11) on a real-world instance from our industrial partner. Then, we discuss how we generated random instances used for the experiment and present the results. Our model was coded in Python 3.8 and solved with Gurobi 9.1.2. on a virtual machine Intel(R) Xeon(R) Gold 6130 with 2.1 GHz and 16 GB of RAM memory, running under Windows Server 2019 Standard. A time limit of 900 seconds per instance was imposed on the solver.

4.1 Results on the Real-World Instance

A real instance for the whole year 2019 including 97 jobs, 27 machines, 7 workers, and 400 working days was solved by the model to optimality in 228 seconds.

We consider 400 working days since there is a delay from the previous year that postpones many of the release dates from January 1, 2019 and jobs are scheduled five days per week. Eight hours per day is the maximum working time for every worker, and different workers' availability is given (e.g., 0 hours of availability during holidays). Machines can process at most one job per day. Release dates are spread during the entire year until day 230 over 400. Compatibility relations are given by the company based on real data. Two types of jobs are defined based on workers' working hours required per day: *Performance tests* require constant supervision of workers (i.e., eight hours per day). *Endurance tests* are usually longer than one week and do not require constant supervision of technicians (i.e., one hour per day as estimated by the company). Four levels of priority are translated into weights (w_j): high (1), medium-high (0.75), medium-low (0.5), and low (0.25) priority. Even though processing times depend on machines in the general problem, we consider equal processing times for every machine in our data as provided by the company. The average processing time is 10 working days. Finally, six precedence relations and 34 contiguity relations must be respected by our solution.

Comparing the solution obtained by the model with the solution implemented manually by the company we obtain an overall improvement of 80% on the weighted tardiness. In our solution, 28% of jobs are in delay against the 63%. The average delay of the model solution is equal to 10 days, while 43 days of average delay result in the real solution. Despite these outstanding results, several factors were not taken into account by our model: for example, the difference in jobs' processing time among machines, and the delayed release dates. Moreover, we overestimated the machine availability at the beginning of the year since we did not consider the delayed tasks from late 2018.

4.2 Random Instance Generation

Because no similar scheduling problems have been solved in the literature to the best of our knowledge, we have designed an ad-hoc data-generator to create random instances with different sizes (but inspired by our real-world case).

We consider one year of scheduling demand for every instance (i.e., time horizon equal to 400 days) varying the number of jobs, machines, and workers. We consider the following factors for every set of instances:

1. Three binary compatibility matrices have been randomly generated to define resource eligibility constraints in the following way: For jobs/machines compatibility, we use the same compatibility distribution as the one observed in our real-world case (i.e., 30% of machines compatible with every job), meaning

that, for each job, we randomly select $0.3 \times |M|$ compatible machines. For jobs/workers and machines/workers compatibility matrices we follow a similar procedure with compatibility 50% and 70%, respectively.
2. Release dates r_j are generated with U(0, 230) (i.e., a random integer uniformly distributed between day 0 and day 230). Due date of job j is generated as $d_j = r_j + U(30, 90)$.
3. Parameters t_j are randomly generated as equal to 8 with 80% of probability, 1 otherwise.
4. Jobs' weights are randomly generated as 1, 0.75, 0.5, 0.25 with probabilities 40%, 30%, 20%, 10%, respectively.
5. Processing times p_{ij} are randomly generated with uniform distributions: U(1, 10) for *performance* tests, U(10, 50) for *endurance* tests.
6. The number of precedence relations is a random integer between 0 and $\lceil \frac{n_{jobs}}{10} \rceil$; the number of contiguity relations is a random integer between 0 and $\lceil \frac{n_{jobs}}{5} \rceil$. We discard contiguity relations that would cause the instance to be infeasible.

Other factors may cause the generation of infeasible instances (e.g., not enough machines or workers to complete all the tasks before the time horizon). Infeasible instances are discarded.

4.3 Results on Randomly Generated Instances

The computational results on our mathematical model are summarised in Table 1. We have considered "small", "medium", "large", and "very large" groups of instances. Each group is called *N. jobs-N.machines-N.workers* based on the number of jobs, benches, and workers it contains. Average results shown in Table 1 are computed for each group on the first five feasible instances solved by the solver. In particular, we show (i) the number of instances solved to optimality (# Opt.), (ii) the average Upper Bound (UB), the average Lower Bound (LB), and the computed Gap ($Gap = 100 \times (UB - LB)/UB\%$), (iii) the average CPU time (i.e. time spent by the solver for the optimization process) expressed in seconds, (iv) the average number of variables, constraints, and non-zero coefficients of the model.

The average gap is equal to 0% for 12 out of 15 groups (i.e., all instances are solved to optimality). Groups with the highest ratio between number of jobs and number of machines (i.e., equal to 20) are the most critical ones to solve with the highest average CPU time and gap. Moreover, from Fig. 1 we derive that CPU time and solution value strongly depend on the instance generation procedure.

To conclude, our model is able to solve the selected groups of instances in most cases to optimality. We plan to work on the model to improve its computational performance as for bigger instances (e.g., with 200 jobs) it is not always able to find a feasible solution within our time limit.

Table 1 Summary of average results for every group of instances

Parameters			AVG results[a]							
Size	Group	Count	# Opt.	UB	LB	Gap (%)	CPU time (s)	nVar	nConstr	nNZ
Small	25-5-5	5	5	37.95	37.95	0%	10.06	32820	8891	1913714
Medium	50-5-5	5	5	100.65	100.65	0%	25.08	66937	10571	2974136
	50-10-5	5	5	23.95	23.95	0%	21.30	100241	15779	3819798
	50-10-7	5	5	33.55	33.55	0%	15.64	118614	12568	3886185
Large	100-5-5	5	1	402.35	306.35	19%	755.86	133904	11123	5116715
	100-10-5	5	5	74.05	74.05	0%	83.40	199678	25559	8704074
	100-10-7	5	5	123.60	123.60	0%	64.17	240824	14324	7993412
	100-20-5	5	5	46.70	46.70	0%	86.77	412546	27952	14020107
	100-20-7	5	5	73.40	73.40	0%	112.50	477280	49605	20582006
	100-20-10	5	5	62.50	62.50	0%	126.80	594467	58826	27137796
Very large	200-10-7	5	0	1766.80	259.30	81%	900.84	479144	18638	14885107
	200-20-7	5	4	169.75	163.25	2%	571.83	944139	62744	37745809
	200-20-10	5	5	147.35	147.35	0%	362.23	1162038	63945	45411608
	200-40-7	5	5	132.05	132.05	0%	587.39	1903131	99617	66621444
	200-40-10	5	5	155.60	155.60	0%	649.22	2344446	200264	118129430

[a]Average results on five instances per group

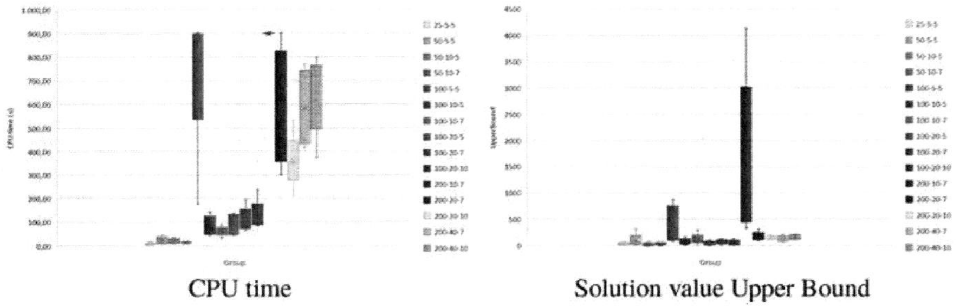

Fig. 1 Boxplots for the different groups of instances

5 Conclusions

In this paper, we have studied a real-world scheduling problem with unrelated parallel machines and workforce occurring in an engineering test laboratory. Eligibility constraints are included in our problem since incompatibilities occur between jobs and resources and between resources (i.e., machines and workforce). We have also considered precedence and contiguity relations between jobs.

We have proposed a Mixed Integer Linear Programming formulation that minimizes the total weighted tardiness, while satisfying all the given constraints. With the presented model, we were able to solve a real-world instance representing the scheduling demand of the whole year 2019 in 228 seconds of CPU time. The solution obtained an improvement of 80% on the total weighted tardiness. The proposed mathematical formulation can be extended to more general versions of scheduling problems (e.g., unrelated parallel machines, additional resources).

Further computational experiments on randomly created instances have shown good scalability of the model on small- and medium-sized instances and a worse performance on large-sized instances up to 200 jobs. Computational time limits were reached when additional larger instances were tested.

Obtaining an optimal solution is important for our industrial partner when scheduling their full annual activity. One direction for our future research will be to improve the performance of the proposed mathematical model.

Moreover, since rescheduling is required in practice on a daily or weekly basis due to several types of perturbation such as delayed release dates and machines' failures, we plan to continue our research focusing on decomposition and metaheuristic procedures able to find good solutions in shorter times both for the scheduling and the related rescheduling problem. We have obtained some preliminary good results on a constructive heuristic algorithm that we plan to improve with local search.

Acknowledgment We thank Dana Motion Systems Italy S.r.l. for describing the real-world problem and providing us with the data.

References

1. Edis, E.B., Oguz, C., Ozkarahan, I.: Parallel machine scheduling with additional resources: notation, classification, models and solution methods. Eur. J. Oper. Res. **230**(3), 449–463 (2013)
2. Pfund, M, Fowler, J.W., Gupta, J.N.D.: A survey of algorithms for single and multi-objective unrelated parallel-machine deterministic scheduling problems. J. Chin. Inst. Ind. Eng. **21**(3), 230–241 (2004)
3. Fanjul-Peyro, L., Perea, F., Ruiz, R.: Models and matheuristics for the unrelated parallel machine scheduling problem with additional resources. Eur. J. Oper. Res. **260**(2), 482–493 (2017)
4. Leung, J.Y.-T., Li, C.: Scheduling with processing set restrictions: a literature update. Int. J. Prod. Econ. **175**, 1–11 (2016)
5. Ventura, J.A., Kim, D.: Parallel machine scheduling with earliness–tardiness penalties and additional resource constraints. Comput. Oper. Res. **30**(13), 1945–1958 (2003)
6. Su, H., Pinedo, M., Wan, G.: Parallel machine scheduling with eligibility constraints: a composite dispatching rule to minimize total weighted tardiness. Nav. Res. Logist. **64**(3), 249–267 (2017)
7. Schaller, J., Valente, J.M.S.: Efficient heuristics for minimizing weighted sum of squared tardiness on identical parallel machines. Comput. Ind. Eng. **119**, 146–156 (2018)
8. Unlu, Y., Mason, S.J.: Evaluation of mixed integer programming formulations for non-preemptive parallel machine scheduling problems. Comput. Ind. Eng. **58**(4), 785–800 (2010)
9. Edis, E.B., Oguz, C.: Parallel machine scheduling with flexible resources. Comput. Ind. Eng. **63**(2), 433–447 (2012)
10. Brucker, P., Drexl, A., Möhring, R., Neumann, K., Pesch, E.: Resource-constrained project scheduling: notation, classification, models, and methods. Eur. J. Oper. Res. **112**(1), 3–41 (1999)
11. Mischek, F., Musliu, N.: A local search framework for industrial test laboratory scheduling. Ann. Oper. Res. **302**, 1–30 (2021)
12. Edis, E.B., Ozkarahan, I.: Solution approaches for a real-life resource-constrained parallel machine scheduling problem. Int. J. Adv. Manuf. Technol. **58**(9), 1141–1153 (2012)
13. Seifi, C., Schulze, M., Zimmermann, J.: A new mathematical formulation for a potash-mine shift scheduling problem with a simultaneous assignment of machines and workers. Eur. J. Oper. Res. **292**(1), 27–42 (2021)

Scheduling K-mers Counting in a Distributed Environment

Lavinia Amorosi, Lorenzo Di Rocco, and Umberto Ferraro Petrillo

Abstract Alignment-free algorithms are used in bioinformatics to efficiently evaluate the similarity between pairs of genomic sequences. They work by extracting and aggregating features from the sequences under study and, then, by comparing them using alignment-free functions. When working on large collections of huge sequences, it is possible to improve the performance of these algorithms by executing the extraction and the aggregation steps, in parallel, across the computing nodes of a distributed system. In this work, we address the problem of finding the optimal schedule to use for assigning to computing nodes the features to be aggregated, in order to minimize the maximum aggregation time. For this purpose, we consider one exact mathematical programming approach and two approximated ones, based on the Longest Processing Time heuristic. These have been implemented using the *Gurobi* solver, and compared to the algorithm used by the Spark distributed computing framework for assigning tasks to computing nodes. The experiments have been performed on some large collections of genomic sequences well-known in literature. The results show that the proposed approaches perform favourably with respect to the scheduling strategy used by Spark, in terms of quality of the solution. In particular the exact approach, run up to the time limit, allows to reduce the makespan up to 77.11%, when considering the largest instance tested. However, the required computational time of the exact approach is not compatible with its online application. The approximated approaches appear more promising in terms of computational time, while providing good quality solutions.

Keywords Identical parallel machines · Job scheduling · Genomic analysis

L. Amorosi · L. Di Rocco (✉) · U. F. Petrillo
Department of Statistical Sciences, Sapienza University of Rome, Rome, Italy
e-mail: lavinia.amorosi@uniroma1.it; lorenzo.dirocco@uniroma1.it;
umberto.ferraro@uniroma1.it

L. Amorosi et al. (eds.), *Optimization in Artificial Intelligence and Data Sciences*,
AIRO Springer Series 8, https://doi.org/10.1007/978-3-030-95380-5_7

1 Introduction

Sequence alignment is a fundamental task in bioinformatics. It is used to arrange genomic or proteomic sequences so to as to identify matching or similar regions. These could be, in turn, due to structural, evolutionary or functional relationships between the considered sequences. The computational cost of sequence alignment algorithms led to the introduction of alternative sequence comparison methods based on an *alignment-free* approach. By this term, we denote those algorithms that determine the similarity between pairs of sequences by, first, extracting a set of features from each sequence and, then, by comparing features instead of the original sequences.

Despite their improved efficiency, even alignment-free algorithms may fail to deliver results in an acceptable time, especially when working with large collections of (possibly huge) genomic sequences. Distributed computing can help to significantly reduce the execution time of these algorithms, by splitting the individual steps they require into independent processes that are executed as tasks on the nodes of a distributed system. This approach has been repeatedly studied in the scientific literature and there are several contributions such as [15, 16] highlighting the potential performance improvement from this approach. However, we note that this speedup is only possible if the running algorithm is able to use all available computational resources, simultaneously, by providing each computing node with approximately the same workload.

In our work, we focus on the optimization of a particular task that exists in distributed alignment-free algorithms. It is the task of choosing which computing node is responsible for collecting and aggregating all occurrences of the same feature. This is a fundamental task as a good allocation allows for an even *workload distribution*, thus exploiting the intrinsic parallelism of a distributed system and reducing the overall completion time.

To this end, we note that the assignment methods commonly used by distributed alignment-free algorithms are those available, by default, from the underlying distributed computing framework. It turns out that these methods do not provide an even distribution of workload, when applied to features extracted by an alignment-free algorithm.

In this paper, we present two optimization approaches and compare their performance with that of the standard assignment method available in the Apache Spark platform [17] and used by some of the main alignment-free distributed algorithms. All experiments have been performed by considering, as datasets, the features extracted from some collection of genomic sequences commonly used for benchmarking alignment-free algorithms.

2 Alignment-Free Algorithms

The term alignment-free (AF, for short) refers to a class of algorithms, that are useful for identifying regions of similarity between different genomic or proteomic sequences, and are an alternative to sequence alignment-based methods. These approaches include, e.g., methods based on exact or approximate k-mer counts, length of common substrings, micro-alignments, and many others (we refer the interested reader to [18] for a thorough review of existing approaches).

The most popular AF approach is, for sure, the one based on k-mer counts [13]. Let Σ be an alphabet and S be a collection of sequence of characters drawn from Σ. We define k-mer as any contiguous substring of length k in S_i, with $S_i \in S$. In a few words, AF algorithms establish the distance/similarity between pairs of sequences in S by comparing their corresponding k-mers frequency tables, using an AF function. There is a vast literature of functions for this purpose, including functions based on the Euclidean distance [18] and on the D2 distance [18].

The whole process of AF evaluation can be summarized in two steps. In the first step, the distinct k-mers that exist in each sequence S_i of S are extracted and counted. As a result, a k-mer frequency table is available for each of the considered sequences. In a second step, k-mer frequency tables of distinct sequences are evaluated pairwise using an AF function. As a result, a similarity/distance matrix is available, which indicates the AF distance/similarity between S_i and S_j at position (i, j).

3 Related Works

The problem we face in this paper can be modeled as a scheduling problem on identical parallel machines, i.e., P||Cmax. This problem is NP-hard and has been extensively studied in combinatorial optimization literature, because of its many different applications. Most of the proposed approaches consist of approximation algorithms like: [7, 8]. The pioneering approximation algorithm to solve this problem is the Longest Processing Time (LPT) rule, proposed by Graham [10]. It consists of sorting an input set of n jobs in non-ascending order by their processing times $p_j (j = 1, \ldots, n)$ and then, given the sorted job set, assigning one job at a time, to the machine whose workload is the lowest so far. Due to its simplicity and good performance, especially when the number of jobs becomes larger, many exact and heuristic algorithms have been developed using this method. Dosa [5] and Dosa and Vizvari [6] are examples of approaches explicitly based on this rule.

As for exact approaches, the literature is smaller. We mention [4], for approaches based on the branch-and-bound algorithm, [14], for an approach based on the cutting-plane algorithm, and [3], for an approach consisting of a scatter search heuristic followed by an exact algorithm based on a specialized binary search and a branch-and-price scheme. The latter uses the relationship between the P||Cmax and the well-known bin packing problem (BPP).

Approximation algorithms have also been designed, that exploit the relation between P||Cmax and BPP. We can mention MULTIFIT [1]. They provide better worst-case performance than the LPT rule, but at the cost of higher running times.

In literature, also polynomial time approximation schemes (PTASs) have been derived for this problem. The most recent one has been proposed in [12].

Finally, in a recent work by Della Croce and Scatamacchia [2], a revisited LPT heuristic rule, called SLACK, has been proposed which improves the approximation ratio derived by Graham, but keeping the same computational complexity.

4 Methods

4.1 Distributed Counting of k-mers

Counting the k-mers present in an input sequence is apparently a very simple task. All that is needed is a linear scan of the input sequence and an associative data structure to store the frequencies for each k-mer. The problem turns out to be quite complex when solved on a distributed environment.

Suppose we have an input collection S of genomic sequences. We are interested in evaluating the matrix D_f, where $D_f(i, j)$ reports the AF distance/similarity between sequences i and j, as evaluated by the AF measure f, with $i, j \leq S$. We also assume that this evaluation is to be performed on a distributed system of n computing nodes, so as to scale the procedure execution time up to $1/n$ the time required to fulfill the same procedure on a non-distributed setting.

The challenge here is to implement the procedure outlined in Sect. 2 by using all n computing nodes simultaneously, in all steps.

A typical distributed layout, followed by distributed algorithms as implemented in [15, 16], assumes that S has been partitioned into n parts, of approximately equal size, each of which is associated with a particular computing node. Furthermore, the range of 4^k possible k-mers is partitioned in m non-overlapping bins of size $\frac{4^k}{m}$, each identified by a unique id number, thanks to a *binning* function. Then, the AF evaluation procedure is organized into the following four distinct and sequential distributed tasks:

- **Task 1: K-mers extraction and partial counting.** Each node extracts and counts k-mers from its part of the input sequences. For each input sequence i, the outcoming counts are recorded in a sequence of m frequency tables (i.e., one for each bin). Upon completion of this task, a set of (key,value) pairs is returned containing, as key, the bin id, and, as value, the table reporting the corresponding sequence id and frequency counts.
- **Task 2: K-mers counts aggregation.** All k-mers frequency tables that have the same bin id are sent to the same computing node where they are aggregated, according to the sequence they refer to.

- **Task 3: AF function partial evaluation** Each node performs a partial evaluation of the AF function of choice, by considering the k-mers frequency tables related to the bins it has been assigned. When finished, this task will return a list of (key, value) pairs having, as key, the id of the two sequences being considered and, as value, their partial AF similarity/distance evaluation.
- **Task 4: AF function complete evaluation** All partial AF similarity/distance evaluations targeting a same pair of input sequences are sent to a same computing node where they will be aggregated thus returning, as a result, their similarity/distance.

Regarding this procedure, provided that input sequences have been partitioned in roughly equal parts, we observe that Task 1 can be executed almost perfectly in parallel, thus has an execution time that is typically about $1/n$ of the time required on a single computing node. We also observe that Tasks 3 and 4 have a very short execution time, since the number of partial distances/similarities to be processed in a typical setting is orders of magnitude smaller than the number of k-mers they refer to. The remaining task, Task 2 is crucial for two reasons. First, the data transfer operation required to send all frequency tables sharing the same bin id to a computing node is typically very time consuming. Second, a wrong assignment of bins to computing nodes, can lead to an unbalanced workload distribution where some nodes are overloaded by a large amount of k-mers frequency counts to process and others have very little work to do.

4.2 Scheduling Models

We model the problem as an identical parallel machines scheduling problem with no preemption, where given a set of n jobs, having an associated processing time p_j $(j = 1, \ldots, n)$, and given a set of m parallel identical machines, each job must be assigned to an exactly one machine so as to minimize the maximum completion time (*makespan*). In our case, jobs refer to the work needed for counting the frequencies of distinct k-mers existing in one particular bin. The time required to process a job is approximated to the number of k-mers in the corresponding bin. Finally, machines represent the computing nodes in charge of processing bins.

The problem, denoted as $P||C_{max}$, is NP-hard [9] and can be mathematically formulated as follows:

$$\min C_{max} \tag{1}$$

$$C_{max} \geq \sum_{j=1}^{N} p_j x_{ij} \quad \forall i \in \{1..m\} \tag{2}$$

$$\sum_{i=1}^{m} x_{ij} = 1 \quad \forall j \in \{1..N\} \tag{3}$$

$$x_{ij} \in \{0, 1\} \quad \forall i \in \{1..m\} \; j \in \{1..N\} \tag{4}$$

where C_{max} is the *makespan*, N is the number of jobs, m is the number of machines, p_j is the processing time of job j, x_{ij} is the decision variable. The value is 1 if the job j is assigned to machine i, otherwise it is 0. We denote this model as W_{opt}.

Due to its complexity, for this problem different approximated and heuristic algorithms have been proposed in literature. Among them, one of the most known is the LPT rule, which consists in sorting jobs with respect to their processing times in non-increasing order and then iteratively assigning the next job to the machine whose current completion time, that is the sum of the processing times of jobs assigned to that machine, is minimum. We denote this model as W_{lpt}.

The worst-case performance ratio of LPT is equal to $\frac{4}{3} - \frac{1}{3}m$. In [2], the authors propose a revised version of this rule, called *slack*. This latter consists, first, in applying the LPT rule for sorting a set of jobs, then, in splitting this set in tuples of m consecutive jobs $(1, \ldots, m; m + 1, \ldots, 2m;$ etc.), to be sorted according to the difference between the longest and the shortest job in the tuple. Then, following this new ordering, each unscheduled job is assigned to the first available machine, that is a list scheduling algorithm is applied. The approximation ratio of this modified version of LPT is $\frac{4}{3} - \frac{1}{3(m-1)}$ for $m \geq 3$ and $\frac{9}{8}$ for $m = 2$. We denote this model W_{lpt}.

5 Experimental Evaluation

In this section, we report the results of the experiments we conducted to evaluate the performance of the scheduling models we presented in Sect. 4.2, in terms of makespan and model computation time. We also give some insights into the impact of the choice of the k-mers length (i.e., parameter k) on the performance of each scheduling model considered.

5.1 Experimental Setup

We consider three datasets, that are commonly used in the literature for benchmarking AF functions (see, e.g., [16]). The first, here denoted **Yersinia**, contains the assembled sequences of 8 bacteria strains. The average length of each sequence is about 4,600,000 characters, giving a total size of about 37,000,000 characters. The second, here denoted *E. coli*, contains the assembled sequences of 29 *E. coli*/Shigella strains. The average length of each sequence is about 5,000,000 characters, giving a total size of about 145,000,000 characters. The third, here denoted **Plants**, contains the unassembled set of about 1,900,000 reads derived from 14 sequences of the *Plants* taxonomic group. The average length of each read is 150 bp, giving a total size of about 290,000,000 characters.

As a preliminary step, we extract the k-mer counts from these datasets using the FADE framework [16]. To investigate the impact of the k-mer length on the performance of the considered models, we consider two possible assignments for k: $11, 25$. Then, we process the outcoming k-mers distributions with the considered models, i.e., W_{opt}, W_{lpt} and W_{slack}. The W_{opt} model has been implemented using the *Gurobi* optimization solver [11]. The remaining models have been coded as Python applications.

5.2 Preliminary Analysis: The Role of k in Determining the k-mers Distributions

The length of the k-mers is a key parameter for many AF methods. Fixed an input genomic sequence, smaller values of k reduce the number of possible distinct substrings, and thus increasing the number of high frequency k-mers. Conversely, larger values of k imply a larger number of unique k-mers, which have a lower frequency.

The binning function typically used by distributed k-mer counting systems is simply based on applying a standard hash function to the numerical encoding of the statistic (i.e., the k-mer), modulo the number of bins. Indeed, this choice could lead to an unbalanced partitioning where multiple large bins are assigned to the same node of the distributed system, affecting the overall completion time. We observe that this data skew is also strongly affected by the length of the k-mers.

To investigate this dependency, we performed a preliminary statistical analysis on the distributions that arise with increasing values of k, when we extract k-mers from the considered datasets. We report in Fig. 1 the results for the (**Plants**) dataset. Smaller values of k determine a histogram distributed over a larger range with a higher value of the standard deviation σ, indicating significant differences between

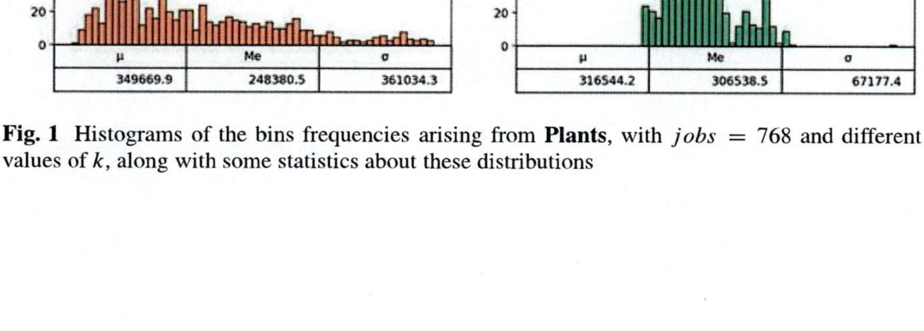

Fig. 1 Histograms of the bins frequencies arising from **Plants**, with $jobs = 768$ and different values of k, along with some statistics about these distributions

bins frequencies. Moreover, the value of μ tends to be larger than the median value, confirming the presence of a large number of bins with a low workload, versus a group of overloaded bins.

On the other hand, as k increases, the mean value of the distribution is more representative since the k-mers are scattered more homogeneously across the bins. Hence, there is a significantly narrower range of variation.

5.3 Experimental Results

We divide the discussion on two parts. First, we present the results of the direct comparison between the performance of the considered models and that of the scheduling approach of distributed k-mer counting frameworks, here denoted W_{std}. We then analyze the behavior of these approaches as a function of k.

5.3.1 Comparison of the Overall Performance

Table 1 reports the results of the experiment described in Sect. 5.1, for the specific case of $k = 11$, in terms of makespan and model computation time. As a first observation, we note that the exact model (i.e., W_{opt}) manages to significantly improve the quality of the solution with respect to W_{std}. However, we note that this significant performance gain requires a very long solution time, which makes this option less attractive for a real use in a distributed k-mer counting framework.

If we instead consider the performance of the heuristic solution (i.e., W_{slack}), we find that its quality is not that far from W_{opt}, but with a much smaller solution time. For the sake of readability, we omit to report the case of W_{lpt}, since its performance

Table 1 Performance improvement provided by W_{opt} and W_{slack}, in terms of makespan percentage reduction with respect to W_{std}, when considering increasing values of m and $jobs$, where the $jobs$ is a function of m. For W_{std}, we report the makespan. For W_{opt} and W_{slack}, we also report the solution time, in seconds

m	**jobs** (w.r.t m)	W_{std}	W_{opt} (%)	Time (s)	W_{slack} (%)	Time (s)
128	1x	894,585	2.60	0.750	0.00	0.006
	2x	–	38.66	2	38.66	0.001
	4x	–	66.72	4	66.72	0.002
	8x	–	69.51	79	69.51	0.004
256	1x	548,769	0.00	3	0.00	0.002
	2x	–	47.46	10	45.74	0.004
	4x	–	75.07	761	74.58	0.011
	8x	–	75.13	1075	75.11	0.007
512	1x	297,753	0.00	45	0.00	0.082
	2x	–	66.27	464	66.27	0.020
	4x	–	67.46	1756	67.46	0.031
	8x	–	77.08	7223	77.05	0.053

is very close to that of W_{slack}. Moreover, as expected, the table shows that setting the number of jobs very close to the number of machines yields a naive scheduling approach, since the makespan coincides with the completion time of the longest job. Increasing the number of jobs, gives better schedules. However, we also observe that, except for the case $jobs = m$, using a larger number of jobs, once the number of machines is fixed, does not yield a significant improvement in the performance of W_{slack} and of W_{opt}. On the other hand, W_{std} does not seem to benefit from a more aggressive binning strategy as its makespan remains roughly the same, regardless of the ratio between $jobs$ and m.

We report a similar behavior for the case of $k = 25$ (data not shown but available upon request).

5.3.2 Performance of Our Methods with Respect to k and to the Dataset Size

Our expectation is that the performance of W_{std} is influenced by the empirical distribution of the k-mers to be counted. Indeed, a heterogeneous distribution leads to an unbalanced jobs assignment that does not allow to fully exploit the distributed computing system. We also expect that this phenomenon to be more pronounced when using small values of k, as anticipated in Sect. 5.2.

To investigate this behavior, we have run a second experiment where we measured the trend of the workload reduction provided by W_{slack} as a function of k. The results, visible in Fig. 2, clearly show that with smaller values of k, W_{slack} achieves a more significant workload reduction as there is more room for improvement. Instead, when considering larger values of k, there is still a significant workload reduction, although less pronounced than in the previous case. This trend can be explained by looking at the standard deviation values (see Fig. 1, $k = 11, 25$).

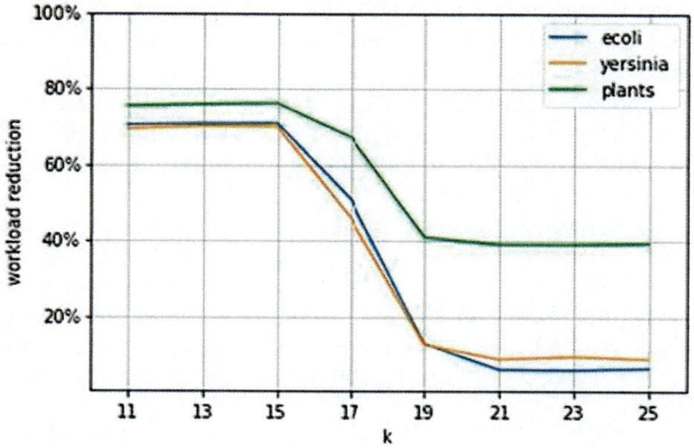

Fig. 2 Makespan improvement, in percentage, provided by W_{slack} with respect to W_{std}, when considering $m = 128$, $jobs = 768$, and increasing values of k

We observe that the greater the reduction in terms of standard deviation is, the higher is the slope of the curve between two values of k. Along the same line, the flat segments of the curve are due to almost constant values of the standard deviation. We also observe that the performance improvements we observe are more important for larger datasets, like **Plants**, likely due to the larger size of bins due to increased redundancy in the input files.

6 Conclusion and Future Directions

In this work, we have investigated the efficiency of the scheduling strategy adopted for counting k-mers during the execution of an alignment-free algorithm on a distributed environment. Our proposed heuristic approach, based on the SLACK algorithm, is compared with the standard one commonly used in these cases, and yields promising results. As expected, the time performances of the exact resolution of the mathematical formulation are not compatible with its online adoption. Instead, the solution obtained by the SLACK algorithm is almost equivalent to the exact solution in most of the cases, but requires only a tiny fraction of its time, while providing a much more balanced assignment of jobs than that of the standard scheduling strategy. As a further direction, we are considering the possibility of developing and experimenting with other scheduling algorithms. Another direction we are pursuing is to effectively integrate our approach with real-world distributed k-mer counting frameworks.

References

1. Coffman, E.G.Jr., Garey, M.R., Johnson, D.S.: An application of bin-packing to multiprocessor scheduling. SIAM J. Comput. **7**, 1–17 (1978)
2. Della Croce, F., Scatamacchia, R.: The longest processing time rule for identical parallel machines revisited. J. Scheduling **23**(2), 163–176 (2020)
3. Dell'Amico, M., Iori, M., Martello, S., Monaci, M.: Heuristic and exact algorithms for the identical parallel machine scheduling problem. INFORMS J. Comput. **20**(3), 333–344 (2008)
4. Dell'Amico, M., Martello, S.: A note on exact algorithms for the identical parallel machine scheduling problem. Eur. J. Oper. Res. **160**, 576–578 (2005)
5. Dosa, G.: Graham example is the only tight one for P || Cmax. Ann. Univ. Sci. Budapest **47**, 207–210 (2004)
6. Dosa, G., Vizvari, A.: The general algorithm LPT(k) for scheduling identical parallel machines. Alkamaz. Mat. Lapok **23**(1), 17–37 (2006)
7. França, P.M., Gendreau, M., Laporte, G., Müller, F.M.: A composite heuristic for the identical parallel machine scheduling problem with minimum makespan objective. Comput. Oper. Res. **21**, 205–210 (1994)
8. Frangioni, A., Necciari, E., Scutellà, M.G.: A multi-exchange neighborhood for minimum makespan machine scheduling problems. J. Combin. Optim. **8**, 195–220 (2004)
9. Garey, M.R., Johnson, D.S.: Computers and Intractability, vol. 174. W. H. Freeman, San Francisco (1979)

10. Graham, R.L.: Bounds on multiprocessing timing anomalies. SIAM J. Appl. Math. **17**(2), 416–429 (1969)
11. Gurobi Optimization, LLC.: Gurobi optimizer version 9.1 (2021). http://www.gurobi.com/
12. Jansen, K., Klein, K.M., Verschae, J.: Improved efficient approximation schemes for scheduling jobs on identical and uniform machines. In: Proceedings of the 13th Workshop on Models and Algorithms for Planning and Scheduling Problems (MAPSP 2017), pp. 77–79 (2017)
13. Luczak, B.B., James, B.T., Girgis, H.Z.: A survey and evaluations of histogram-based statistics in alignment-free sequence comparison. Brief. Bioinform. **20**(4), 1222–1237 (2017)
14. Mokotoff, E.: An exact algorithm for the identical parallel machine scheduling problem. Eur. J. Oper. Res. **152**, 768–769 (2004)
15. Petrillo, U.F., Sorella, M., Cattaneo, G., Giancarlo, R., Rombo, S.E.: Analyzing big datasets of genomic sequences: fast and scalable collection of K-mer statistics. BMC Bioinf. **20**(4), 138 (2019)
16. Petrillo, U.F., Palini, F., Cattaneo, G., Giancarlo, R.: Alignment-free genomic analysis via a big data spark platform. Bioinformatics **37**(12), 1658–1665 (2021)
17. Zaharia, M., Chowdhury, M., Franklin, M.J., Shenker, S., Stoica, I.: Spark: cluster computing with working sets. 10:10 (2010)
18. Zielezinski, A., Girgis, H.Z., Bernard, G., Leimeister, C.-A, Tang, K., Dencker, T., Lau, A.K., Röhling, S., Choi, J.J., Waterman, M.S., Comin, M., Kim, S.-H., Vinga, S., Almeida, J.S., Chan, C.X., James, B.T., Sun, F., Morgenstern, B., Karlowski, W.M.: Benchmarking of alignment-free sequence comparison methods. Genome Biol. **20**(1), 144 (2019)

Workload Balancing at Cross-Docking Platforms

Nicolas Zufferey, Marc-Antoine Coindreau, Olivier Gallay, and Gilbert Laporte

Abstract We consider the management of cross-docking platforms (CPs). Every day, suppliers deliver products to the CP by trucks (inbound flow). The same day or during another day, the products are then loaded into containers (outbound flow). At the end of the planning horizon (one week in this work), boats ship the containers to offshore production plants. Workers are required to perform the involved loading/unloading tasks. Constrained by the fact that a container can only be loaded if its entire content is available at the CP, most of the loading is currently done during the last days of the week. A mixed-integer linear program (MILP) is proposed to smooth the workload. It allows to simultaneously manage the inbound and outbound flows. Results are proposed for real instances and highlight the benefit of the proposed approach when compared to a common practice rule, where inbound and outbound flows are managed independently.

Keywords Workload balancing · Cross-docking platforms · Integrated logistics

1 Introduction

In this work, we model and solve a problem faced by CM (a car manufacturer that cannot be named because of a non-disclosure agreement). CM consolidates the product flows of its European suppliers to its offshore plants (in America, Africa, or

N. Zufferey (✉) · M.-A. Coindreau
GSEM, University of Geneva - Uni Mail, Geneva, Switzerland
e-mail: n.zufferey@unige.ch; marc-antoine.coindreau@unige.ch

O. Gallay
Faculty of Business and Economics (HEC Lausanne), Department of Operations, University of Lausanne, Lausanne, Switzerland
e-mail: olivier.gallay@unil.ch

G. Laporte
HEC Montréal, Montréal, QC, Canada
e-mail: gilbert.laporte@hec.ca

L. Amorosi et al. (eds.), *Optimization in Artificial Intelligence and Data Sciences*, AIRO Springer Series 8, https://doi.org/10.1007/978-3-030-95380-5_8

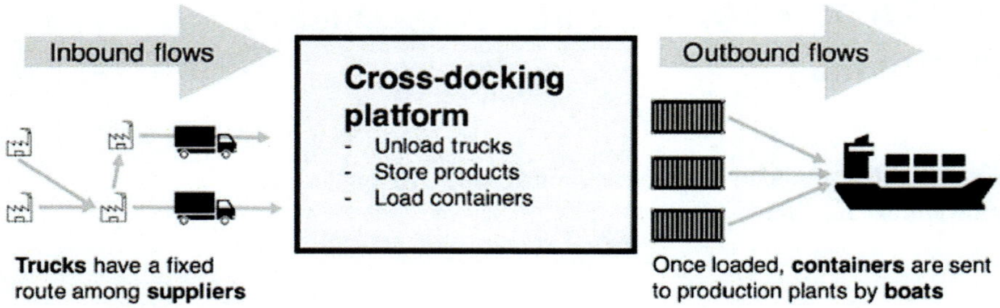

Fig. 1 Operations related to a CP

Asia) by routing these products through cross-docking platforms (CPs). Over one week (from Monday to Friday), products are delivered by trucks to the CPs. These products are then sorted, reconditioned (to satisfy the constraints of sea transport, in particular) and loaded into containers which are finally shipped by boats to the production plants. The reader interested in having more information on the related cross-docking literature is referred to [3, 12, 15, 16], whereas papers on integrated logistics include [6, 11, 13]. The flows related to the CP are depicted in Fig. 1.

CM seeks to smooth the workload over the week. The workload of a day is proportional to the volume of products handled (i.e., the sum of the volumes of products unloaded from trucks and loaded into containers). To do this, CM has determined the content of the trucks, the content of the containers, and the loading day of the containers. We denote this problem as the CM problem.

For each CP, CM solves several complex optimization problems. First of all, CM defines in advance the truck routes to collect the products from the suppliers (e.g., [2]). Truck routes cannot be modified (i.e., the truck arrival day and the list of visited suppliers), but CM can nevertheless choose the products that are collected from each supplier in each route (as long as the truck loading constraints are satisfied). Next, CM minimizes the number of containers needed to ship all the products. Finally, during the loading phase, CM must wait until all the products composing a container are present in the CP before starting to load the container (the number of loading doors being limited).

Currently, CM independently solves the optimization problems related to the content of trucks and containers. Once these problems have been solved, CM decides on the loading day of each container. As the entire content of the containers must be available in the CP at the time of loading, it appears that CM must wait until the end of the week to carry out most of the loading operations. This creates unbalanced workload within the CP over the week (the workload is 2–3 times larger in the last days of the week than in the first days).

In this work, we propose to jointly optimize the content of trucks and containers to smooth the workload over the week. Changing the content of trucks and containers involves taking into account complex loading constraints (size, weight and position of the transported products). A detailed description of these constraints can be found in [14]. The full consideration of loading constraints would lead to an untractable optimization problem. However, these loading problems can be simplified as the products are loaded into boxes, and the loading constraints only apply to the boxes regardless of the products they contain. In the data provided by CM, more than 70% of the boxes can transport different products with weight variations of less than 10 kg. Thus, for a given loading of boxes into containers (resp. trucks), we can evaluate a large number of box-to-product assignments without violating the loading constraints.

The contributions of this work are the following. First, we model the CM problem with a mixed-integer linear program (MILP). Second, from a managerial point of view, we quantify the gain brought by the joint optimization of truck and container contents in the CPs by comparing our results with current industrial practice on real instances provided by CM.

2 Model

In a preliminary work [5], we considered a simplified version of the CM problem detailed above, where we did not take into account the truck contents. The reader is referred to that work for a description of the associated literature review and for the justification of the $\mathcal{NP}$-hardness of the problem considered here.

2.1 Considered Sets, Parameters and Variables

The following sets, parameters and variables are considered, where T denotes the time horizon (i.e., days). The exponents $^{(in)}$ (resp. $^{(out)}$) refer to the parameters related to trucks (resp. containers).

The following sets are considered.

- C is the set of CP clients (i.e., production plants)
- P is the set of product types
- B is the set of box types
- S is the set of suppliers
- I is the set of trucks
- I_t is the set of trucks arriving on day $t \in T$
- O is the set of containers
- O_c is the set of containers that are sent to the client $c \in C$

The following parameters are given.

- $g_p \in \mathbb{N}$: number of units of product type $p \in P$ already available in the inventory at the beginning of the week

- M_{op}: largest amount of product of type $p \in P$ that can be transported in container $o \in O$
- $d_{cp} \in \mathbb{N}$: demand of client $c \in C$ for product type $p \in P$ (in units)
- $l_{pb} \in \mathbb{R}^+$: weight of a box of type $b \in B$ when filled with product type $p \in P$ (in kg)
- $q_{pb} \in \mathbb{N}$: number of units of product type $p \in P$ that can be transported in box type $b \in B$
- $n_{ib}^{(in)} \in \mathbb{N}$: number of units of boxes of type $b \in B$ transported in truck $i \in I$
- $n_{ob}^{(out)} \in \mathbb{N}$: number of units of boxes of type $b \in B$ transported in container $o \in O$
- $l^{(in)} \in \mathbb{R}^+$: maximum allowed weight that can be transported by a truck (in kg)
- $l^{(out)} \in \mathbb{R}^+$: maximum allowed weight that can be transported by a container (in kg)
- $h_p \in \mathbb{R}^+$: volume of a product of type $p \in P$ (in m^3)
- $\pi_{pi} = 1$ if truck $i \in I$ visits the supplier that can provide product type $p \in P$ ($\pi_{pi} = 0$ otherwise)

  The following variables have to be determined.

- $u_{pt} \in \mathbb{N}$: number of units of product type $p \in P$ in stock on day $t \in T$ before loading the containers
- $v_{pt} \in \mathbb{N}$: number of units of product type $p \in P$ in stock on day $t \in T$ after loading the containers
- $r_{pt} \in \mathbb{N}$: number of units of product type $p \in P$ received on day $t \in T$
- $s_{pt} \in \mathbb{N}$: number of units of product type $p \in P$ sent on day $t \in T$
- $z_{ibp} \in \mathbb{N}$: number of boxes of type $b \in B$ assigned to product type $p \in P$ in truck $i \in I$
- $x_{obp} \in \mathbb{N}$: number of boxes of type $b \in B$ assigned to product type $p \in P$ in container $o \in O$
- $w_{opt} \in \mathbb{N}$: number of units of product type $p \in P$ sent by container $o \in O$ on day $t \in T$
- $m_t \in \mathbb{R}$: workload performed on day $t \in T$ (in m^3)
- $f \in \mathbb{R}$: workload difference between the most loaded day and the least loaded one
- $y_{ot} = 1$ if container $o \in O$ is loaded on day $t \in T$ ($y_{ot} = 0$ otherwise)

2.2 Presentation of the MILP

The MILP is described as follows.

$$\text{Minimize } f \tag{1}$$

Subject to

$$f \geq m_{t_1} - m_{t_2} \quad t_1, t_2 \in T \tag{2}$$

$$m_t = \sum_{p \in P} h_p \cdot r_{pt} + \sum_{o \in O} \sum_{p \in P} h_p \cdot w_{opt} \quad t \in T \tag{3}$$

$$v_{pt} = u_{pt} - s_{pt} \quad p \in P, t \in T \tag{4}$$

$$u_{pt} = v_{p,t-1} + r_{pt} \quad p \in P, t \in T \tag{5}$$

$$v_{p0} = g_p \quad p \in P \tag{6}$$

$$r_{pt} = \sum_{b \in B} \sum_{(i \in I_t | \pi_{pi} > 0)} q_{pb} \cdot z_{ibp} \quad p \in P, t \in T \tag{7}$$

$$s_{pt} = \sum_{o \in O} w_{opt} \quad p \in P, t \in T \tag{8}$$

$$\sum_{t \in T} y_{ot} = 1 \quad o \in O \tag{9}$$

$$w_{opt} \leq M_{op} \cdot y_{ot} \quad t \in T, o \in O, p \in P \tag{10}$$

$$w_{opt} \leq \sum_{b \in B} q_{pb} \cdot x_{obp} \quad t \in T, o \in O, p \in P \tag{11}$$

$$\sum_{o \in O_c} \sum_{t \in T} w_{opt} = d_{cp} \quad c \in C, p \in P \tag{12}$$

$$\sum_{b \in B} \sum_{p \in P} l_{pb} \cdot x_{obp} \leq l^{(out)} \quad o \in O \tag{13}$$

$$\sum_{p \in P} x_{obp} \leq n_{ob}^{(out)} \quad o \in O, b \in B \tag{14}$$

$$\sum_{b \in B} \sum_{p \in P} l_{pb} \cdot z_{ibp} \leq l^{(in)} \quad i \in I. \tag{15}$$

$$\sum_{p \in P | \pi_{pi} > 0} z_{ibp} \leq n_{ib}^{(in)} \quad i \in I, b \in B \tag{16}$$

Constraints (2) set the difference between the most loaded working day and the least loaded one. Constraints (3) calculate for each day the value of the workload. Constraints (4) (resp. (5)) compute the available inventory in the CP at the end (resp. at the beginning) of the day. Constraints (6) fix the initial inventory in the CP at the beginning of the planning horizon (i.e., the products not obtained in the planning horizon already belong to the inventory from the first day of the week). For each day, constraints (7) determine the amount of products obtained at the CP, whereas constraints (8) compute the number of units sent for each product type.

Constraints (9) ensure that no container is loaded more than once. Constraints (10) restrict each product to be sent on the loading day of a container. For each container, constraints (11) bound the amount of products sent. Constraints (12) satisfy the demand of each client. The loading constraints of the containers (resp. trucks) are in constraints (13) and (14) (resp. (15) and (16)). More precisely, constraints (13) (resp. (15)) limit the weight of the transported products to the container (resp. the truck) capacity, and constraints (14) (resp. (16)) ensure that the number of boxes transported in a container (resp. in a truck) satisfy the allowed upper bound.

3 Experiments

We use CPLEX 12.10.0.0 to solve the MILPs. The employed computer has the following configuration: 2.2 GHz Intel Core i7 with 16 Go 1600 MHz DDR3 of RAM memory. CPLEX is used in deterministic parallel mode with 8 threads. Default parameters were used to perform the computation (preprocessing presolve and aggregator with primal and dual reduction).

3.1 Presentation of the Instances

We consider the data for two different CPs provided by CM, denoted as V and G. Table 1 presents the characteristics of the considered instances regarding the number of containers, trucks, suppliers and clients. It also indicates the number of different product types and box types.

Table 1 Characteristics of the test instances

Instance	$\|O\|$	$\|I\|$	$\|P\|$	$\|B\|$	$\|S\|$	$\|C\|$
V1	28	48	326	206	151	17
V2	51	78	358	290	171	20
V3	49	67	424	315	190	21
V4	59	82	454	334	191	20
G1	67	98	1181	616	544	8
G2	71	112	1199	644	554	7
G3	68	89	1353	575	572	8
G4	88	112	1401	718	606	8
G5	80	122	1548	605	646	8
G6	85	136	1676	748	678	7

Table 2 Performances of CPLEX preprocessing step

	Preprocessing							Branch and cut		
Instance	Time (s)	Best int.	Best bound	Nb. rows	Nb. cols	Nb. bin.	Nb. gen.	Time (s)	Nb. nodes	Opt. value
V1	<1	777	723	499	606	94	506	0.37	1587	777
V2	<1	1597	1515	831	761	155	600	5	5354	1626
V3	<1	1780	1780	339	472	124	342	0	1	1780
V4	<1	2071	1866	1162	1175	220	949	3	3469	2024
G1	2	2795	2643	5213	4914	388	4520	0	1	2795
G2	65	883	774	12,913	10,846	560	10,280	780	4899	881
G3	14	3806	3587	12,876	11,324	593	10,765	40	3716	3700
G4	148	2352	2260	11,887	10,520	650	9864	620	3389	2352
G5	180	2042	1772	23,792	21,521	753	20,762	>3600	4804	2022
G6	21	2739	2473	19,298	18,278	969	17,303	212	5612	2717

3.2 *CPLEX Performances*

Table 2 presents the output of the CPLEX preprocessing and branch-and-cut steps. Columns "Time (s)" give the time spent (in seconds) for both these steps. Column "Best int" (resp. "Best bound") indicates the best integer solution (resp. the best bound) obtained after the preprocessing step, whereas column "Opt. value" gives the value of the optimal solution. Columns "Nb. rows", "Nb. cols", "Nb. bin." and "Nb. gen." present the characteristics of the reduced MIP obtained after the preprocessing step (i.e., the number of rows and columns of the matrix describing the constraints, and the number of binary and integer variables). Column "Nb. nodes" gives the number of subproblems explored by CPLEX during the branch-and-cut phase. Overall, this table shows that the preprocessing step is rather efficient as it provides a feasible solution that is on average 1% above the optimal solution returned at the end of the branch-and-cut part, within a computation time representing on average 5% of the computation time associated with the branch-and-cut phase.

3.3 *Benefit of the Integrated Approach*

In Table 3, the following solutions are compared.

- The solutions of the configuration where both the truck and container contents are optimized are displayed under column Q (it is the most generic problem).
- The solutions of the configuration where the container contents are fixed and set to the contents used by CM are displayed under column Q_x. In this case, the values of the x_{obp} variables are known and do not change during the resolution of the MILP.
- The solutions of the configuration where the truck contents are fixed and set to the contents used by CM are displayed under column Q_z. In this case, the values

Table 3 Results of the different configurations for the V and G instances

Instance	$Q_{x,z}$		Q_z			Q_x			Q				
	Value	Time	Value	Time	% CM	Value	Time	% CM	Value	Time	% CM	% Q_z	% Q_x
V1	1037	<1	1037	<1	0.0%	780	<1	−24.8%	778	<1	−25.0%	−25.0%	−0.3%
V2	1863	<1	1757	<1	−5.7%	1728	<1	−7.2%	1626	<1	−12.7%	−7.5%	−5.9%
V3	1946	<1	1841	<1	−5.4%	1781	<1	−8.5%	1778	<1	−8.6%	−3.4%	−0.2%
V4	2528	<1	2304	<1	−8.9%	2237	<1	−11.5%	2024	<1	−19.9%	−12.2%	−9.5%
G1	3964	<1	3317	<1	−16.3%	3701	<1	−6.6%	2774	<1	−30.0%	−16.4%	−25.0%
G2	2492	<1	1593	8	−36.1%	1818	<1	−27.0%	881	13	−64.6%	−44.7%	−51.5%
G3	5014	<1	3977	<1	−20.7%	4661	<1	−7.0%	3700	<1	−26.2%	−7.0%	−20.6%
G4	3565	<1	2633	5	−26.1%	3173	<1	−11.0%	2352	10	−34.0%	−10.7%	−25.9%
G5	4859	<1	3703	31	−23.8%	3328	<1	−31.5%	2022	>60	−58.4%	−45.4%	−39.2%
G6	4915	<1	3697	<1	−24.8%	4087	<1	−16.8%	2717	4	−44.7%	−26.5%	−33.5%

of the z_{ibp} variables are known and do not change during the resolution of the MILP.

- The solutions of the configuration where both the container and the truck contents are fixed are displayed under column $Q_{x,z}$. It captures the current practice at CM, where the truck and container contents are built at an earlier stage with two different optimization systems and, ultimately, the decision maker builds "by hand" the loading day of each container (in a step-by-step fashion). In this case, the values of the x_{obp} and z_{ibp} variables are known and do not change during the resolution of the MILP.

The optimal solution value is given in column "Value", whereas column "Time" indicates the time needed for CPLEX to find optimality (in minutes). Columns "% CM", "% Q_z" and "% Q_x" give the improvement percentage with respect to CM, Q_z and Q_x, respectively. For instance, the improvement of configuration Q over Q_z is given in column "% Q_z" and is computed as $\frac{f(Q)-f(Q_z)}{f(Q_z)}$, where $f(Q_z)$ (resp. $f(Q)$) is the obtained workload gap when considering configuration Q_z (resp. Q).

As already highlighted in [5], reworking the container contents leads to significant improvements with respect to the current practice (on average, an improvement of 6.2% for the V instances, and of 30% for the G instances). Similar improvements are obtained here when reworking the truck contents (on average, an improvement of 13% for the V instances and of 19% for the G instances). The main achievement is obtained when reworking the truck and container contents in an integrated manner (together with the loading day of the containers). Indeed, when compared to [5], the additional average improvement obtained by solving Q instead of Q_z increases to 11.8% (resp. 30%) for the V (resp. G) instances. When compared to the current practice at CM, the average improvement is of 19% for the V instances, and 70% for the G instances. Moreover, for the V and G instances, the largest execution time to find the optimal solutions with CPLEX is 31 minutes.

4 Conclusions and Future Works

Considering various operations in cross-docking platforms, we have proposed models and solution techniques for a real problem using real data. The workload has to be smoothed over the planning horizon (a week), which is obtained through the minimization of the gap between the most loaded working day and the least loaded one. We compare our results with current practice at CM, that acts as a non-integrated solution method where truck and container contents are optimized independently. Computational experiments showed that allowing product reassignment from one container to another and from a truck to another leads to improvements up to 70%.

A next step of this project would consist in considering larger instances (i.e., larger CPs), for which the MILP approach cannot be used. In such a context, the use of a fix-and-optimize matheuristic [4, 7, 9] would be relevant. It consists in iteratively solving sub-problems generated by temporarily fixing the value of certain variables in the MILP (e.g., fix some truck or container contents or both). This differs from the literature on filtering techniques [10], where non-promising solutions are definitely discarded from the search process. Note that the size of the fixed parts of the problem (and as a consequence, the size of the remaining problem to solve with the MILP) could be managed in a nested and well-controlled fashion, as it is the case for variable-neighborhood-search metaheuristics [1, 8].

References

1. Bierlaire, M., Thémans, M., Zufferey, N.: A heuristic for nonlinear global optimization. INFORMS J. Comput. **22**(1), 59–70 (2010)
2. Boctor, F., Laporte, G., Renaud, J.: Heuristics for the traveling purchaser problem. Comput. Oper. Res. **30**(4), 491–504 (2003)
3. Boysen, N., Fliedner, M.: Cross dock scheduling: classification, literature review and research agenda. Omega **38**(6), 413–422 (2010)
4. Chen, H.: Fix-and-optimize and variable neighborhood search approaches for multi-level capacitated lot sizing problems. Omega **56**, 25–36 (2015)
5. Coindreau, M.A., Gallay, O., Zufferey, N., Laporte., G.: Integrating workload smoothing and inventory reduction in three intermodal cross-docking platforms of a European car manufacturer. Comput. Oper. Res. **112**, 104762 (2019)
6. Darvish, M., Coelho, L.: Sequential versus integrated optimization: production, location, inventory control, and distribution. Eur. J. Oper. Res. **268**, 203–214 (2018)
7. Dorneles, Á.P., de Araújo, O.C., Buriol, L.S.: A fix-and-optimize heuristic for the high school timetabling problem. Comput. Oper. Res. **52**, 29–38 (2014)
8. dos Santos, J.P.Q., de Melo, J.D., Neto, A.D.D., Aloise, D.: Reactive search strategies using reinforcement learning, local search algorithms and Variable Neighborhood Search. Expert Syst. Appl. **41**, 4939–4949 (2014)
9. Helber, S., Sahling, F.: A fix-and-optimize approach for the multi-level capacitated lot sizing problem. Int. J. Prod. Econ. **123**(2), 247–256 (2010)
10. Hertz, A., Schindl, D., Zufferey, N.: Lower bounding and tabu search procedures for the frequency assignment problem with polarization constraints. 4OR **3**(2), 139–161 (2005)
11. Respen, J., Zufferey, N., Wieser, P.: Three-level inventory deployment for a luxury watch company facing various perturbations. J. Oper. Res. Soc. **68**(10), 1195–1210 (2017)
12. Serrano, C., Delorme, X., Dolgui, A.: Scheduling of truck arrivals, truck departures and shop-floor operation in a cross-dock platform, based on trucks loading plans. Int. J. Prod. Econ. **194**, 102–112 (2017)
13. Thevenin, S., Zufferey, N., Widmer, M.: Order acceptance and scheduling with earliness and tardiness penalties. J. Heuristics **22**(6), 849–890 (2016)
14. Toffolo, T., Esprit, E., Wauters, T., Vanden Berghe, T.: A two-dimensional heuristic decomposition approach to a three-dimensional multiple container loading problem. Eur. J. Oper. Res. **257**(2), 526–538 (2017)
15. Van Belle, J., Valckenaers, P., Cattrysse, D.: Cross-docking: state of the art. Omega **40**(6), 827–846 (2012)
16. Yu, W., Egbelu, P.J.: Scheduling of inbound and outbound trucks in cross docking systems with temporary storage. Eur. J. Oper. Res. **184**(1), 377–396 (2008)

Knowledge Before Solutions: Some Reflections on a Successful O.R. Case Study

Giovanni Righini and Pier Giorgio Villani

Abstract This paper describes how a capacity planning problem arising in health care services design and optimization was successfully tackled with mathematical programming techniques. What made the project successful was not the design of a sophisticated algorithm providing optimal solutions, but rather the iterative development of an integer linear programming model of the problem, solved by a general-purpose MILP solver. This approach was made possible by the characteristics of the mathematical model itself and the user-friendly tools that were used. As a result, the problem expert could autonomously challenge and improve the model and the data in a countless number of iterations with little or no intervention of the O.R. expert. This allowed to reduce the development cost to zero and the development time to a few days.

Keywords Decision science · Mathematical modelling · Capacity planning

1 Introduction

On 18/02/2020 a 37-year-old man in apparent good health and with no pathological history came to the emergency room of Codogno hospital for fever, dyspnea and productive cough, on X-ray evidence of right basal pneumonia. Excluding the most common causes of pneumonia, the alarm bell, which lead to the execution of the swab test for COVID19, was his distant connection with China, linked to the visit of an acquaintance who had recently returned from the East. On February 20th at 9:30 pm confirmation of the positivity at the swab test for Sars-Cov-2 arrived [3]. The regional administration of Lombardy gave the crisis unit of Lodi hospital complete

G. Righini (✉)
Department of Computer Science, University of Milan, Milan, Italy
e-mail: giovanni.righini@unimi.it

P. G. Villani
Department of Emergency and Critical Care, Hospital of Cremona, Cremona, Italy
e-mail: piergiorgio.villani@asst-cremona.it

L. Amorosi et al. (eds.), *Optimization in Artificial Intelligence and Data Sciences*, AIRO Springer Series 8, https://doi.org/10.1007/978-3-030-95380-5_9

power in decision-making for managing the emergency [2]. On February 21st, the crisis unit decided to close the Codogno hospital to concentrate resources in a single hospital, that of Lodi. For three months the operating room activity in Codogno and Lodi was interrupted and all physicians of the anesthesia and resuscitation service were engaged in facing the health emergency in Lodi.

At the end of May 2020 the heads of the two hospitals decided to reopen the first aid and intensive care service in Codogno together with the operating rooms of Lodi and Codogno, at the beginning of June. Therefore, the problem arose of ensuring coverage of all work shifts in the two hospitals, while at the same time guaranteeing annual leave periods in the summer for all physicians of the anesthesia and intensive care service.

The problem was tackled and solved thanks to a mathematical optimization model, developed in collaboration between a physician and an O.R. expert. The aim of this short paper is to illustrate and discuss the use of mathematical optimization models for decision-support in a real situation, highlighting the importance of knowledge generation compared with optimal solution computation as well as the flexibility and ease-of-use of models explicitly described in mathematical terms.

2 The Problem

The problem was a capacity planning problem: a given number of physicians with non-identical skills have to be assigned to six different types of work shifts for a given number of weeks, allowing some of them to be on annual leave in each week. The goal was not to define a complete work schedule for the physicians, taking into account, for instance, preferences, additional activities, balance requirements in work shifts assignments in holidays and other details (a problem of this kind is described in [1]). The goal was rather to understand how the strategic decision of reopening the health services could be implemented. A formal description of the problem is given hereafter.

Data The problem data are the following:

1. a set P of physicians;
2. a set T of types of work shifts; namely: Morning, Afternoon, Night, On-call availability, Operating room, Clinic service, Rest (compulsory day of rest following a night work shift);
3. a set D of days of the week (1=Monday, 7=Sunday);
4. a set W of weeks (the planning horizon);
5. a set S of hospitals (Lodi and Codogno).

Additional data appear only in some constraints and therefore they are described when these constraints are introduced in the remainder.

Variables The main variables of the model are the following:

1. Binary variables x representing the selected assignments have five indices: $x_{twdsp} = 1$ if and only if physician $p \in P$ is assigned a work shift of type $t \in T$ in day $d \in D$ of week $w \in W$ in hospital $s \in S$.
2. Additional binary variables y represent non-working days besides days on leave: $y_{wdp} = 1$ if and only if physician $p \in P$ does not work in day $d \in D$ of week $w \in W$.
3. Variables $h' \geq 0$ and $h'' \geq 0$ count the number of work shifts not covered by the available staff of physicians, but assigned to external resources, which implies an extra cost for the administration: h'_{wd} is the number of operating room work shifts assigned to external resources in day $d \in D$ of week $w \in W$; h''_{wd} is the number of night work shifts assigned to external resources in day $d \in D$ of week $w \in W$. Owing to the integrality of the x variables and the right-hand-sides of the assignment constraints (see below), it is not necessary to impose integrality requirements on these variables.

The model also includes other variables that appear only in some constraints and are thus described when needed in the remainder.

Constraints

Assignment Constraints for Work Shifts in Lodi ($s = 1$)

1. Five work shifts must be covered every day: morning, afternoon, night, on-call availability and rest.

$$\sum_{p \in P} x_{twd1p} = 1 \; \forall w \in W, d \in D, t \in \{1, 2, 3, 4, 7\}$$

2. A given number ω_w of operating rooms ($t = 5$) must be active in Lodi in the working days ($d = 1, \ldots, 5$) of each week. Operating rooms are not active in the week-ends.

$$\sum_{p \in P} x_{5wd1p} + h'_{wd} = \omega_w \quad \forall w \in W, d \in \{1, \ldots, 5\}$$

$$\sum_{p \in P} x_{5wd1p} = 0 \;\; \forall w \in W, d \in \{6, 7\}$$

3. A clinic service work shift ($t = 6$) is required every Tuesday and Thursday ($d = 2, 4$) and not in the other days of the week.

$$\sum_{p \in P} x_{6wd1p} = 1 \;\; \forall w \in W, d \in \{2, 4\}$$

$$\sum_{p \in P} x_{6wd1p} = 0 \;\; \forall w \in W, d \in \{1, 3, 5, 6, 7\}$$

Assignment Constraints for Work Shifts in Codogno ($s = 2$)

1. Two work shifts must be covered every day: morning and on-call availability ($t = 1, 4$).

$$\sum_{p \in P} x_{twd2p} = 1 \ \forall w \in W, d \in D, t \in \{1, 4\}$$

2. Afternon and clinic service work shifts ($t = 2, 6$) are not required.

$$\sum_{p \in P} x_{twd2p} = 0 \ \forall w \in W, d \in D, t \in \{2, 6\}$$

3. A night work shift ($t = 3$) must be covered every day, either by the available staff or by external resources.

$$\sum_{p \in P} x_{3wd2p} + h''_{wd} = 1 \ \forall w \in W, d \in D$$

4. A single work shift for the operating room ($t = 5$) must be covered on Wednesdays and Fridays ($d = 3, 5$). The operating room is not active in the other days of the week.

$$\sum_{p \in P} x_{5wd2p} = 1 \quad \forall w \in W, d \in \{3, 5\}$$

$$\sum_{p \in P} x_{5wd2p} = 0 \quad \forall w \in W, d \in \{1, 2, 4, 6, 7\}$$

Skills Not all physicians can be assigned to each work shift: incompatibilities are easily forbidden by fixing the corresponding binary variables x to 0.

Compulsory Pairings Every night work shift ($t = 3$) must be immediately followed by a rest day ($t = 7$).

$$x_{3,w,d,s,p} = x_{7,w,d+1,s,p} \forall w \in W, d \in \{1, \ldots, , 6\}, s \in S, p \in P$$

$$x_{3,w,7,s,p} = x_{7,w+1,1,s,p} \forall w \in W, s \in S, p \in P.$$

Similar constraints were also introduced to force the correct pairing of the rest day in day 1 of week 1 with the last night work shift of the previous planning period in each hospital.

Forbidden Pairings By contrast, in some other cases pairs of work shifts were declared incompatible, thus forbidding their assignment to the same person. For instance, on-call availability work shifts are incompatible with morning and afternoon shifts in the next day.

$$\sum_{s \in S} \left(x_{4,w,d-1,s,p} + x_{1,w,d,s,p} \right) \leq 1 \ \ \forall w \in W, d \in D, p \in P$$

$$\sum_{s \in S} \left(x_{4,w,d-1,s,p} + x_{2,w,d,s,p} \right) \leq 1 \ \ \forall w \in W, d \in D, p \in P$$

Similar constraints were introduced at the boundaries of the planning horizon.

Joint Work Shifts Some work shifts can be joined, i.e. they can be assigned to the same physician in the same day. The constraint that forbids multiple assignments of shifts to physicians has the form $\sum_t x_{twdsp} \leq 1 \ \forall w, d, s, p$. The possibility of joining two work shifts was introduced by assigning them coefficient $1/2$ in the left-hand-side of the constraint. These constraints were one of the main issues that were examined, to explore the boundary between feasible and infeasible instances. Here is a sample set of constraints among the many that were tested.

1. In the working days ($d = 1, \ldots, 5$) in Lodi $s = 1$ it is allowed to join morning and afternoon shifts $t = 1, 2$ as well as clinic service and on-call availability shifts ($t = 4, 5$).

$$\frac{1}{2} \sum_{t \in \{1,2,4,5\}} x_{t,w,d,1,p} + \sum_{t \in \{3,6,7\}} x_{t,w,d,1,p} + y_{w,d,p}$$
$$\leq 1 \ \ \forall w \in W, d \in \{1, \ldots, 5\}, p \in P$$

2. On Saturdays ($d = 6$) in Lodi ($s = 1$) the morning shift and the on-call availability shift ($t = 1, 4$) can be joined.

$$\frac{1}{2} \sum_{t \in \{1,4\}} x_{t,w,6,1,p} + \sum_{t \in \{2,3,7\}} x_{t,w,6,1,p} + y_{w,6,p} \leq 1 \ \ \forall w \in W, p \in P$$

3. On Sundays ($d = 7$) in Lodi ($s = 1$) the morning shift and the afternoon shift ($t = 1, 2$) can be joined.

$$\frac{1}{2} \sum_{t \in \{1,2\}} x_{t,w,7,1,p} + \sum_{t \in \{3,4,7\}} x_{t,w,7,1,p} + y_{w,7,p} \leq 1 \ \ \forall w \in W, p \in P$$

4. When operating rooms are active in Codogno ($s = 2$) on Wednesday and Friday ($d = 3, 5$), the operating rooms shifts $t = 5$ can be joined with on-call availability shifts ($t = 4$).

$$\frac{1}{2} \sum_{t \in \{4,5\}} x_{t,w,d,2,p} + \sum_{t \in \{1,3,7\}} x_{t,w,d,2,p} + y_{w,d,p} \leq 1 \ \ \forall w \in W, d \in \{3, 5\}, p \in P$$

5. In the other days in Codogno no work shifts can be joined.

$$\sum_{t\in\{1,3,4,7\}} x_{t,w,d,2,p} + y_{w,d,p} \le 1 \;\; \forall w \in W, d \in \{1,2,4,6,7\}, p \in P$$

6. No two work shifts can be joined if they belong to different hospitals.

$$x_{t',w,d,1,p} + x_{t'',w,d,2,p} \le 1 \;\; \forall t' \in T, t'' \in T, w \in W, d \in D, p \in P$$

Forced Joined Work Shifts In some cases two work shifts are mandatorily joined, i.e. they must be assigned to the same physician in the same day.

1. On Sundays ($t = 7$) in Lodi ($s = 1$) morning and afternoon work shifts are joined.

$$x_{1,w,7,1,p} = x_{2,w,7,1,p} \;\; \forall w \in W, p \in P$$

2. On Saturdays ($t = 6$) in both hospitals morning and on-call availability work shifts ($t = 1, 4$) are joined.

$$x_{1,w,6,s,p} = x_{4,w,6,s,p} \;\; \forall w \in W, s \in S, p \in P$$

3. In working days ($d = 1, \ldots, 5$) in Lodi ($s = 1$) the physician who is available on-call ($t = 4$) is also assigned an operating room work shift (but not necessarily vice versa).

$$x_{4,w,d,1,p} \le x_{5,w,d,1,p} \;\; \forall w \in W, d \in \{1,\ldots,5\}, p \in P$$

4. On Wednesdays and Fridays ($d = 3, 5$) in Codogno ($s = 2$) the physician who is available on-call ($t = 4$) is also assigned an operating room work shift (but not necessarily vice versa).

$$x_{4,w,d,2,p} \le x_{5,w,d,2,p} \;\; \forall w \in W, d \in \{3,5\}, p \in P.$$

Days Off A complicating feature of the model is the presence of days off. An integer variable r_{wp} indicates how many days off a physician $p \in P$ must have in week $w \in W$. Days off in week w are assigned to physicians who have been assigned demanding work shifts, such as a joint morning+afternoon shift on Sunday of week $w - 1$ or a night shift on Saturday or Sunday in week $w - 1$.

$$r_{w,p} = x_{1,w-1,7,1,p} + \sum_{s\in S}(x_{3,w-1,6,s,p} + x_{3,w-1,7,s,p}) \;\; \forall w \in W, p \in P$$

Similar constraints are used to make the plan in week 1 consistent with the last work shifts assigned in the previous days.

Annual Leaves All physicians must be on leave for two weeks every year. This period is usually concentrated in two consecutive weeks in the summer. The need for this study was triggered by the question whether leave periods were compatible with the need of covering all services in the two hospitals and it proved that actually this would have been impossible without changing the constraints of the problem.

A binary variable $f_{wp} = 1$ indicates that physician $p \in P$ is on leave in week $w \in W$. These variables occur in several constraints.

1. Days off cannot be taken in leave weeks

$$r_{wp} + f_{wp} \leq 1 \;\; \forall w \in W, p \in P.$$

   Suitable boundary constraints ensure that in the last days of the planning horizon not too many demanding work shifts are assigned to physicians in the subset of those who still have to be assigned leave weeks. This is done to make these constraints feasible in the next planning period.
2. Physicians on leave cannot be assigned any work shift, apart from the rest day following a night shift (a rest day can occur at the beginning of a leave week).

$$x_{t,w,d,s,p} + f_{wp} \leq 1 \;\; \forall t \in \{1, \ldots, 6\}, w \in W, d \in D, p \in P, s \in S.$$

3. The number of non-working days for each physician during a non-leave week is equal to 1 plus the number of required days-off.

$$\sum_{d \in D} y_{wdp} \geq 1 - f_{wp} + r_{wp} \;\; \forall w \in W, p \in P$$

4. A suitable constraint was introduced to force consecutive leave weeks for each physician.

$$f_{w,p} = f_{w+1,p} \;\; \forall w \in W : w \pmod 2 = 1, p \in P$$

5. A different number ϕ_w of physicians on leave was decided for each week $w \in W$, according to the forecasted needs of the hospitals, and it was imposed by suitable constraints.

$$\sum_{p \in P} f_{w,p} \geq \phi_w \;\; \forall w \in W.$$

   It was used as a lower bound to give the model more flexibility.
6. A maximum number of leave weeks is given for each physician in each planning period. Subset P' includes physicians that have not been assigned leave weeks in the previous planning periods.

$$\sum_{w \in W} f_{w,p} \leq 2 \;\; \forall p \in P'.$$

For those who have already been assigned leave weeks, variables f are forced to 0.

Objective The model was initially formulated just to check the satisfiability of all constraints, with no objective. Then it was formulated to minimize the number of work shifts to be assigned to external physicians.

$$\text{minimize} \sum_{w \in W, d \in D} (h'_{w,d} + h''_{w,d}).$$

Finally, it was used in a multi-objective fashion to perform a parametric analysis to explore the trade-off between the number of external work-shifts and some indicators of the quality of service in the two hospitals from the viewpoint of patients and physicians. For instance, one of these indicators was the number of joint morning+afternoon shifts in Lodi (to be minimized) apart from the week-ends (when they are explicitly forced to occur). Such an objective can be expressed as follows:

$$\text{minimize} \sum_{w \in W, d \in \{1,\ldots,5\}} \delta_{w,d}$$

with the additional constraints

$$x_{1,w,d,1,p} + x_{2,w,d,1,p} \leq 1 + \delta_{w,d} \quad \forall w \in W, d \in \{1, \ldots, 5\}, p \in P,$$

where $\delta_{w,d} \geq 0$ is a non-negative auxiliary variable that is forced to 1 every time the same person p is assigned both the morning and the afternoon shifts ($t = 1$ and $t = 2$) in a working day ($d \in \{1, \ldots, 5\}$) of a week w in Lodi ($s = 1$).

2.1 *The Solution Process*

Since time constraints did not allow for the development of a customized mathematical programming algorithm and since there was no budget to carry out the analysis, it was mandatory to rely on a free MILP solver. The solver *glpsol* with its *Gusek* interface was selected, both because it is free and because of its ease of use.

Real instances with 12 physicians and a time horizon of 14 weeks turned out out to be by far out of reach for *glpsol*. For this reason the model was solved in a rolling horizon fashion, two weeks at a time. Provably optimal solutions were not always found, depending on the activated and deactivated constraints. However, a five minutes timeout for each run was enough to provide the necessary insight into the problem complexity and explore the boundary between feasibility and infeasibility.

The model, written in MathProg and including many comments, was about two hundreds lines long, plus an additional hundred lines for the data and the commands to produce an easy-to-understand output in a text file. The user-friendliness of the Gusek interface and the MathProg language turned out to be instrumental for the success of the study, because it made possible to the problem expert to directly use the model written and commented by the O.R. expert, thus speeding up the process. The problem expert necessarily had to learn the MathProg language, which is a standard in mathematical programming and is well documented on the web. In this way he could grasp the meaning of each instruction, becoming able to add, remove or modify constraints and objectives autonomously. Hence he modified and run the mathematical model countless times, since each solution (including the answer "No feasible solution found") was used as a starting point to modify either the model (the possible decisions, the constraints to be enforced, the objectives to be optimized) or the data or both. The main model parameters to act upon were the type and number of required services, the different possible definitions of allowed joined shifts, the use of external resources in specific services in either hospital, and the rules to assign days-off.

3 Discussion and Conclusions

The mindset of O.R. experts is instinctively oriented to the computation of optimal solutions through efficient algorithms that suitably exploit the mathematical properties of the models representing critical decision problems. However, one of the main lessons that can be learned from this study concerns the generation of knowledge that comes well before the computation of an optimal solution and can even be treated as an objective by itself.

In natural sciences, knowledge is generated by iteratively comparing abstract models with empirical observations. Every model is challenged by new observations, triggering the search for more general or more refined models. In a similar way, when the object of the study is not a natural phenomenon but rather a complex decision problem, knowledge can be generated by continuously improving the mathematical model of the problem: each solution round provides a feedback that challenges the model and the input data, possibly triggering the development of a more detailed model or the collection or observation of more reliable and precise data. When used in this way, mathematical programming is a powerful tool to generate knowledge, well before providing optimal solutions.

It is worth remarking that algebraic modeling languages, relying on mathematics as a universal and unambiguous language, allow any user endowed with a sufficient mathematical education to understand the model and to use it as a tool to investigate the problem, not necessarily to solve it. Furthermore, the separation between the logical structure of the model and the numerical values of the data, placed in two separate files or in two separate sections of the same file is instrumental in making the problem expert autonomous in evaluating alternative models.

This remark is especially important in an age in which "solutionism" is heavily criticized (not without some good reason) [4], "artificial intelligence" is often presented as *the* way to solve complex problems and strong emphasis is placed on data, especially "big" data. However, when using an "artificial intelligence" tool, one could only examine solutions, often without any clue about why they have been suggested and how they depend on data, because the model is not explicit. By contrast, mathematical optimization puts emphasis on models, that are represented in an algebraic language. This allows the user to examine the effects of the changes he himself has introduced into the model. In this way mathematical optimization and decision science aim at empowering human intelligence and ability to understand complex problems, in order to formulate them better and better. Solutions come later, almost as a side effect.

This project was no exception. Its main outcome was not a best possible solution, but first of all a good model. This should be remarked, because in general problem experts do not know what is the right model of their problems at the beginning; they perfectly know their needs, but in general this is not enough to translate them into a model. The search for the model should obviously precede the search for the solutions and not rarely when a solution is provided after countless efforts in algorithm development, it turns out that the model is wrong, incomplete, or flawed for some reason. It may be the case that some "constraints" are not constraints but decisions and the same holds for some "data". Similarly it may be the case that the initially assumed objective turns out not to be the main objective, because different performance indicators have priority. This is why the definition of a model must be challenged by a critical examination of the solutions (not necessarily the optimal ones) obtained from it.

Making the problem expert autonomous in managing this knowledge generation process was extremely beneficial to the development of the project. The feedback from the solution back to the model and the data typically requires the intervention of both the problem expert and the technical expert. On the contrary, in this case after the development and documentation of an initial ILP model, the iterative feedback was completely managed by the problem expert, with just a limited support from the O.R. expert for major changes, such as the steps from constraint satisfaction to optimization and then to multi-objective optimization. This significantly reduced the effort, the time and the amount of interaction needed to carry the study to a positive end under very strict time requirements (about ten days overall).

Acknowledgments The authors acknowledge the role of Enrico Storti, who strongly supported the adoption of a mathematical optimization approach in the hospitals of Lodi and Codogno and encouraged the writing and presentation of this contribution.

The detailed comments of an anonymous referee were very helpful.

An informative article on the same subject recently appeared on OR/MS Today [5].

References

1. Cervesato, E., Righini, G., Rellini, G.L., Cassin, M., Piazza, R., Nicolosi, G.L.: Optimization of shifts and on-call coverage of cardiologists working in a hospital complex structure by using free software. Comput. Cardiol. **41**, 477–480 (2014)
2. Gagliano, A., Villani, P.G., Co', F.M., Manelli, A., Paglia, S., Bisagni, P.A.G., Perotti, G.M., Storti, E., Lombardo, M.: Epidemic in the central province of Northern Italy: Impact, logistics and strategy in the front-line hospital. Disaster Med. Publ. Health Preparedness **14**(3), 372–376 (2020). https://doi.org/10.1017/dmp.2020.51
3. Micheli, V., Mancon, A., Malara, A., Mileto, D., Villani, P.G., Rizzo, A., Pagani, C., Alquati, O., Gismondo, M.R.: What was behind the first recognition and characterization of the indigenous transmission of SARS-CoV-2 in Italy: the impact on European scenario. Clinical Case Reports, Wiley (2021). https://doi.org/10.1002/ccr3.4154
4. Morozov, E.: To Save Everything, Click Here: The Folly of Technological Solutionism. PublicAffairs, New York (2013)
5. Villani, P.G., Righini, G.: O.R. reopens ORs in Italy. OR/MS Today, INFORMS (April 2021)

A Genetic Algorithm to Optimize Dynamics of Supply Chains

Luigi Rarità

Abstract This paper focuses on a model for supply chains, based on partial and ordinary differential equations, that model, respectively, densities of parts on suppliers and queues between consecutive arcs. An optimization approach is discussed via a cost functional that, in consideration of a wished outflow, weights queues of materials by variations of processing velocities for suppliers. The minimization of the cost functional is achieved via a genetic algorithm that, as for the processing velocities, considers mechanisms of selection, crossover and mutation. A simulation example is discussed for the optimization procedure.

Keywords Genetic algorithms · Supply chains · Simulations

1 Introduction

Managing supply systems is an important issue, as particular phenomena, such as dead times and bottlenecks, represent serious matters within industrial contexts. Various mathematical approaches are useful in this regard. Some of them are based on Discrete Event Simulations (DES) [1], while others refer to Ordinary and/or Partial Differential Equations (ODEs, PDEs) [2–4]. In this paper we consider a continuous model that, based on differential equations for the dynamics of goods on arcs and queues among them, is introduced in [5], further analyzed in [6, 7], and solved numerically in [8]. Notice that the proposed model is different from others based on mixed integer linear programming with possible issues about combinatorial optimization, see for instance [9]. On the other hand, the used numerical approach is similar to ones described, for instance, in [10–12].

L. Rarità (✉)
Department of Management and Innovation Systems, University of Salerno, Fisciano, Italy

Department of Science and Technology, University of Sannio, Benevento, Italy
e-mail: lrarita@unisa.it

L. Amorosi et al. (eds.), *Optimization in Artificial Intelligence and Data Sciences*,
AIRO Springer Series 8, https://doi.org/10.1007/978-3-030-95380-5_10

In the proposed research, once the PDE-ODE model is solved numerically by an upwind scheme for the density of parts over suppliers, and by an explicit Euler method for queues of goods between consecutive arcs, an optimization procedure is described. A correct definition of the optimal performances is useful to improve the productivity and often involves different questions. For instance, in [6], two different optimal control problems arise: the first considers the minimization of queues in terms of a pre-defined outflow; the second focuses on possible values of distributions rates that minimize queues for a supply network with vertices of dispersing type. In [13], the authors describe a procedure that, by adjusting a piecewise constant input flow, aims to minimize queues and to approximate the wished supply chain outflow. In particular, [6] focuses on a rich numerical investigation, based on the software Matlab; [13], on the other hand, provides a correct analysis for an analytical optimal solution, but only for particular cases of input flows.

In this paper, a procedure, based on a genetic algorithm (GA), allows, from one side, to compute optimal solutions for a generic supply system that could have an input flow of various shapes, unlike the case presented in [13]. On the other hand, the adoption of a GA ensures a suitable theoretical basis for optimization issues, that are considered only in terms of simulations in [6]. Finally, various analysis of the approach proposed in this paper confirmed the results of [6] and [13], thus showing a robust approach of resolution. Such features provide the key elements of novelty for the following paper: the possibility of adapting classical numerical schemes in order to simulate networks of medium/big dimensions with reduced computational times; the definition of a robust optimization approach, based on a GA, for a supply system modelled by ODEs and PDEs. Notice that the adoption of GAs is only a starting point as other possible simulation schemes, based for instance on particle swarm optimization as well as ant/bee colony dynamics, could be proposed. Indeed, unlike other suitable optimization procedures, that are still under investigation, GAs already present a complete analysis of various properties, see [14–18], while applications of GAs in the context of supply systems are reported in [19] and [20]: the former describes an integrated model for a supplier selection model of both multi-item and multi-supplier frameworks via a two-level GA that decides about selections of suppliers and splitting of demands; in the latter, a GA works as a decision support system for dynamics of an integrated inventory control in case of backlogged shortage. Here, a GA is used in a different way. Precisely, a cost functional (see [6]), that weights the amount of queues and a wished outflow, is minimized in terms of processing velocities of suppliers. The different iterations of the GA allow variations of the velocities of suppliers by mechanisms of selection, crossover and mutation.

Some numerical simulations are also discussed. In particular, a possible supply chain with twenty arcs is considered. The queues show an evident dependence on processing velocities and maximal capacities of suppliers. A further investigation allows a possible optimization. Different iterations of the genetic algorithm are considered and it is shown the queue decrease in successive steps.

The paper is structured as follows. Section 2 focuses on the ODE–PDE model and numerical approaches. Section 3 describes a possible optimal control problem for the chosen model. Section 4 focuses on a test case and its optimization. The paper ends with conclusions and future research activities in Sect. 5.

2 Model and Numerics for Supply Chains

We describe a model for supply chains, characterized by ODEs and PDEs ([5, 6]), on the basis of a reformulated approach proposed in [2].

A supply chain has a set of vertices $\mathscr{V} = \{1, \ldots, M-1\}$ and a set of arcs $\mathscr{A} = \{1, \ldots, M\}$. Each arc $m \in \mathscr{A}$ is a supplier, indicated by an interval $[\alpha_m, \beta_m]$. For each vertex, one incoming arc is connected to one outgoing arc and the various arcs are consecutively labelled, namely arc m connects arc $m+1$ with $\beta_m = \alpha_{m+1}$. For the first and the last arc, $\alpha_1 = -\infty$ and $\beta_M = +\infty$, respectively, with suitable boundary data.

For each supplier $m \in \mathscr{A}$, we have: length $L_m > 0$; processing time $T_m > 0$, and hence a processing velocity $V_m := L_m/T_m$; the highest processing capacity $\mu_m > 0$; the density of parts at point x and time t, represented by the continuous function $D_m(t, x) \in \left[0, D_m^{\max}\right]$. Finally, for each supplier $m \in \mathscr{A} \setminus \{1\}$, at $x = \alpha_m$ the function $Q_m(t)$ represents a time dependent queue of goods, that travel between consecutive arcs.

Then, for densities $D_m(t, x)$ and queues $Q_m(t)$, the model obeys the equations:

$$\frac{\partial D_m(t,x)}{\partial t} + \frac{\partial \phi_m(D_m(t,x))}{\partial x} = 0, \quad \forall\, x \in [\alpha_m, \beta_m], \quad t > 0, \tag{1}$$

$$D_m(0,x) = D_{m,0}(x) \geq 0, \quad D_m(t, \alpha_m) = \frac{\phi_{m,inc}(t)}{V_m}, \tag{2}$$

$$\frac{d}{dt} Q_m(t) = \phi_{m-1}(D_{m-1}(\beta_{m-1}, t)) - \phi_{m,inc}(t), \quad m \in \mathscr{A} \setminus \{1\}, \tag{3}$$

$$Q_m(0) = Q_{m,0} \geq 0, \tag{4}$$

where: $\phi_m(D_m(t,x)) := \min\{\mu_m, V_m D_m(t,x)\}$ is the flux function; $D_{m,0}(x)$ is the initial datum (to assign); $\phi_{m,inc}(t)$ is the flux on the outgoing arc m, namely:

$$\phi_{m,inc}(t) := \begin{cases} F(t), & m = 1, \\ \min\{\phi_{m-1}(D_{m-1}(\beta_{m-1}, t)), \mu_m\}, & Q_m(t) = 0, m \in \mathscr{A} \setminus \{1\}, \\ \mu_m, & Q_m(t) > 0, m \in \mathscr{A} \setminus \{1\}, \end{cases} \tag{5}$$

whose interpretation is as follows: if $m = 1$ (first arc of the supply chain), $\phi_{m,inc}(t)$ is $F(t)$, assigned input profile on the left boundary $\{(\alpha_1, t) : t \in \mathbb{R}\}$. If $m \in \mathscr{A} \setminus \{1\}$,

$\phi_{m,inc}(t)$ is dependent on the queue: if $Q_m(t) = 0$, inflow to supplier m and outflow from supplier $m-1$ are equal; otherwise, we get the maximal inflow.

Remark 1 Notice that $D_m(t,x) \geq 0$, $Q_m(t) \geq 0$ for every $m \in \mathscr{A}$, $t \geq 0$ and x, see [8] for details.

Now, consider suitable numerical schemes to approximate $D_m(t,x)$, $m \in \mathscr{A}$, and $Q_m(t)$, $m \in \mathscr{A} \setminus \{1\}$.

For an arc $m \in \mathscr{A}$, denote by N_m and η_m, respectively, the number of grid points for a partition of $[0, L_m] \times [0, T]$. Consider a fixed time mesh Δt and varying space meshes $\Delta x_m = V_m \Delta t$. Then, the grid points are $(x_i, t^n)_m = (i\Delta x_m, n\Delta t_m)$, $i = 0, \ldots, N_m$, $n = 0, \ldots, \eta_m$.

The upwind scheme, useful to define the parts density of arc m, reads as:

$$\frac{{}^m D_i^{n+1} - {}^m D_i^n}{{}^m D_{i-1}^n - {}^m D_i^n} \Delta x_m = \Delta t V_j, \tag{6}$$

where ${}^m D_i^n$ is the approximation of D_m at $(x_i, t^n)_m$, see (1), $\forall\, m \in \mathscr{A}$, $i = 0, \ldots, N_m$, $n = 0, \ldots, \eta_m$, while the Courant-Friedrich-Levy (CFL) condition is satisfied since:

$$\Delta t = \min\left\{\frac{\Delta x_m}{V_m} : m \in \mathscr{A}\right\}. \tag{7}$$

The proposed numerical approach allows advantageous computational times, as well as properties of convergence and stability, as described carefully in [8].

If $\alpha_j < -\infty$, the explicit Euler method, that allows to construct queues, reads as:

$$Q_m^{n+1} - Q_m^n + \Delta t\, \phi_{m,inc}^n = \Delta t\, \phi_{m-1}^n({}^m D_{N_m}^n), \quad n = 0, \ldots, \eta_m, \tag{8}$$

where $\phi_{m,inc}^n$ is defined by using (5) while details for numerical corrections are in [8]. Notice that, if $\alpha_j = -\infty$, boundary data are used by ghost cells.

3 Optimization

Now, we consider a possible optimal control problem for the model of Sect. 2. Fix a time horizon $[0, T]$ and define the cost functional:

$$G(V_1, V_2, \ldots, V_M) = \sum_{k=2}^{M} \int_0^T Q_k(t)\, dt + \int_0^T [V_M D_M(\beta_M, t) - \delta(t)]^2\, dt, \tag{9}$$

where Q_k refers to (3), $V_M D_M (\beta_M, t)$ is the outflow of the supply chain with density level lower than μ_M, while $\delta(t) \in L^\infty((0,T), [0,+\infty[)$ is a pre-assigned flow. The second integral of (9) represents a sort of measure between the effective outflow of the supply chain and a reference output $\delta(t)$. Notice that the solution of (1), D_m, is implicitly part of (9), hence the numerical solution of (1) and (3) represents a priority for the optimization issue.

We analyze the minimization problem:

$$\min_{(V_1,\ldots,V_M)} G(V_1, V_2, \ldots, V_M), \tag{10}$$

with $V_m^{\min} \le V_m \le V_m^{\max}$, $m = 1, \ldots, M$. Hence, the aim is the minimization of the queues and the distance between the effective outflow and $\delta(t)$ by referring to the velocities V_m, $m = 1, \ldots, M$.

A solution to problem (10) is sought via a Genetic Algorithm (GA). Such an approach is deeply considered [14] for numerical optimization, while convergence details are widely analyzed in [17, 18].

For a maximal number of iterations Λ, the algorithm works as follows: at the iteration 0, generate an initial population $V^0 = \left(V_1^0, V_2^0, \ldots, V_M^0\right)$ and compute the value $\Gamma_0 := G\left(V_1^0, V_2^0, \ldots, V_M^0\right)$ of the fitness function (9).

In general, indicating by $\Gamma_k := G\left(V_1^k, V_2^k, \ldots, V_M^k\right)$ the value of (9) at the iteration k, $k \ge 1$, the steps are:

Step 1 Via selection, crossover and mutation, get $\overline{V^k} = \left(\overline{V_1^k}, \overline{V_2^k}, \ldots, \overline{V_M^k}\right)$ from $V^{k-1} = \left(V_1^{k-1}, V_2^{k-1}, \ldots, V_M^{k-1}\right)$;

Step 2 compute $\overline{\Gamma_k} := G\left(\overline{V_1^k}, \overline{V_2^k}, \ldots, \overline{V_M^k}\right)$ of (9);

Step 3 if $\overline{\Gamma_k} < \Gamma_{k-1}$, set $V^k := \overline{V^k}$ and go to step 4; otherwise, come back to step 1;

Step 4 set $k := k + 1$ and come back to step 1 if $k \le \Lambda$; otherwise, stop.

Remark 2 The just described procedure uses a maximal number of iterations Λ as stop criterion. Indeed, further optimization schemes could be addressed, also considering different ways to stop iterations, see for instance [14, 15].

4 Simulations

For simulations, we deal with a test supply chain of twenty arcs, see Fig. 1 for a possible structure. The number of arcs is purely indicative as the aim is to simulate a network of medium dimensions, also considering the possibilities due to the used numerical approaches. For the analysis of different supply networks, as well as for computational times, see [8]. The supply chain has the following characteristics: for arcs, $L_k = T_k = 1$, $m = 1, \ldots, 20$; $\mu_1 = 550$; $\mu_{20} = 10$; $\mu_m = 50 - 2m$,

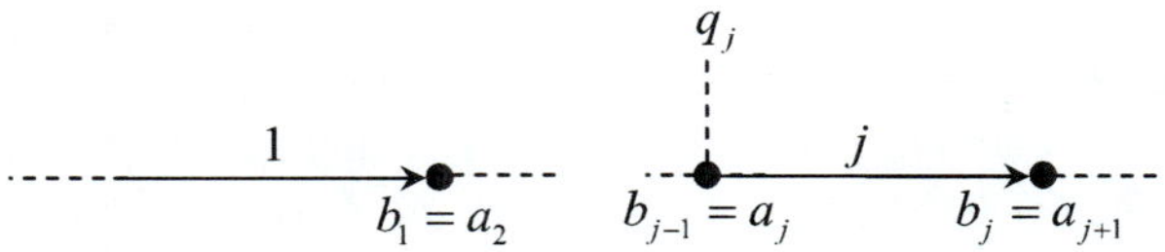

Fig. 1 An example of supply chain, see [8]

$m = 2, \ldots, 19$; $D_m(0, x) = 0$, $m = 1, \ldots, 20$; $Q_m(0) = 0$, $m = 2, \ldots, 20$; total simulation time $\overline{T} = 1800$; input profile given by:

$$F(t) = \begin{cases} t, & 0 \leq t \leq 60, \\ 60, & 60 < t \leq 130, \\ 216 - \frac{6}{5}t, & 130 < t \leq 180, \\ 0, & t > 180. \end{cases} \tag{11}$$

As for the initial conditions, the supply chain is simulated in case of empty arcs and queues. The processes velocities are all equal to one in order to simulate a homogeneous starting situation when optimization criteria are used. The processing capacities of all arcs $m = 2, \ldots, 20$, are chosen in order to create queues among arcs. In fact, following the model described in Sect. 2, in case of equal processing velocities among consecutive arcs, queues occur if $\mu_{m+1} > \mu_m$, $m = 1, \ldots, 19$. The processing capacity of the first arc is chosen so that the inflow $F(t)$ is not cut and totally directed to the first arc, as foreseen from the model of Sect. 2. In order to simulate dynamics that are typical of industrial realities, function $F(t)$ is provided in order to simulate an inflow of this type: strong injection (increasing profile), constant injection, light injection (decreasing profile). Finally, the input of the system equals zero, and phenomena on the test supply chain are only due to possible dynamics on the last arcs.

According to the numerical schemes described in Sect. 2, we used $\Delta t = 0.025$. Figure 2 presents various queues.The behaviour of queues is a direct consequence of the choice of $F(t)$, considering that conservation laws have flux functions that could, in some cases, be constant. Richer phenomena, that deal with further profiles for queues, are widely described in [6, 8, 13]. In the case of the presented paper, the slopes of $Q_m(t)$, $m = 17, 18, 19$, are quite different due to the values of μ_m, $m = 1, \ldots, 20$. Moreover, although (11) is zero $\forall\, t > 180$, queues dynamics is very slow. This is confirmed by $Q_{19}(t)$ that vanishes at $t \simeq 350 > 180$.

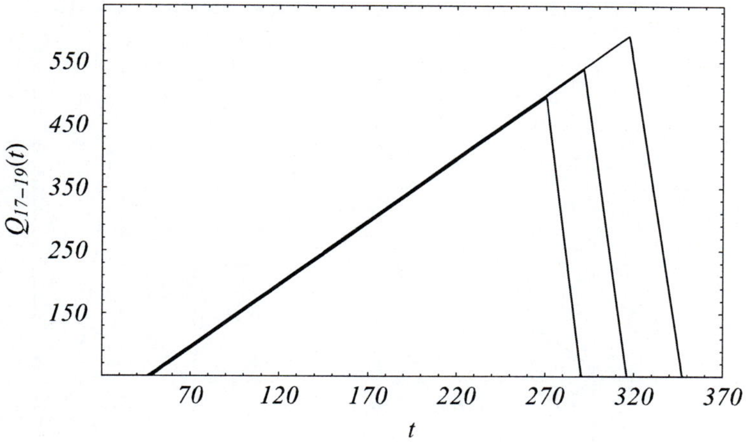

Fig. 2 Queues $Q_m(t)$, $m = 17, 18, 19$; $Q_{17}(t)$ is the first on the left, $Q_{18}(t)$ the second on the left, and so on

Table 1 Iterations, values of some velocities at various iterations and corresponding values of (9)

Iteration k	(V_5^k, V_6^k, V_7^k)	Γ_i	Iteration k	(V_5^k, V_6^k, V_7^k)	Γ_i
1	(1.05, 1.08, 1.21)	356603	10	(1.55, 1.73, 1.54)	315021
2	(1.07, 3.11, 1.71)	350021	11	(1.61, 1.74, 1.22)	312212
3	(1.12, 2.45, 1.54)	349121	12	(1.71, 1.87, 1.17)	309989
4	(1.24, 2.31, 1.13)	348112	13	(1.61, 1.85, 1.89)	308874
5	(1.44, 2.18, 1.29)	339212	14	(1.58, 1.88, 2.17)	306721
6	(1.52, 2.14, 1.17)	334121	15	(1.59, 1.81, 2.19)	305361
7	(1.78, 1.79, 1.16)	317719	16	(1.59, 1.77, 2.21)	303218

For the optimization, we fix $V_m^{\min} = 0.35$, $V_m^{\max} = 2.75$, $m = 1, \ldots, 20$, $\delta(t) = 155$ and $\Delta t = 0.05$. The initial population is defined by V^0 with entries $V_m^0 = 1$, $m = 1, \ldots, 20$. In this case, the fitness function (9) equals $\Gamma_0 = 356603$. Fixing $\Lambda = 16$ iterations, Table 1 reports the various values of (9) and some processing velocities.

All queues decrease at the various iterations. Figure 3 presents $Q_{19}(t)$ for the first and the last iteration. For iteration 0, $Q_{19}(t)$ has a maximum $M \simeq 580$, and vanishes at $t_v \simeq 350$; for the last iteration, $M \simeq 400$, and vanishes at $t_v \simeq 230$.

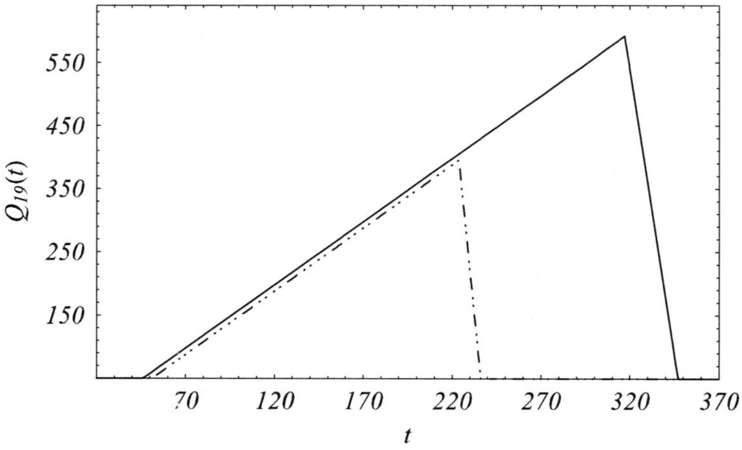

Fig. 3 Evolution of $Q_{19}(t)$ for iteration 0 (continuous line) and iteration 16 (dot dashed line)

5 Conclusions

This paper has described possible dynamics of supply chains modeled by PDEs and ODEs, and solved numerically by an upwind scheme for densities and an explicit Euler method for queues. A genetic algorithm has been described and tested to optimize the performances of a supply system via minimization of a cost functional that considers either queues or a pre-defined outflow. The aim of future research issues is to study different evolutionary algorithms for the optimization.

References

1. Daganzo, C.: A Theory of Supply Chains. Springer, New York (2003)
2. Armbruster, D., Degond, P., Ringhofer, C.: A model for the dynamics of large queueing networks and supply chains. SIAM J. Appl. Math. **66**, 896–920 (2006)
3. Armbruster, D., Degond, P., Ringhofer, C.: Kinetic and fluid models for supply chains supporting policy attributes. Bull. Inst. Math. Acad. Sin. **2**, 433–460 (2007)
4. Armbruster, D., Marthaler, D., Ringhofer, C., Kinetic and fluid model hierarchies for supply chains. Multiscale Model. Simul. **2**, 43–61 (2003)
5. Göttlich, S., Herty, M., Klar, A.: Network models for supply chains. Commun. Math. Sci. **3**, 545–559 (2005)
6. Göttlich, S., Herty, M., Klar, A.: Modelling and optimization of supply chains on complex networks. Commun. Math. Sci. **4**, 315–330 (2006)
7. Herty, M., Klar, A., Piccoli, B.: Existence of solutions for supply chain models based on partial differential equations. SIAM J. Math. An. **39**, 160–173 (2007)
8. Cutolo, A., Piccoli, B., Rarità, L.: An Upwind-Euler scheme for an ODE-PDE model of supply chains. SIAM J. Sci. Comput. **33**(4), 1669–1688 (2011)

9. Merchant, D.K., Nemhauser, G.L.: A model and an algorithm for the dynamic traffic assignment problem. Transp. Sci. **12**(3), 183–199 (1978)
10. Tomasiello, S.: Q based methods: theory and application to engineering and physical sciences. In: Leng, J., Sharrock, W. (eds.) Handbook of Research on Computational Science and Engineering: Theory and Practice, pp. 316–346. Hershey, IGI Global (2012). https://doi.org/10.4018/978-1-61350-116-0.ch014
11. Macías-Díaz, J.E., Tomasiello, S.: A differential quadrature-based approach àla Picard for systems of partial differential equations associated to fuzzy differential equations. J. Comput. Appl. Math. **299**, 15–23 (2016)
12. Tomasiello, S.: A functional network to predict fresh and hardened properties of self-compacting concretes. Int. J. Numer. Methods Biomed. Eng. **27**(6), 840–847 (2011)
13. D'Apice, C., Manzo, R., Piccoli, B.: Optimal input flows for a PDE-ODE model of supply chains. Commun. Math. Sci. **10**(4), 1225–1240 (2012)
14. Michalewicz, Z., Janikow, C.Z.: Genetic algorithms for numerical optimization. Stat. Comput. **1**(2), 75–91 (1991)
15. Berthiau, G., Siarry, P.: Etat de l'art des methodes d' "optimisation globale". RAIRO Oper. Res. **35**(3), 329–365 (2001)
16. Kar, A.K.: Bio inspired computing - a review of algorithms and scope of applications. Exp. Syst. Appl. **59**, 20–32 (2016)
17. Barrios, D., Malumbres, L., Rios, J.: Convergence conditions of genetic algorithms. Int. J. Comput. Math. **68**(3–4), 231–241 (1998)
18. Cerf, R.: Asymptotic convergence of genetic algorithms. Adv. Appl. Prob. **30**(2), 521–550 (1998)
19. Sana, S.S., Chedid, J.A., Navarro, K.S.: A three layer supply chain model with multiple supplier, manufacturers and retailers for multiple items. Appl. Math. Comput. **229**, 139–150 (2014)
20. Pourakbar, M., Farahani, R.Z., Asgari, N.: A joint economic lot-size model for an integrated supply network using genetic algorithm. Appl. Math. Comput. **189**, 583–596 (2007)

Closed-Loop Supply Chain Network Equilibrium with Online Second-Hand Trading

Georgia Fargetta and Laura Scrimali

Abstract This paper studies a closed-loop supply chain network equilibrium problem with online second-hand trading of high-uniqueness products. The closed-loop supply chain network consists of manufacturers, retailers, demand markets, and one online second-hand platform engaging in both horizontal and vertical competition. The optimal behaviors of all the decision-makers are modeled as variational inequality problems, and the governing closed-loop supply chain network equilibrium conditions are given.

Keywords Variational inequality · Reverse logistics · Second-hand market · Sustainable consumption

1 Introduction

Reverse logistics aims at improving the exploitation of used products through recycling or re-manufacturing and leads to a reduction in environmental damage. Products may reverse direction in the supply chain for several motivations, such as manufacturing returns, product recalls, warranty or service returns, end-of-use returns, and end-of-life returns. Reverse logistics and, in particular, second-hand trading have received large interest for the opportunity of sustainable consumption, extending the life span of products, and reducing adverse environmental impacts due to the purchase of new goods. Recently, the increasing use of the Internet and trading platforms, such as eBay or Vinted, has completely changed the market conditions, see [1, 15]. In the U.S., the e-commerce sales have reached \$876 billion in the first quarter of 2021, up 38% year-over-year. Moreover, the speculative buying of limited edition goods has become a real business. In fact, people may gain profit

G. Fargetta · L. Scrimali (✉)
Department of Mathematics and Computer Science, University of Catania, Catania, Italy
e-mail: georgia.fargetta@phd.unict.it; laura.scrimali@unict.it

L. Amorosi et al. (eds.), *Optimization in Artificial Intelligence and Data Sciences*, AIRO Springer Series 8, https://doi.org/10.1007/978-3-030-95380-5_11

from selling high-uniqueness goods at a price much higher than the original one to potential consumers when the goods get scarce over time.

Due to several advantages, closed-loop supply chains (CLSCs) have been extensively studied in recent years. Many researchers have investigated the network structure of the CLSC, which includes competitive manufacturers, competitive retailers and consumer markets. For example, Nagurney and Toyasaki [4] develop a network model for supply chain decision-making with environmental criteria. In [5], the authors explore a reverse supply chain network model using variational inequality. In [10], Shen et al. examine the sale of second-hand products through an online platform on a supply chain consisting of contributors, one second-hand online platform, and one supplier. Different scenarios in terms of CLSC structure and block-chain use are considered. In [13], Wang et al. study the waste of electrical and electronic equipment and provide a variational inequality to model the CLSC network. In [16], the authors examine a CLSC network equilibrium problem in multiperiod planning horizons, with consideration to product lifetime and carbon emission constraints. By variational inequalities and complementary theory, the governing CLSC network equilibrium model is established. Other valuable contributions can be found in [3, 9, 11, 12, 14].

Motivated by all the above analysis, this paper establishes a CLSC network equilibrium model with online second-hand trading of high-uniqueness products. We consider a CLSC network consisting of manufacturers, retailers, demand markets, and one online platform, in which the consumers purchase new products and collect them. Then, collectors sell the goods to consumers through the online platform. By variational inequalities, the optimal behaviors of all the decision-makers are modeled, and, in turn, the governing CLSC network equilibrium model is given. The main contributions of this paper are: the modeling of the second-hand market in a reverse logistics setting, and the study of the horizontal competition among the members of the same tiers as well as the vertical one between adjacent tiers. We describe the forward and the reverse logistics, taking into account capacity constraints of manufacturers and retailers, as well as consumers' risk-aversion to purchasing second-hand goods, and platform's risk-aversion to transacting with collectors.

The paper is organized as follows. In Sect. 2, we develop the CLSC model by describing the manufacturers' and the retailers' competitive behavior, and the interactions with the demand markets and the online platform. We then provide a variational inequality formulation of the optimal behavior of decision-makers. In Sect. 3, we state that the governing equilibrium conditions of the CLSC network. We summarize our results and present our conclusions in Sect. 4.

2 The Closed-Loop Supply Chain Network

We consider a CLSC network consisting of multiple manufacturers, multiple retailers, and multiple demand markets, in which the consumers purchase new products and collect them. Collectors distribute the used goods through an online

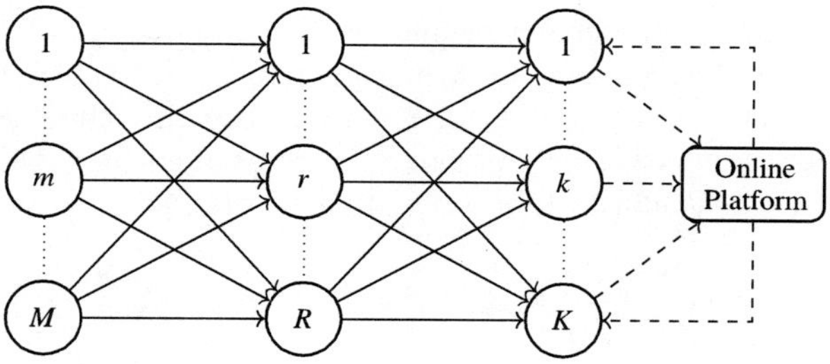

Fig. 1 The closed-loop supply chain network

platform for the aim of gaining profit on the resale. The criterion of each player in the network is the total profit maximization.

Let M be the set of manufacturers (we denote by m the typical manufacturer), R be the set of retailers (we denote by r the typical retailer), K be the set of demand markets (we denote by k the typical demand market) and we consider a single online platform like eBay, Marketplace by Facebook, Vinted etc... Abusing notation, without risk of confusion, we use the same symbols here to denote the sets M, R, K and their cardinalities. Furthermore, we introduce the set of collectors K^c, with $|K^c| \leq K$, that represents the set of consumers who decide to resell their collectibles. The network can be divided into two parts: the forward chain, formed by manufacturers, retailers and consumers, and the reverse chain, formed by collectors, the online platform and consumers. The collectors and the online platform make it possible to connect the forward and the reverse chains and form the closed-loop network. We consider two different types of items: the new ones denoted by index $n = 1, \dots N$ and the used ones indicated by index $u = 1 \dots U$. The model network can be represented as in Fig. 1. The solid lines represent the forward transactions and the dashed lines refer to the reverse ones.

We first focus on the manufacturers. We then turn to the retailers, to the consumers, and, finally, to the platform. The complete equilibrium model is then constructed as a variational inequality.

2.1 The Optimal Behavior of the Manufacturers

Let x_{mr}^n be the quantity of new item n sold by manufacturer m to retailer r. We group all the n and r elements into the vector $x_m \in \mathbb{R}^{NR}$, and then we group all the vectors (x_m) for all m into the vector $x^M \in \mathbb{R}^{NMR}$. We denote by x_m^{max} the production capacity of manufacturer m.

In the forward logistics, a manufacturer incurs production costs and transaction costs. In order to maximize his own profit, each manufacturer m must decide the quantity x_{mr}^n of new item n to be sold to retailer r. We associate with each manufacturer the production cost, c_m, and assume that it can depend, in general, on the entire vector of production outputs, namely, $c_m = c_m(x^M)$. We denote by $t_{mr}(x^n_{mr})$ the transaction cost from manufacturer m to retailer r. Moreover, we assume that $c_m(x^M)$ and $t_{mr}(x^n_{mr})$ are continuously differentiable and convex functions. Finally, we consider p^n_{mr} as the selling price of a new product. Given the above notation, each manufacturer m wishes to maximize the profit as follows:

$$\max \sum_{n \in N} \sum_{r \in R} p^n_{mr} x^n_{mr} - c_m(x^M) - t_{mr}(x^n_{mr}) \tag{1}$$

$$\sum_{n \in N} \sum_{r \in R} x^n_{mr} \leq x^{max}_m, \quad x^n_{mr} \geq 0, \forall n, r. \tag{2}$$

The objective function (1) maximizes the profit, which equals sales revenue minus costs associated with production and transaction. The first constraint in (2) expresses the production capacity of manufacturer m. All the manufactures compete in a non-cooperative fashion, and each manufacturer seeks to maximize his profit given other manufacturers' decisions. Thus, the optimality conditions of all the manufacturers can be described by the following variational inequality, see [2]:

$$\sum_{n \in N} \sum_{m \in M} \sum_{r \in R} \left(\frac{\partial c_m(x^{*M})}{\partial x^n_{mr}} + \frac{\partial t_{mr}(x^{*n}_{mr})}{\partial x^n_{mr}} - p^n_{mr} \right) (x^n_{mr} - x^{*n}_{mr}) \geq 0, \forall x^M \in S^M, \tag{3}$$

$$S^M = \left\{ x^M \in \mathbb{R}^{NMR}_+ : \sum_{n \in N} \sum_{r \in R} x^n_{mr} \leq x^{max}_m, \forall m \in M \right\}. \tag{4}$$

2.2 *The Optimal Behavior of Retailers*

The retailers interact with manufacturers and consumers. Specifically, they decide the amount of products to order from the manufactures, so as to transact with the demand markets, while seeking to maximize their profit. The product shipment of new good n between retailer r and consumer k is denoted by x^n_{rk}; the product shipments x^n_{rk} for all n and k are then grouped into the column vector $x_r \in \mathbb{R}^{NK}$ and, further, into the vector $x^R \in \mathbb{R}^{NRK}$.

Each retailer r has associated management cost c_r, which may include, for example, the storage cost associated with the products in stock. For the sake of generality, and to enhance the modeling of competition, see [6], we allow the function to depend also on the amounts of the products held by other retailers, that is, $c_r = c_r(x^M)$. Let $\hat{c}^n_{mr}(x^n_{mr})$, be the transportation cost from m for new items and let p^n_{rk} be the sale price associated with a new item. Moreover , retailers incur transaction costs $t^n_{rk}(x^n_{rk})$, when selling new products to consumers. Finally, we assume that $c_r(x^n_{mr})$, $\hat{c}^n_{mr}(x^M)$, and $t_{rk}(x^N_{rk})$ are continuously differentiable and convex functions.

Each retailer r seeks to maximize his profit function as follows:

$$\max \sum_{n\in N}\left(\sum_{k\in K} p^n_{rk}x^n_{rk} - \sum_{m\in M}\hat{c}^n_{mr}(x^n_{mr}) - \sum_{k\in K} t^n_{rk}(x^n_{rk}) - \sum_{m\in M} p^n_{mr}x^n_{mr}\right) - c_r(x^M) \tag{5}$$

$$\sum_{n\in N}\sum_{k\in K} x^n_{rk} \le \sum_{n\in N}\sum_{m\in M} x^n_{mr}, \quad x^n_{mr}\ge 0,\ x^n_{rk}\ge 0, \forall n, m, k. \tag{6}$$

Objective function (5) expresses that the profit of the retailer is equal to sales revenues minus costs associated with the management, the transportation, the transaction and the payout to the manufacturers. The first constraint in (6) states that consumers cannot purchase more from a retailer than is held in stock. Since all the retailers compete in a non-cooperative fashion, the optimality conditions for all retailers can be expressed as the variational inequality:

$$\sum_{n\in N}\sum_{r\in R}\sum_{k\in K}\left(\frac{\partial t_{rk}(x^{*n}_{rk})}{\partial x^n_{rk}} - p^n_{rk}\right)(x^n_{rk} - x^{*r}_{rk})$$

$$+\sum_{n\in N}\sum_{m\in M}\sum_{r\in R}\left(p^n_{mr} + \frac{\partial c_r(x^{*n}_{mr})}{\partial x^n_{mr}} + \frac{\partial \hat{c}^n_{mr}(x^{*M})}{\partial x^n_{mr}}\right)(x^n_{mr} - x^{*r}_{mr}) \ge 0,$$

$$\forall (x^R, x^M) \in S^R, \tag{7}$$

$$S^R = \left\{(x^R, x^M) \in \mathbb{R}^{NRK+NMR}_+ : \sum_{n\in N}\sum_{k\in K} x^n_{rk} \le \sum_{n\in N}\sum_{m\in M} x^n_{mr}, \forall r\in R\right\}. \tag{8}$$

2.3 *The Optimal Behavior of the Consumers*

The consumers at demand markets transact with the retailers as well as the online platform. Specifically, in the forward supply chain, consumers purchase new products; in the reverse supply chain, consumers act as collectors and sell their goods on the online platform, that are then purchased by consumers at demand markets. We analyze these situations separately.

2.3.1 The Consumers in the Forward Logistics

Let $\hat{c}^n_{rk}(x^n_{rk})$ be the transportation cost for new product n sold by retailer r to demand market k. Moreover, let d^n_k be the demand of new item n at demand market k and be

p_k^n the price of new product n at demand market k. The equilibrium conditions for consumers at demand market k are (see [4–8]):

$$p_{rk}^{*n} + \hat{c}_{rk}^n(x_{rk}^{*n}) \begin{cases} = p_k^{*n} & \text{if } x_{rk}^{*n} > 0, \\ \geq p_k^{*n} & \text{if } x_{rk}^{*n} = 0, \end{cases} \quad \forall r, k, n. \tag{9}$$

$$d_k^n(p_k^{*n}) \begin{cases} = \sum_{r \in R} x_{rk}^{*n} & \text{if } p_k^{*n} > 0, \\ \leq \sum_{r \in R} x_{rk}^{*n} & \text{if } p_k^{*n} = 0, \end{cases} \quad \forall k, n. \tag{10}$$

Inequality (9) states that if the consumers at demand market k purchase the products from retailer r, then the price charged by the retailer for the product plus the transportation cost undertaken by the consumers does not exceed the price that the consumers are willing to pay. Equation (10) states that if the equilibrium price that the consumers are willing to pay for the new products at the demand market is positive, then the quantities purchased of new goods from the retailers will be exactly equal to the demand. These conditions correspond to the well-known spatial price equilibrium conditions, see [2].

2.3.2 The Consumers in the Reverse Logistics

In the reverse supply chain, some consumers resell collectible items to the demand market through the online platform. The product shipment of second-hand good u between collector k_c and consumer k, using the platform, is denoted by $x_{k_c k}^u$. The product shipments $x_{k_c k}^u$, for all u and k, are then grouped into the column vector $x_{k_c} \in \mathbb{R}^{UK}$ and, further, into the vector $x^U \in \mathbb{R}^{UK^cK}$. We set Q_{k_c} as the amount of items in the collection of collector k_c. Let $p_{k_c}^u$ be the price charged by the collector k_c for second-hand items. We note that selling on the online platform can give higher visibility to the products and, as a consequence, it can be more profitable, even if the platform retains a portion of the sale price. For instance, on eBay the transaction price amounts to the 10% of the selling price, indicated by the coefficient γ. Let $c_{k_c}(x_{k_c})$ be the maintenance and restoring cost of the collector k_c, depending on the amount of items that he resells on the online platform. Let $\hat{c}_{rk_c}^n(x^R)$, be the transportation cost from r for new item n to collector k_c. We assume that $c_{k_c}(x_{k_c})$ and $\hat{c}_{rk_c}^n(x^R)$ are continuously differentiable and convex functions. We denote by $\mu_{k_c} \in (0, 1]$ the portion of second-hand goods that collector $k_c \in K^c$ decides to sell on the platform.

Each collector $k_c \in K^c$ seeks to maximize his profit function as follows:

$$\max \sum_{u \in U} \sum_{k \in K} (1 - \gamma) p_{k_c}^{*u} x_{k_c k}^u - c_{k_c}(x_{k_c}) - \sum_{n \in N} \sum_{r \in R} \left(p_{rk_c}^n x_{rk_c}^n - \hat{c}_{rk_c}^n(x^R) \right) \tag{11}$$

$$\sum_{u \in U} \sum_{k \in K} x_{k_c k}^u \leq \mu_{k_c} Q_{k_c}, \quad x_{k_c k}^u, x_{rk_c}^n \geq 0, \quad \forall n, u, r, k. \tag{12}$$

Objective function (11) expresses that the profit of the collector is equal to sales revenues minus costs associated with restoring, purchasing and transportation. The first constraint in (12) states that the amount of products collector k_c decides to sell should be less than or equal to the amount of collectibles in k_c's collection.

We now examine the transactions between the platform and the demand market k. Let $\hat{c}^u_{k_c k}(x^U)$ be the transportation cost from collector k_c to consumer k for used product u purchased on the platform. Furthermore, let d^u_k be the demand of second-hand items at demand market k, and ρ^u_k be the willingness to pay second-hand items at demand market k. We group all these ρ^u_k into a column vector $\rho_k \in \mathbb{R}^U$, and then into the vector $\rho^U \in \mathbb{R}^{UK}$. We also consider a risk associated with purchasing second-hand items from the trading platform. Therefore, each consumer exhibits risk aversion that may be dependent on flows controlled by other demand markets. Hence, the risk aversion function can be expressed as the continuous function $\pi_k(x^U)$ [8]. The equilibrium conditions for consumers at demand market k in the reverse supply chain are

$$p^{*u}_{k_c} + \hat{c}^u_{k_c k}(x^{*U}) + \pi_k(x^{*U}) \begin{cases} = \rho^{*u}_k & \text{if } x^{*u}_{k_c k} > 0, \\ \geq \rho^{*u}_k & \text{if } x^{*u}_{k_c k} = 0, \end{cases} \quad \forall k_c, u. \tag{13}$$

$$d^u_k(\rho^{*U}) \begin{cases} = x^{*u}_{k_c k} & \text{if } \rho^{*u}_k > 0, \\ \leq x^{*u}_{k_c k} & \text{if } \rho^{*u}_k = 0, \end{cases} \qquad p^{*n}_{rk} - \rho^{*u}_k \begin{cases} < 0 & \text{if } x^{*n}_{rk} = 0, \\ \geq 0 & \text{if } x^{*n}_{rk} > 0, \end{cases} \quad \forall r, k, u, n. \tag{14}$$

Equality (13) states that if the consumers at demand market k purchase the product on the online platform, then the price charged by the collector k_c for second-hand items plus the transportation cost plus the risk undertaken by the consumer is equal to the price that the consumer is willing to pay. The first condition in (14) states that if the equilibrium price the consumers at demand market k are willing to pay for the second-hand product is positive, then the amount purchased of second-hand product should exactly be equal to the demand of this second-hand item. The second condition in (14) means that the unitary price of a second-hand collectible is higher than the unitary price of a new collectible that is totally sold out.

2.3.3 The Consumers' Equilibrium Conditions

Combining consumer behaviors in both forward and reverse supply chain, and assuming that the tranposrtation costs, the demand functions and the risk function are continuous, the equilibrium conditions for all the demand markets can be expressed as the following variational inequality, see [4–8]:

$$
\sum_{n\in N}\sum_{k\in K}\left(\sum_{r\in R} x^{n}_{rk} - d^{*n}_{k}(p^{*n}_{k})\right)(p^{n}_{k} - p^{*n}_{k})
$$

$$
+\sum_{u\in U}\sum_{k_c\in K^c}\left(\frac{\partial c_{k_c}(x_{k_c})}{\partial x^{u}_{k_c k}} + \gamma p^{*u}_{k_c} + \hat{c}^{u}(x^{*U}_{k_c k}) + \pi_k(x^{*U}) - \rho^{*u}_{k}\right)(x^{u}_{k_c k} - x^{*u}_{k_c k})
$$

$$
+\sum_{u\in U}\sum_{k\in K}\left(x^{*u}_{k_c k} - d^{u}_{k}(\rho^{*U})\right)(\rho^{u}_{k} - \rho^{*u}_{k})
$$

$$
+\sum_{n\in N}\sum_{r\in R}\sum_{k\in K}\left(p^{*n}_{rk} + \hat{c}^{n}_{rk}(x^{*n}_{rk}) - p^{*n}_{k} + p^{*n}_{rk} - \sum_{u\in U}\rho^{*u}_{k} + \frac{\partial \hat{c}^{n}_{rk_c}(x^{R})}{\partial x^{n}_{rk}}\right)(x^{n}_{rk} - x^{*n}_{rk}),
$$

$$
+\sum_{n\in N}\sum_{r\in R} p^{n}_{rk_c}(x^{n}_{rk_c} - x^{*n}_{rk_c}) \geq 0, \quad \forall (p^{N}, x^{UK}, \rho^{U}, x^{U}, x^{R}) \in S^{K}, \tag{15}
$$

$$
S^{K} = \left\{(p^{N}, x^{UK}, \rho^{U}, x^{U}, x^{R}) \in \mathbb{R}_{+}^{KN+2UK+UK^cK+NRK} : \sum_{u\in U}\sum_{k\in K} x^{u}_{k_c k} \leq \mu_{k_c} Q_{kc}\right\}.
$$

2.4 *The Behavior of the Online Platform*

Now, we present the behavior of the online platform as an intermediary that matches consumers and collectors. As an intermediary, the platform is involved in transactions both with the collectors, as well as with the consumers at the demand markets.

Collectors resell items on the platform and determine the unitary price p^{u}_{k} of second-hand goods. Let $C^{u}(x^{U})$ be the management costs of second-hand product u, including processing and advertisement, and let $\hat{t}^{u}_{k}(x^{u}_{k})$ be the transaction cost function between the platform and demand market k, where $x^{u}_{k} = \sum_{k_c} x^{u}_{k_c k}$. Since the platform has no decision-making power on the choice of products that will be sold, it takes the risk of owning false objects or with descriptions that do not correspond to the real conditions of the item. As a consequence, the intermediary may have risk associated with transacting with the various collectors and with the demand markets. Let $\pi(x^{U})$ denote the risk function associated with online platform. We assume that $C(x^{U})$, $\hat{t}^{u}_{k}(x^{u}_{k})$ and $\pi(x^{U})$ are continuously differentiable and convex. Let μ_{k_c} the portion of second-hand goods that collector k_c decides to sell on the platform, and satisfies $\mu_{k_c} \in (0, 1]$. We define $Q^{u} \in R^{U}$ as the total amount of item u on the online platform. Each online platform makes his optimal decisions based on maximizing the following profit function:

$$\max \sum_{u\in U}\Big(\sum_{k_c\in K^c}\sum_{k\in K}\gamma p^u_{k_c} x^u_{k_c k} - C^u(x^U) - \sum_{k\in K}\hat{t}^u_k(x^u_k) - \pi(x^U)\Big). \tag{16}$$

$$\sum_{k\in K} x^u_k \le Q^u, \qquad x^u_k, x^u_{k_c k} \ge 0,\ \forall u,\ \forall k. \tag{17}$$

Objective function (16) expresses that the profit of the online platform is equal to a percentage of the profit of sale of the product minus the management, transaction costs and the risk. The first constraint in (17) states that the total amount of each second-hand item bought by all consumers k on platform should be less or equal than the availability of item u.

Under our assumptions, the optimality conditions for the online platform can be expressed as the variational inequality:

$$\sum_{u\in U}\sum_{k_c\in K^c}\sum_{k\in K}\left(\frac{\partial C^u(x^{*U})}{\partial x^u_{k_c k}} - \frac{\partial \pi^u(x^{*U})}{\partial x^u_{k_c k}} - \gamma p^{*u}_k\right)(x^u_{k_c k} - x^{*u}_{k_c k}) \tag{18}$$

$$+\sum_{u\in U}\sum_{k\in K}\left(\frac{\partial \hat{t}^u_k(\bar{x}^u_k)}{\partial x^u_k}\right)(x^u_k - x^{*u}_k) \ge 0,\ \forall (x^U, x^{K^c}) \in S^P \tag{19}$$

$$S^P = \Big\{(x^U, x^{K^c}) \in \mathbb{R}^{UK^cK+UK}_+ : \sum_{k\in K} x^u_k \le Q^u, \forall u \in U\Big\}. \tag{20}$$

3 The Equilibrium Conditions of the CLSC Network

In equilibrium state, the optimality conditions for all suppliers, manufacturers, retailers, demand markets and online platform must be satisfied simultaneously. We now define the CLSC network equilibrium and give an equivalent variational inequality formulation.

Definition 1 The CLSC network is at equilibrium if the forward and reverse flows between the tiers of the decision-makers coincide and the product flows and prices satisfy the sum of optimal conditions in (3), (7), (15), and (19).

Using standard arguments, it can be proved that the equilibrium conditions governing the CLSC network model with competition are equivalent to solve a single variational inequality problem. We can establish the following theorem:

Theorem 1 *The equilibrium conditions governing the CLSC network model with competition are equivalent to solve a single variational inequality problem, given by the sum of problems* (3), (7), (15), *and* (19).

The equilibrium conditions presented can be used by policy-makers to anticipate the effects of the second-hand business on the market. Second-hand economy is an increasingly important phenomenon because it is a sustainable way for manufacturers and retailers to operate, and also because it is a convenient system for users. In fact, it allows consumers to put a used or unwanted product back into the supply chain and gain money.

4 Conclusions

This paper presents an equilibrium model of a CLSC network consisting of manufacturers, retailers, demand markets, and one online platform, in which the consumers purchase new products and collect them. Then, the collectors sell the goods to consumers through the online platform. We take into consideration capacity constraints of manufacturers and retailers, as well as consumers' risk-aversion to purchasing second-hand goods, and platform's risk-aversion to transacting with collectors. We model the optimal behaviors of all the decision-makers as variational inequality problems and provide the governing CLSC network equilibrium conditions.

Our study can provide an analytical tool for investigating the market equilibrium when collectors engage in the second-hand business. We emphasize that adopting circular business models can be an effective system to extend the life span of products and to inspire sustainable consumption. The entire society will benefit from the lengthened use of available resources.

Future research can explore the equilibrium problem in multi-period planning horizons, and examine some random factors in the demand functions.

Acknowledgments The research was partially supported by the research project "Programma ricerca di ateneo UNICT 2020–22 linea 2-OMNIA" University of Catania. This support is gratefully acknowledged.

References

1. Hu, S., Henninger, C.E., Boardman, R., Ryding, D.: Challenging current fashion business models: entrepreneurship through access-based consumption in the second-hand luxury garment sector within a circular economy. In: Gardetti M., Muthu S. (eds.) Sustainable Luxury. Environmental Footprints and Eco-design of Products and Processes. Springer, Singapore (2019). https://doi.org/10.1007/978-981-13-0623-5_3
2. Nagurney, A.: Network Economics: A Variational Inequality Approach (2nd edn. (rev.)). Kluwer Academic, Boston (1999)
3. Nagurney, A., Ke, K.: Financial networks with intermediation. Quant. Finance **1**(4), 441–451 (2001)
4. Nagurney, A., Toyasaki, F.: Supply chain supernetworks and environmental criteria. Transp. Res. D **8**, 185–213 (2003)

5. Nagurney, A., Toyasaki, F.: Reverse supply chain management and electronic waste recycling: a multi-tiered network equilibrium framework for ecycling. Transp. Res. E **41**, 1–28 (2005)
6. Nagurney, A., Dong, J., Zhang, D.: A supply chain network equilibrium model. Transp. Res. E **38**, 281–303 (2002)
7. Nagurney, A., Loo, J., Dong, J., Zhang, D.: Supply chain networks and electronic commerce: a theoretical perspective. Netnomics **4**(2), 187–220 (2002)
8. Nagurney, A., Cruz, J., Dong, J., Zhang, D.: Supply chain networks, electronic commerce, and supply side and demand side risk. Eur. J. Oper. Res. **164** (1), 120–142 (2005)
9. Qiang, Q.: The closed-loop supply chain network with competition and design for remanufacture ability. J. Cleaner Product. **105**, 348–356 (2014)
10. Shen, B., Xu, X., Yuan, Q.: Selling secondhand products through an online platform with blockchain. Transp. Res. E Logist. Transp. Rev. **142**, 102066 (2020)
11. Toyasaki, F., Daniele, P., Wakolbinger, T.: A variational inequality formulation of equilibrium models for end-of-life products with nonlinear constraints. Eur. J. Oper. Res. **236**, 340–350 (2014)
12. Wakolbinger, T., Nagurney, A.: Dynamic supernetworks for the integration of social networks and supply chains with electronic commerce: modeling and analysis of buyer-seller relationships with computations. Netnomics **6**, 153–185 (2004)
13. Wang, W., Zhang, P., Ding, J., Li, J., Sun, H., He, L.: Closed-loop supply chain network equilibrium model with retailer-collection under legislation. J. Ind. Manage. Optim. **15**(1), 199–219 (2019)
14. Yang, G.F., Wang, Z.P., Li, X.Q.: The optimization of the closed-loop supply chain network. Transp. Res. E **45**, 16–28 (2009)
15. Yrjölä, M., Hokkanen, H., Saarijärvi, H.: A typology of second-hand business models. J. Market. Manage. **37**(7–8), 761–791 (2021)
16. Zhang, G., Sun, H., Hu, J., Dai, G.: The closed-loop supply chain network equilibrium with products lifetime and carbon emission constraints in multiperiod planning horizon. Discrete Dyn. Nat. Soc. **2014**, 784637, 16pp. (2014). https://doi.org/10.1155/2014/784637

Design and Optimization of a Regional Buffalo Milk Supply Chain: A Case Study

Andrea Bettinelli, Giovanni Righini, and Fabrizio Venturelli

Abstract A feasibility study was carried out to assess the economic viability of a regional supply chain of buffalo milk mozzarella in Lombardy, Italy. The design and optimization of the supply chain required the solution of several combinatorial optimization problems at a strategic and tactical level: location, generalized assignment, transportation and inventory/routing. Some of them could be easily solved with a spreadsheet add-in, while others required mixed-integer programming solvers like *glpsol* and ILOG CPLEX.

Keywords Supply chain management · Location · Routing

1 A New Regional Supply Chain

Buffalo milk mozzarella is an appreciated typical Italian fresh cheese. Its protected designation of origin (D.O.P.) requires its production to occur in Campania, in Southern Italy, where specialized dairies are located. However, a very large fraction of Italian production of buffalo milk occurs in Lombardy, in the North of the country; moreover, Lombardy, the most populated Italian region, is also the largest national market for mozzarella. More details can be found in [3]. This geographically unbalanced scenario implies the transportation of significant North-to-South flows of milk and South-to-North flows of cheese. The cost has been estimated in 1200 Euros for each truck traveling forth and back. Moreover every

A. Bettinelli
OptIt, Bologna, Italy
e-mail: andrea.bettinelli@optit.net

G. Righini (✉)
Department of Computer Science, University of Milan, Milan, Italy
e-mail: giovanni.righini@unimi.it

F. Venturelli
FDL Servizi, Breno, Italy

L. Amorosi et al. (eds.), *Optimization in Artificial Intelligence and Data Sciences*, AIRO Springer Series 8, https://doi.org/10.1007/978-3-030-95380-5_12

other travel is done with an empty vehicle, because different types of vehicles are required to trasport milk and fresh cheese. This motivated a feasibility study to assess the economic viability of a regional supply chain for the production of buffalo milk mozzarella, not endowed with the D.O.P. certification but possibly more profitable than the current national supply chain.

2 Strategic Level: Location and Allocation

At a strategic level two scenari were considered: scenario A implies the construction of a new specialized dairy, to be optimally located and sized in order to support the whole regional supply chain; scenario B is based on the re-conversion of two existing dairies of given production capacity. For each scenario it was also required to study how milk transport operations should be organized to minimize the total transportation cost.

Scenario A: Locating and Sizing a New Dairy
The strategic problem of optimally locating a new specialized dairy was formulated as a weighted 1-median problem (see [5] for a comprehensive treatment of discrete location problems) on the road network of South-Lombardy: distances were weighted with the (known) average daily production of each of the twenty-eight buffalo farms involved.

The optimization concerned only milk transportation cost, i.e. the first echelon of the supply chain; the transportation cost of cheese to retailers, i.e. the second echelon, was not considered, because of the high variability in the spatial distribution of the demand and the heterogeneity of the destinations, including stores, supermarkets, restaurants and hotels.

Two additional requirements were stated: (1) the location of the new dairy had to be easily accessible to trucks, i.e. along some main road or highway; (2) it also had to be close to a city, in order to have a nearby set of potential customers large enough to make a local shop sustainable. Such a local shop is meant for selling fresh mozzarella with no final packaging and possibly additional products characterized by relatively small production volumes.

The optimal solution was computed on a weighted digraph obtained from digital maps of the road network of South Lombardy, with a software tool based on MapWinGIS [4]. To comply with the two additional constraints a GIS-based decision-support system was designed to compute the set of points for which the value of the objective function is within a user-defined small percentage of the optimum, as shown in Fig. 1.

The optimal solution occurs along one of the main internal streets of Soncino, close to the intersection of the borders between the three provinces of Cremona, Brescia and Bergamo. This point is a few hundred meters from the ring road surrounding the town, which could be a perfectly suitable location satisfying both the above mentioned requirements (see Fig. 2).

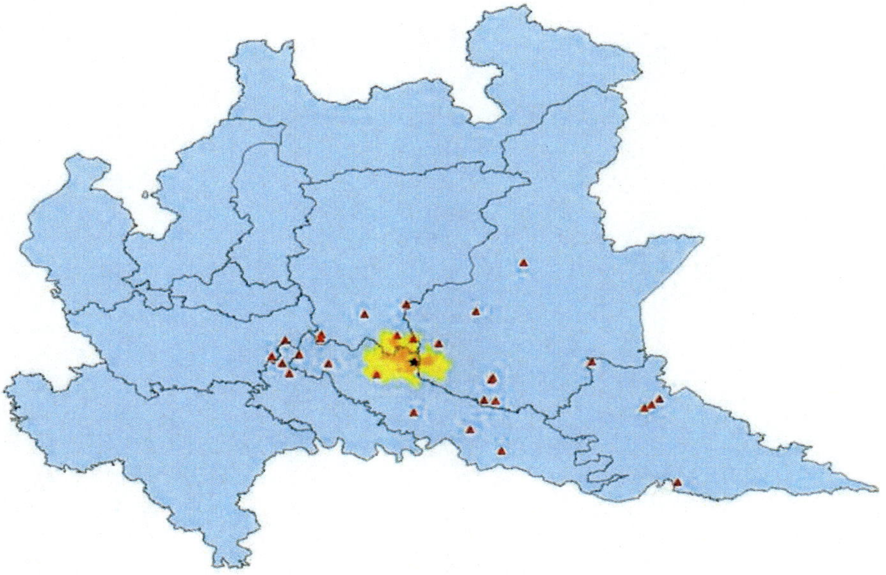

Fig. 1 Scenario A: quasi-optimal zones within 1% (green), 5% (orange) and 10% (yellow) from the optimum

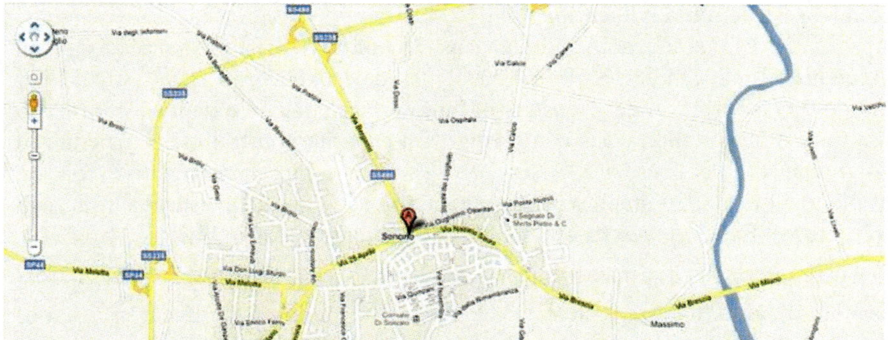

Fig. 2 Scenario A: the optimal location and the ring road just North of it

In scenario A the sizing problem is trivial: the needed capacity should be at least equal to the total estimated production capacity of the twenty-eight buffalo farms of the supply chain (18,477 L of milk per day).

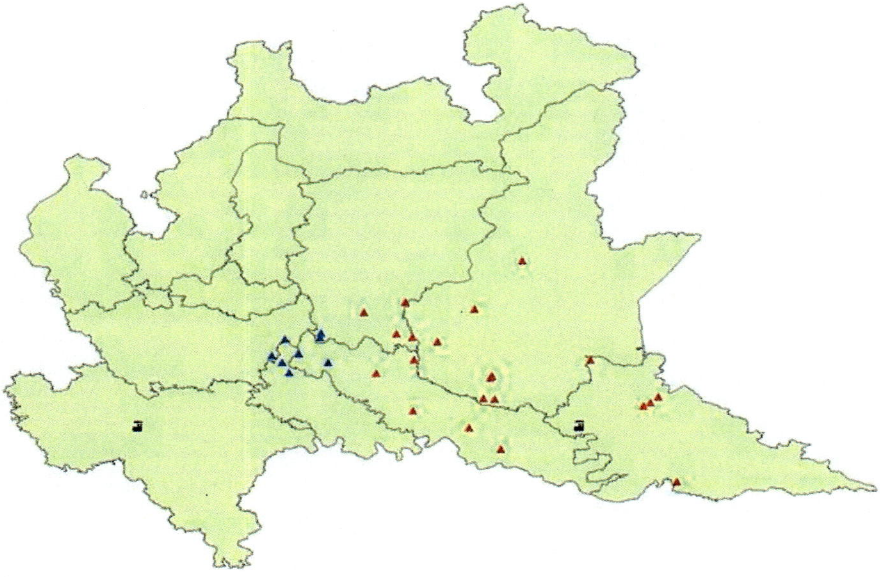

Fig. 3 Optimal allocation of farms (blue and red triangles) to dairies (black icons)

Scenario B: Demand Allocation
In scenario B the location and the capacity of two existing dairies is given (Acquanegra sul Chiese (MN), 20,000 L per day and Dorno (PV), 3000 L per day) and two corresponding clusters of buffalo farms must be defined in order to minimize transportation costs complying with capacity constraints. A generalized assignment problem instance arises (see [2] for a comprehensive survey on the problem). Owing to its small size, it could be solved with a spreadsheet add-in such as Microsoft Excel Solver. Its optimal solution is represented in Fig. 3.

All farms turn out to be allocated to the closest dairy with a single exception: Casirate d'Adda (the break item in the optimal solution of the linear relaxation) is 91 km far from Acquanegra sul Chiese and 76 km far from Dorno, but it is assigned to the former one. The closest dairy (Dorno) has enough residual capacity to accommodate about 75% of the milk production from Casirate d'Adda. However, the option of splitting the Casirate d'Adda supply, allowing for a fractional allocation, was discarded in favor of the more expensive integer allocation, because of three main reasons: (1) integer allocation implies lower administrative costs and simpler administrative procedures; (2) it allows for more reliable forward tracing, which is especially important in food supply chains; (3) the resulting solution is more robust, because no dairy is saturated and this is useful to absorb fluctuations of the daily milk production.

3 Tactical Level: Optimal Pick-Up Frequencies

For the whole set of suppliers in scenario A and for each farm cluster in scenario B one has to decide how to organize the visits of vehicles to pick-up milk and carry it to the dairies. We indicate with $\mathcal{P}$ the set of farms to be visited.

Since milking occurs twice a day, the time slots considered in this problem correspond to half-days. For each time slot the forecasted milk production at each farm is known. An additional datum is the maximum amount s_p of milk that can be stored at each farm $p \in \mathcal{P}$. This amount is not larger than the total production of 7 time slots, because buffalo milk cannot be stored for more than three-and-a-half days. The available storage is further reduced by the production of the last time slot, because it is a good practice to leave fresh milk "resting" for about twelve hours before transporting it.

The possibility of storing milk at the farms allows not to visit every farm in every time slot, which would be very expensive from the viewpoint of transportation costs. Instead, one wants to define a cyclic schedule of visits that repeats every T time slots, where T is a user-defined parameter. We indicate with $\mathcal{T}$ the set of time slots for each problem instance.

The mathematical model includes binary variables v_{pt}, representing whether farm $p \in \mathcal{P}$ is visited in time slot $t \in \mathcal{T}$ or not, as well as continuous non-negative variables z_{pt} and y_{pt}, representing respectively the amounts of milk picked-up at farm $p \in \mathcal{P}$ in time slot $t \in \mathcal{T}$ and the amount of milk stored at farm $p \in \mathcal{P}$ at the end of time slot $t \in \mathcal{T}$. The integer linear programming model includes the dairy capacity constraint (a datum m indicates the capacity of the dairy in every time slot), the farms storage capacity constraint, the typical flow balance constraints $y_{pt-1} + q_p = z_{pt} + y_{pt}$ that link the stored amount y, the production q and picked-up amount z in each time slot for each farm, and constraints stating that pick-up operations can occur only when a farm is visited (i.e. $z_{pt} > 0$ implies $v_{pt} = 1$).

In principle the objective function of this problem should be the total travel cost, i.e. the total distance traveled in the T time slots of the planning horizon. This would generate a vehicle routing problem (actually, a periodic inventory-routing problem) which cannot be solved by general-purpose ILP solvers. Instead of implementing an ad hoc optimization algorithm, it was decided to minimize the total number of visits to the farms. Intuitevely, this is a good proxy for the solution that minimizes the total distance traveled, with the disadvantage that several solutions can exist implying the same total number of visits but different transportation costs.

The model is as follows:

$$\text{minimize } f = \sum_{p \in \mathcal{P}} \sum_{t \in \mathcal{T}} v_{pt}$$

$$\text{subject to } \sum_{p \in \mathcal{P}} z_{pt} \leq m \qquad \forall t \in \mathcal{T}$$

$$y_{pt} = y_{pt-1} - z_{pt} + q_p \qquad \forall p \in \mathcal{P}\ \forall t \in \mathcal{T} : t \neq 1$$

$$
\begin{aligned}
& y_{p1} = y_{pT} - z_{p1} + q_p && \forall p \in \mathcal{P} \\
& y_{pt} \leq s_p && \forall p \in \mathcal{P}\, \forall t \in \mathcal{T} \\
& z_{pt} \leq v_{pt} s_p && \forall p \in \mathcal{P}\, \forall t \in \mathcal{T} \\
& z_{pt} \geq 0 && \forall p \in \mathcal{P}\, \forall t \in \mathcal{T} \\
& y_{pt} \geq 0 && \forall p \in \mathcal{P}\, \forall t \in \mathcal{T} \\
& v_{pt} \text{ binary} && \forall p \in \mathcal{P}\, \forall t \in \mathcal{T}
\end{aligned}
$$

The optimization was repeated for several values of T. It is worth noting that in order to allow farmers to easily plan the visits, it is not recommendable to use large values of T.

In spite of the simplified objective function, the resulting ILP model turned out to be too difficult for the Microsoft Excel Solver and also for the free solver *glpsol*, even for the smallest instance (Dorno dairy in scenario B, with only 7 allocated farms). Using ILOG CPLEX, on the contrary, it was possible to solve all instances of interest to optimality. Moreover, it is worth observing that even values of T are preferable to odd values, because their effect is that visits to each farm always occur in the morning or in the afternoon without annoying alternations.

The results reported here refer to the case where all farms are assumed to be equipped with sufficent capacity to store a 7 time slots production. Two different cases were studied, indicated by "constrained" and "unconstrained". In the constrained case, the farm refrigerator is assumed to be completely emptied when a visit occurs, while the unconstrained case allows for partial pick-up. The unconstrained case is more difficult to implement in practice, because it requires separated refrigerators and an additional information to be communicated to the driver (i.e. the exact amount of milk to pick-up), which may be a source of errors. However the unconstrained case was considered as a benchmark to evaluate the costs of the constrained case. Tables 1, 2, and 3 report the results.

Table 1 Optimal visit frequencies (scenario A, Soncino dairy)

	Unconstrained			Constrained		
T	Visits	Comp. time	Visits/slot	Visits	Comp. time	Visits/slot
1	28	0.000	28.00	28	0.000	28.00
2	28	0.005	14.00	28	0.000	14.00
3	28	0.008	9.33	28	0.015	9.33
4	28	0.022	7.00	28	0.037	7.00
5	28	0.270	5.60	28	0.244	5.60
6	28	0.296	4.67	28	0.199	4.67
7	28	0.008	4.00	28	0.007	4.00
8				56	0.156	7.00
9				56	0.253	6.22
10				56	5.733	5.60

Table 2 Optimal visit frequencies (scenario B, Acquanegra dairy)

	Unconstrained			Constrained		
T	Visits	Comp. time	Visits/slot	Visits	Comp. time	Visits/slot
1	21	0.000	21.00	21	0.000	21.00
2	21	0.003	10.50	21	0.000	10.50
3	21	0.007	7.00	21	0.007	7.00
4	21	0.011	5.25	21	0.031	5.25
5	21	0.309	4.20	21	0.196	4.20
6	21	0.321	3.50	21	0.186	3.50
7	21	0.005	3.00	21	0.006	3.00
8				42	0.132	5.25
9				42	2.048	4.67
10				42	114.048	4.20

Table 3 Optimal visit frequencies (scenario B, Dorno dairy)

	Unconstrained			Constrained		
T	Visits	Comp. time	Visits/slot	Visits	Comp. time	Visits/slot
1	7	0.001	7.00	7	0.000	7.00
2	7	0.001	3.50	7	0.000	3.50
3	8	0.004	2.67	8	0.002	2.67
4	8	0.004	2.00	8	0.003	2.00
5	9	0.023	1.80	8	0.090	1.60
6	10	0.071	1.67	10	0.047	1.67
7	11	0.026	1.57	10	0.002	1.43
8	16	133.075	2.00	16	0.102	2.00
9	17	406.767	1.89	16	0.133	1.78
10	18	203.080	1.80	16	1.407	1.60
11	19	50.519	1.73	18	412.731	1.64
12	20	8.504	1.67	18	21.644	1.50
13	20	0.092	1.54	19	1.241	1.46
14	22	0.215	1.57	19	0.045	1.36

The results show that the constraint does not yield any significant increase of the number of visits per time slot, with only a few exceptions of the smallest dairy (Dorno) for some values of T.

The average number of visits per slot decreases when T increases, because farms storage capacity is better and better exploited. When the time threshold of 7 slots is reached, the average number of visits per slot suddenly doubles. This effect would not be so concentrated on a specific value of T if the farm storage capacities were assumed to be different from one another.

4 Routing

The last step in this feasibility study was the evaluation and minimization of milk transportation costs. This would require to solve some instances of the capacitated vehicle routing problem (see [6] for a comprehensive survey) with some additional constraints. However, owing to budget and time limits, in this preliminary feasibility study it was not required to develop an exact solution algorithm (typically a branch-and-price or branch-and-cut-and-price algorithm), but rather to compute a reliable estimate of the transportation cost.

For this purpose, three assumptions were made. First, a single vehicle was assumed to be sufficient for each time slot; this is not restrictive, because the capacity of the available vehicles is larger than the amount of milk to be transported in every scenario and for every value of T. Second, no constraints were imposed on the maximum distance and the maximum duration for each vehicle route; this is also unlikely to be restrictive in a real situation, because farms are not far from one another. Third, it was assumed that routes were determined by the optimal farm-to-slot assignment computed in the previous phase of the study, where the number of visits had been minimized. This is the most restrictive assumption, because the composition of the farms subsets assigned to the same time slots does not take into account farm-to-farm distances; therefore, from the viewpoint of travel cost minimization such clusters are likely to be sub-optimal. This effect is mitigated by the decentralized position of the dairies in scenario B: this means that the total distance travelled depends more on the number of routes than on their composition.

Owing to these assumptions, it is possible to obtain an estimate of the total transportation costs by solving some instances of the asymmetric traveling salesman problem (see [1] for a comprehensive survey), one for each scenario, for each dairy, for each value of T and for each time slot. The size of these instances was between 2 and 29, including the dairy which is assumed to act as the vehicle depot. The results obtained with ILOG CPLEX for T up to 7 are reported in Table 4.

The reason why the constrained cost is sometimes lower than the unconstrained one is the third assumption mentioned above. This effect is possible when one

Table 4 Average distance [Km] travelled in each time slot for different values of T

	Soncino (A)		Acquanegra (B)		Dorno (B)	
T	Constr.	Unconstr.	Constr.	Unconstr.	Constr.	Unconstr.
1	485.35	485.35	387.58	387.58	180.43	180.43
2	348.98	304.46	281.38	277.22	153.00	160.99
3	280.89	275.24	259.50	259.91	144.88	144.88
4	247.77	244.29	213.73	223.69	141.72	143.38
5	215.50	244.24	199.77	212.35	140.46	113.81
6	217.98	193.17	198.91	194.20	136.01	117.17
7	190.92	191.70	181.35	164.38	134.95	136.02

estimates costs using clusters of farms that are not guaranteed to be optimal from the viewpoint of distance minimization.

5 Extensions and Conclusions

The aim of the preliminary study was to evaluate the feasibility and economic sustainability of a new regional supply chain. For this reason some assumptions and approximations were done to pursue a trade-off between reliability of estimates on one side and development time and cost on the other side.

Several extensions are possible from this starting point. A first possible extension consists of formulating and solving a vehicle routing problem, replacing the many instances of the asymmetric traveling salesman problem, in order to obtain more accurate estimates of the milk transportation cost. In such a vehicle routing prooblem one can allow for the use of two types of vehicles (with capacity 17,000 and 35,000 L). Another extension consists of using tighter (and more realistic) constraints on the available storage capacity at the farms. However this would change the results but not the structure of the mathematical models.

An important possible extension concerns the production and distribution of cheese. Taking into account the milk-to-cheese conversion factor (typically 25–28% for mozzarella), one should determine the optimal production mix of mozzarella and other types of cheese. For this purpose one should also consider seasonal effects both in production and consumption. On one side, the monthly production of milk is different along the year and it also depends on the buffalos reproduction cycle. On the other side, the market demand is not uniform along the year; in winter, demand for mozzarella is lower and demand for other types of cheese is higher than in summer. A critical decision is how to manage production surplus in winter, i.e. deciding whether to sell the surplus milk (for instance in the Southern Italy market) or to freeze it to meet demand peaks in summer.

Finally it should be pointed out that this feasiblity study has shown that a new regional supply chain of buffalo milk mozzarella in Lombardy would allow to transport every liter of milk for about 33 km in average, instead of the 700 km currently travelled to supply dairies in South Italy.

Acknowledgments The authors acknowledge the support by Ilaria Massari and Alessandra Ginelli (Crema Ricerche), who proposed and started the project, and by Silvio Massari, owner and manager of the buffalo farm "Gerardo da Izano" and president of ARAL—Associazione Regionale Allevatori Lombardia, who described the problem and provided the relevant data. They also acknowledge the invaluable help of Andrea Pinciroli for his tool that was used to obtain weighted graphs from digital maps of road networks.

References

1. Applegate, D.L., Bixby, R.E., Chvátal, V., Cook, W.J.: The Traveling Salesman Problem: A Computational Study. Princeton University Press, Princeton (2006)
2. Kundakcioglu, O.E., Alizamir, S.: Generalized assignment problem. In: Floudas, C., Pardalos, P. (eds.) Encyclopedia of Optimization. Springer, Boston (2008)
3. L'allevamento delle bufale in Lombardia (Breeding Buffalos in Lombardy). A.R.A.L. e Regione Lombardia (2009)
4. MapWinGIS Reference Manual. A function guide for the free MapWindow GIS ActiveX map component (2007)
5. Mirchandani, P.B., Francis, R.L.: Discrete Location Theory. Wiley, New York (1990)
6. Toth, P., Vigo, D. (eds.): Vehicle Routing: Problems, Methods, and Applications. MOS-SIAM Series on Optimization, 2nd edn. SIAM, Philadelphia (2014)

Truck-and-Drone Parcel Delivery in the Alps

Olivier Gallay, Marc-Antoine Coindreau, and Nicolas Zufferey

Abstract In this paper, we consider a parcel-delivery situation encountered in practice by a logistics provider in France, involving a truck and a drone. Whereas the majority of the deliveries take place in the valley, some parcels have to be delivered to a remote place located in a mountainous area with poor accessibility. These remote deliveries can be performed by a drone launched from the truck located in the valley. We study this parcel-delivery configuration and extend it by allowing the truck to launch and retrieve the drone from multiple positions in the valley. We propose a strengthened Mixed-Integer-Linear-Programming model, and we solve it for instances with up to 23 parcels. We compare our results with those of a classical delivery framework involving a truck only, and we highlight the practical relevance of using drones in this mountainous context.

Keywords Vehicle routing in mountains · Drones · Mixed-integer linear program

1 Introduction

In this paper, we address a parcel-delivery framework encountered in practice by the logistics provider DPD in France [9], and we refer to it as the DPD Problem. Since September 2019, DPD is operating, in a mountainous region close to Grenoble, a delivery solution involving a truck and a drone. Following the DPD current practice, the truck first reaches a dedicated zone located at the bottom of the mountainous area. Next, a drone (carrying the parcel) flies up to the remote destination, drops the parcel, and comes back to the dedicated zone where the truck waits. This bi-modal delivery framework has several advantages. First, whereas the round trip

O. Gallay (✉)
HEC Lausanne, University of Lausanne, Lausanne, Switzerland
e-mail: olivier.gallay@unil.ch

M.-A. Coindreau · N. Zufferey
GSEM, University of Geneva, Geneva, Switzerland
e-mail: marc-antoine.coindreau@unige.ch; n.zufferey@unige.ch

L. Amorosi et al. (eds.), *Optimization in Artificial Intelligence and Data Sciences*, AIRO Springer Series 8, https://doi.org/10.1007/978-3-030-95380-5_13

from the valley to the uphill destination would take 30 minutes by truck, the drone requires only 8 minutes to perform the delivery. Second, the mountain road is narrow and thus unsafe (in particular in winter conditions), the drone delivery reduces the accident risk. Finally, the energy consumption, as well as the associated environmental footprint, can be drastically reduced by favoring drone instead of truck delivery. It is interesting to note that the implementation of the DPD solution was probably facilitated in practice as the drone avoids flying above urban areas. Indeed, the drone is launched outside the city and most of the flight takes place above a forest. In that regard, legislation aspects and noise pollution constitute smaller obstacles towards the implementation of drone delivery.

Since the seminal paper by Murray and Chu [8], delivery fleets made of both trucks and drones have received substantial interest from the research community. In [6], 63 papers focusing on this research trend are reviewed (with publishing date between 2015 and 2020). The various papers in this field differ in the employed configurations (e.g., number of trucks, number of drones per truck, number of parcels per drone), in the considered objectives and constraints (e.g., minimization of the makespan or costs, time windows constraints, drone autonomy), and also in the location at which the drones are allowed to be launched and retrieved. In terms of solution methods, Mixed-Integer-Linear-Programming (MILP) is used for small instance sizes, and larger instances require the use of meta/heuristics.

In this paper, we study the opportunity of using a truck-and-drone pair to deliver a set of parcels (called jobs), some of them being located in a poorly accessible area (denoted as remote jobs). We compare the three following configurations:

- The traditional truck delivery (TTD), where one single truck is used to deliver all the parcels.
- The DPD current configuration (DPDC), where the truck driver launches and retrieves the drone from a dedicated position located at the bottom of the mountainous area. The drone can transport one parcel at a time and can be launched multiple times. The driver waits until the drone has completed its delivery before recharging it for another delivery or continuing its tour.
- The DPD extended configuration (DPDE), where the drone can be launched and retrieved at any job location in the valley. When the drone is flying, the driver can make other deliveries and retrieve the drone at another job location (instead of simply waiting at the launch point). The time between the launch and the retrieval of the drone is limited by its flight autonomy.

With respect to the increasing interest in routing problems with drones (see [6] for a recent review), the present article proposes the following contributions. First, we consider a real application currently in operation and which addresses the use of drones to deliver parcels in poorly accessible areas. The latest review papers indicate that this application has not been considered in the literature (see [2] for an exhaustive overview of delivery practices using drones). Second, we model and solve realistic instances based on this new application, and we highlight the limitation of current practices (i.e., truck waiting during the drone delivery), in particular when the number of remote jobs increases. Third, we show that these limitations can be narrowed down by: increasing the level of synchronization between the trucks and the drones; allowing the trucks to move during the drone deliveries. Last, we propose a new and efficient MILP formulation that is able to manage several types of synchronization constraints between the truck and the drone. Using strengthening constraints, the MILP is able to find competitive solutions for instances involving up to 23 parcels in less than one hour of execution time. An overview of exact methods developed for similar types of problems can be found in [7]. In comparison, the MILP proposed by Murray and Chu [8] on a related problem requires hours of computation for instances involving 10 parcels.

2 Models

2.1 Considered Sets, Parameters and Variables

Let $J_V = \{1, \dots, n\}$ be the set of jobs located in the valley and $J_R = \{n+1, \dots, n+m\}$ the set of remote jobs located in the mountainous area (only the jobs in J_R can be served by drone). $J = J_V \cup J_R$ denotes the set of all the jobs. For job $j \in J$: $p_j \in \mathbb{N}$ (resp. $\tilde{p}_j \in \mathbb{N}$) denotes its processing time in minutes when served by a truck (resp. by a drone). Between two jobs $(i, j) \in J \times J$, the traveling (resp. flying) distance is given by $d_{ij} \in \mathbb{R}$ (resp. $\tilde{d}_{ij} \in \mathbb{R}$), and the driving (resp. flying) time is denoted by $\tau_{ij} \in \mathbb{R}$ (resp. $\tilde{\tau}_{ij} \in \mathbb{R}$). The drone autonomy (i.e., maximum flying time) is $\mathcal{E}$.

To handle the necessary synchronization of the truck and the drone, we create virtual nodes for each job to distinguish: the time at which the drone is launched, the time at which it is retrieved, and the job service time (i.e., the time at which the parcel is delivered at the remote location). Accordingly, a drone is launched and retrieved at virtual nodes. We introduce the following extended sets and variables. $J^- = \{1^-, \dots, n^-\}$ is the set of virtual entry nodes, whereas $J^+ = \{1^+, \dots, n^+\}$ is the set of virtual exit nodes. In our modeling framework, when the truck serves job j, it first visits j^-, then j, and j^+ (j^- and j^+ are two potential launch and retrieval nodes for the drone). Moreover, 0 denotes the node representing the starting depot, and 0^+ represents the terminal depot. We introduce $V = J^- \cup J^+$ as the set of virtual nodes, $V^- = V \cup \{0\}$, and $V^+ = V \cup \{0^+\}$. $A_1 = \{(i, j) \in (V^- \times V^+)$, with $i \neq j$, such that: if $i = 0$ then $j \in J^-$, if $i \in J^-$ then $j = i^+$, if $i \in J^+$ then $j \in \{J^- \cup 0^+\}\}$ is the set of arcs that can be used by the truck. $A_2 = \{(i, j, k) \in (V^- \times J_R \times V^+)$, such that $\tilde{\tau}_{ij} + \tilde{\tau}_{jk} + \tilde{p}_j \leq \mathcal{E}\}$ is the set of arcs that can only be used by the drone.

The following decision variables are used.

- $x_{ij} = 1$ if the truck travels from $i \in V^-$ to $j \in V^+$; $x_{ij} = 0$ otherwise.
- $y_{ijk} = 1$ if the drone visits $j \in J_R$ in a flight route starting from $i \in V^-$ and arriving at $k \in V^+$; $y_{ijk} = 0$ otherwise.
- $z_{ij} = 1$ if the truck transports the drone from $i \in V^-$ to $j \in V^+$; $z_{ij} = 0$ otherwise.

- $u_j \in \mathbb{R}$: time at which the truck leaves $j \in V^-$.
- $u_{0^+} \in \mathbb{R}$: time at which the truck returns to the depot.
- $s_j \in \mathbb{R}$: service start time of job $j \in J$.
- $\tilde{w}_{ijk} \in \mathbb{R}$: flying time corresponding to the flight route where the drone is launched at $i \in V^-$, serves job $j \in J$, and is retrieved at $k \in V^+$ ($\tilde{w}_{ijk} = 0$ if such a flight does not exist).

2.2 MILP Model for the DPDE Configuration

Objective (1) minimizes the truck completion time (i.e., the time at which the truck comes back to the depot).

$$\min u_{0^+} \tag{1}$$

Job-satisfaction constraints:

$$\sum_{(i,j^-)\in A_1} x_{i,j^-} + \sum_{(i,j,k)\in A_2} y_{ijk} = 1, \;\; j \in J \tag{2}$$

Constraints (2) ensure that each job is performed exactly once by either the truck or the drone. The left component indicates that if a truck enters a virtual entry node ($j^- \in J^-$), it serves the corresponding job. The right component checks whether the considered job belongs to a drone flight route.
Vehicle-flow constraints:

$$\sum_{(j,i)\in A_1} x_{ji} = \sum_{(i,j)\in A_1} x_{ij}, \;\; i \in V \tag{3}$$

$$\sum_{(0,i)\in A_1} x_{0i} = \sum_{(i,3\cdot n+1)\in A_1} x_{i0^+} \tag{4}$$

$$\sum_{(0,i)\in A_1} x_{0i} \le 1 \tag{5}$$

$$\sum_{(k,j,i)\in A_2} y_{kji} + \sum_{(k,i)\in A_1} z_{ki} = \sum_{(i,j,k)\in A_2} y_{ijk} + \sum_{(i,k)\in A_1} z_{ik}, \;\; i \in V \tag{6}$$

$$\sum_{(k,j,i)\in A_2} y_{kji} + \sum_{(k,i)\in A_1} z_{ki} \le 1, \;\; i \in V \tag{7}$$

$$\sum_{(i,j,k)\in A_2} y_{ijk} + \sum_{(i,k)\in A_1} z_{ik} \le 1, \;\; i \in V^- \tag{8}$$

$$z_{ij} \le x_{ij}, \;\; (i,j) \in A_1 \tag{9}$$

Constraints (3) ensure that when the truck arrives at a virtual node, it will ultimately leave this node. Constraints (4) ensure that the truck, after having left the depot, will return to the depot at the end of its tour. Constraints (5) state that the truck can only exit the depot using one arc. If the drone is retrieved or launched at node $i \in V$, the truck route must pass by i. Constraints (6) ensure that when the drone arrives at $i \in V$ by flying or being transported in a truck, it must exit the node by either flying or being transported by a truck. Similarly, Constraints (7) and (8) prevent the drone from arriving at (or exiting from) a node $i \in V$ by two different paths. Constraints (9) ensure the en-route synchronization between the drone and the truck.

Temporal Constraints

$$u_j \geq u_i + \tau_{ij} - M \cdot (1 - x_{ij}), \quad (i, j) \in A_1 \tag{10}$$

$$s_j \geq u_{j^-}, \quad j \in J, \tag{11}$$

$$s_j \geq u_i + \tilde{\tau}_{ij} - M \cdot (1 - \sum_{(i,j,k) \in A_2} y_{ijk}), \quad j \in J, i \in V^- \tag{12}$$

$$u_{j^+} \geq s_j + p_j, \quad j \in J \tag{13}$$

$$u_k \geq s_j + \tilde{p}_j + \tilde{\tau}_{jk} - M \cdot (1 - \sum_{(i,j,k) \in A_2} y_{ijk}), \quad j \in J, k \in V^+ \tag{14}$$

$$\tilde{w}_{ijk} \geq u_k - u_i - M \cdot (1 - y_{ijk}), \quad i \in V^-, j \in J, k \in V^+ \tag{15}$$

$$\tilde{w}_{ijk} \leq \mathcal{E}, \quad i \in V^-, k \in V^+, j \in J \tag{16}$$

At a node $j \in J^- \cup J^+$, the truck can: retrieve or launch the drone, perform both of these tasks, or do none of them. Furthermore, the truck must leave the corresponding node after all these operations take place. We set $M = \sum_{(i,j) \in (J \times J)} \tau_{ij} + \sum_{j \in J} p_j$, which is a sufficiently large number that is strictly greater than the total en-route time of every possible solution. Constraints (10) require the truck to leave node $j \in V$ after its arrival. Constraints (11) (resp. (12)) ensure that service occurs after the truck (resp. the drone) arrival. Constraints (13) force the truck to leave the node after completing the associated service. Constraints (14) allow the truck to leave the node after the drone has been retrieved. Constraints (15) compute the flight time of the drone. Constraints (16) forbid any drone flight from having a longer duration than the drone autonomy.

Preliminary experiments showed that the MILP described by Eqs (1)–(16) suffered from having a poor lower bound. To overcome this issue, we propose the following strengthening constraints to improve the quality of the lower bound on the time at which the driver returns the depot (i.e., u_{0^+}).

Strengthening Constraints

$$u_{0^+} \geq \sum_{(i,j)\in A_1} \tau_{ij} x_{ij} + \sum_{j\in J_V} p_j \tag{17}$$

$$\sum_{(i,j,k)\in A_2} y_{ijk} \geq 1, \quad \forall j \in J_R \tag{18}$$

$$s_{j_1} \leq s_{j_2} \tag{19}$$

$$u_{0^+} \geq \max_{(i,j)\in J} \tau_{0^-,i} + \tau_{i,j} + \tau_{j,0^+} + \sum_{j\in J} p_j \tag{20}$$

Constraints (17) state that the truck driver en-route time is greater than the time spent driving plus the time spent delivering the parcels in the valley. Constraints (18) force the drone to visit the remote jobs (in order to avoid solutions in which the truck visits one of the remote jobs). Constraint (19) allows cutting some symmetries by forcing one of the job j_1 in the valley to be served before another job j_2 (j_1 and j_2 being randomly selected). Constraint (20) is a lower bound for the considered objective.

2.3 Models for the TTD and the DPDC Configurations

The TTD is an occurrence of the Traveling Salesman Problem (TSP). It aims at finding the route for one single truck visiting all delivery locations while minimizing the en-route time (see for instance [1]). Such configurations also occur in production planning when the makespan to perform all the jobs has to be minimized, while satisfying various constraints [12].

To solve the DPDC configuration, we first transform it into a TSP problem. To do it, we create an instance involving: all the jobs that are located in the valley; one aggregated job that contains all the information related to the drone deliveries. The aggregated job is positioned at the dedicated launching position located in the valley. Its duration is equal to the time needed for the drone to perform all the necessary round trips and services to deliver all the parcels. Accordingly, the duration of this aggregated job is equal to $m \times t_f$, where m is the number of parcels to be delivered in the remote area, and t_f denotes the total time needed for the drone to send a parcel uphill (it includes loading and flying times).

TSP problems have been broadly studied over the last decades (see [5] for an overview of efficient MILP formulations for the TSP). We thus use a generic MILP version of the TSP to find the optimal solutions of the instances of the TTD and DPDC configurations.

3 Experiments

CPLEX 12.10 was used to solve the MILPs. The employed computer has the following characteristics: 2.2 GHz Intel Core i7 with 16 Go 1600 MHz DDR3 of RAM memory.

3.1 Instances

We generate instances based on data representing the valley currently delivered by DPD. Figure 1 displays a map of the considered region, which is located 5km in the north west of Grenoble. Fontanil-Cornillon (FC) is the city where DPD currently launches and retrieves the drone, and Mont-Saint-Martin (MSM) is the village equipped with the parcel retrieval platform. The drone flies over the platform to drop the parcel. The parcel is secured by a locking system, and it is then available to be picked up by its recipient (at any time after the delivery has taken place).

The valley containing the jobs is modeled by a rectangle grid of 10 × 4 km. The start of the uphill to MSM (the drone launching position) is located at position (5, 4 km), and the depot at position (0, 0 km) on the grid. MSM is located at 10 km from FC by road, and at 3 km when flying. The average truck speed in the valley is 25 km/h, and the average drone speed is 45 km/h (these values can be found in [9]). Hence, the flying time for a round trip between FC and MSM is equal to 8 minutes. Jobs are randomly generated in the valley, following a uniform distribution. We use the Manhattan distance between jobs that are located in the valley (road network), but the Euclidean distance between a job in the valley and a remote job. The road distance between a job in the valley and a remote job is computed as the distance to the beginning of the uphill in FC plus 10 km (the round trip for a truck is 20 km and

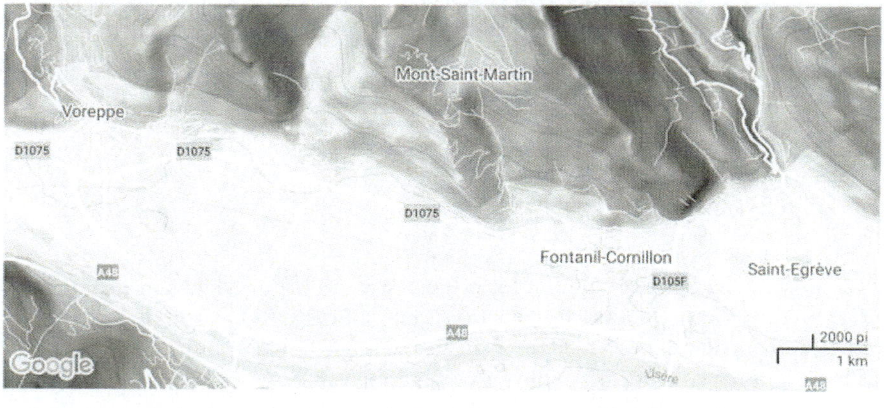

Fig. 1 Geographical characteristics of the territory under study. Source: Google Maps

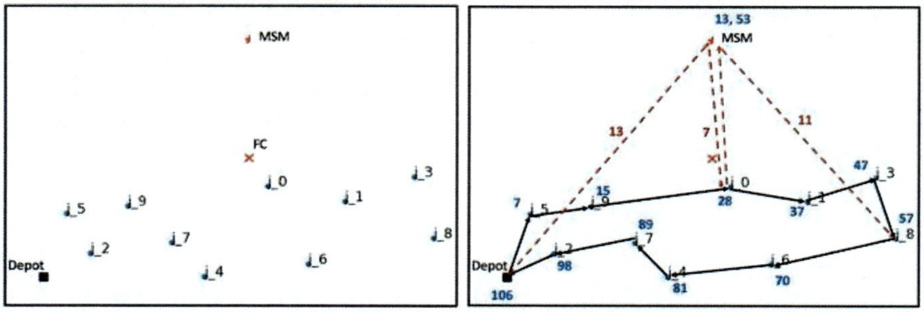

Fig. 2 Spatial distribution of the job locations for an instance involving $|J_V| = 10$ jobs in the valley and $|J_R| = 2$ remote jobs (left part). Associated optimal solution (right part)

lasts 30 minutes). The service time is set to 3 minutes when the parcel is delivered by the truck driver, and to 1 minute when the parcel is dropped by the drone in the parcel-retrieval platform at MSM. The time required to load a parcel to the drone is equal to 1 minute. As a result, considering the DPD problem, the total time to deliver one parcel from FC to MSM by drone is equal to 10 minutes.

We will compare the performance of the three considered configurations (i.e, TSP, DPDC and DPDE) with respect to the number of remote jobs to be performed. We propose 5 instance structures (3 structures involving $|J_V| = 10$ jobs, and 2 structures involving $|J_V| = 20$ jobs, which corresponds to the workload of half a day). For each instance structure, the number of remote jobs varies, with $|J_R| \in \{1, 2, 3\}$. Instances having the same "Id" involve jobs located at the same positions in the valley.

As an illustration, the left part of Fig. 2 shows how jobs are spread for one representative instance. The square at the bottom left corner designates the depot, the red cross "FC" indicates the start of the uphill to MSM, and "MSM" shows the position of the remote jobs. The right part of Fig. 2 illustrates the associated optimal DPDE solution obtained by the MILP. The straight (resp. dashed) lines indicate the paths used by the truck (resp. drone). The value next to each job location is the time at which the delivery occurs, and the drone flight times are indicated next to the corresponding dashed lines.

3.2 Performance of CPLEX

CPLEX found the optimal solution for instances with $|J_V| = 10$ jobs in 13 minutes (on average). Within 1 hour of execution time, CPLEX found solutions with an average gap to optimality of 14% for instances with $|J_V| = 20$ jobs. Constraints (17) was beneficial, as its reduces the objective-function value by 1% (resp. 15%) for the instances with $|J_V| = 10$ (resp. $|J_V| = 20$). Moreover, as u_{0^+} is poorly constrained in the original MILP (even after 1 hour of execution time for the smallest instances,

CPLEX returned a lower bound equal to 0), adding these constraints is mandatory to obtain a good lower bound on the results, and to prove the optimality of some solutions. Finally, adding constraint (19) (cutting symmetries) helps in reducing the execution time (to find the optimal solution) by 45% for the instances with $|J_V| = 10$. The use of more advanced filtering techniques [3, 4] to discard non-promising solutions is left here as an avenue of research.

3.3 Results and Managerial Implications

Table 1 compares the results of the three considered configurations (TTD, DPDC and DPDE). Instance characteristics are depicted in the first three columns. Next, for each configuration, we display the en-route driver time (in minutes) in column "ERDT", and the truck driving distance (in km) in column "TDD". Columns "%(TTD)" is the percentage improvement (of the objective-function value) of DPDC (or DPDE, depending on the column) with respect to the TTD objective-function value. Similarly, "%(DPDC)" is the percentage improvement achieved by DPDE over DPDC.

We can observe that DPDC reduces the truck driving distance (TDD) (and hence the associated greenhouse gas emissions) when compared to TTD, which highlights the benefit of using a drone. On average, TDDs are reduced by 39.5% (i.e., indeed, 20 km of driving on mountainous roads is removed from the solution). However, regarding the en-route driver time (ERDT), which is the main cost for the company, the superiority of DPDC (over TTD) decreases when the number of remote jobs increases. Indeed, the more jobs have to be delivered at the remote location, the more round trips must be achieved by the drone, and hence the more the driver has to wait at the drone launching position. As a result, when the number of remote jobs is equal to 3 (or more), DPDC is not more efficient than TTD regarding the driver's salary. In contrast, DPDE reduces both ERDT and TDD, regardless of the number of remote jobs to be served. More precisely, on average, TDD is reduced by 41.2% (resp 2.9%) when compared to TTD (resp. DPDC). At the same time, on average, ERDT is decreased by 22.5% (resp. 15.6%) when compared to TTD (resp. DPDC). This gain increases with the number $|J_R|$ of remote jobs. Compared to DPDC, when $|J_R|$ is equal to 1, 2, and 3, this average improvement is equal to 9.5, 15.9, and 21.4%, respectively. Such ERDT improvements are possible for DPDE thanks to the fine-tuned synchronization between the driver and the drone. Indeed, for DPDE, the drone can be launched and retrieved at any job location, and not only at a single launching position (as for DPDC).

Table 1 Comparison of the TTD, DPDC and DPDE configurations

Instance			TTD		DPDC				DPDE					
Id	$\|J_V\|$	$\|J_R\|$	ERDT	TDD	ERDT	%(TTD)	TDD	%(TTD)	ERDT	%(TTD)	%(DPDC)	TDD	%(TTD)	%(DPDC)
1	10	1	137	49.6	116	15.3%	29.6	40.3%	104	24.1%	10.3%	28.8	41.9%	2.7%
1	10	2	138	49.6	126	8.7%	29.6	40.3%	104	24.6%	17.5%	28.8	41.9%	2.7%
1	10	3	139	49.6	136	2.2%	29.6	40.3%	104	25.2%	23.5%	29.2	41.2%	1.5%
2	10	1	142	50.9	121	14.8%	30.9	39.3%	106	25.4%	12.4%	29.0	43.0%	6.2%
2	10	2	143	50.9	131	8.4%	30.9	39.3%	106	25.9%	19.1%	29.0	43.0%	6.2%
2	10	3	144	50.9	141	2.1%	30.9	39.3%	106	26.4%	24.8%	29.0	43.0%	6.2%
3	10	1	130	46.9	109	16.2%	26.9	42.7%	97	25.4%	11.0%	25.8	45.1%	4.2%
3	10	2	131	46.9	119	9.2%	26.9	42.7%	97	26.0%	18.5%	25.8	45.1%	4.2%
3	10	3	132	46.9	129	2.3%	26.9	42.7%	97	26.5%	24.8%	25.8	45.1%	4.2%
4	20	1	186	55.1	165	11.3%	35.1	36.3%	154	17.2%	6.7%	34.5	37.4%	1.8%
4	20	2	187	55.1	175	6.4%	35.1	36.3%	154	17.6%	12.0%	34.8	36.9%	1.0%
4	20	3	188	55.1	185	1.6%	35.1	36.3%	155	17.6%	16.2%	35.6	35.4%	−1.3%
5	20	1	178	51.5	157	11.8%	31.5	38.8%	146	18.0%	7.0%	31.1	39.6%	1.4%
5	20	2	179	51.5	167	6.7%	31.5	38.8%	146	18.4%	12.6%	31.1	39.6%	1.4%
5	20	3	180	51.5	177	1.7%	31.5	38.8%	146	18.9%	17.5%	31.4	39.1%	0.4%

4 Conclusions and Future Works

In this paper, we model a realistic situation involving bi-modal parcel delivery using a truck and a drone. More precisely, we address a truck-and-drone framework currently operated by the French logistics provider DPD in a mountainous area where some parcels have to be delivered in a poorly accessible location. In the current DPD configuration (DPDC), a single launch-and-retrieve position (located in the valley) is used, and the driver has to wait while the drone flies to drop the parcels at the remote location. Keeping the realistic hypotheses of DPDC (i.e., most of the drone flights are outside the populated areas, hence reducing accident risks), we propose an extended configuration (DPDE) for which the drone can be launched and retrieved from various locations in the valley, and the driver can deliver one or more parcels while the drone is on flight. We develop a strengthened MILP that is able to tackle instances involving up to 23 jobs while managing the complex synchronization among the driver and the drone. We measure the benefit of DPDE with respect to DPDC and to TTD. Indeed, the en-route driver time and the truck driving distance can be both significantly reduced (and hence the associated greenhouse gas emissions, as no more round trips by truck are needed in the mountains).

The superiority of DPDE should hold when multiple remote areas have to be served within the same routes, and mountainous areas with denser population should open the door for even greater improvement potential. Among future works, and in line with some truck allocation problems [10], we could consider a fleet of trucks and drones, where the trucks can carry several drones and are hence used as drone hubs. Another avenue of research would be to daily solve an order-acceptance-and-scheduling problem (i.e., the deliveries to be performed have to be selected first, and performed second), as it is often the case in production planning [11].

References

1. Bektas, T.: The multiple traveling salesman problem: an overview of formulations and solution procedures. Omega **34**(3), 209–219 (2006)
2. Chung, S.H., Sah, B., Lee, J.: Optimization for drone and drone-truck combined operations: a review of the state of the art and future directions. Comput. Oper. Res. **123**, 105004 (2020)
3. Dupont, A., Alvernhe, E., Vasquez, M.: Efficient filtering and tabu search on a consistent neighbourhood for the frequency assignment problem with polarisation. Ann. Oper. Res. **130**(1–4), 179–198 (2004)
4. Hertz, A., Schindl, D., Zufferey, N.: Lower bounding and tabu search procedures for the frequency assignment problem with polarization constraints. 4OR **3**(2), 139–161 (2005)
5. Letchford, A.N., Lodi, A.: The traveling salesman problem: a book review. 4OR **5**, 315–317 (2007)
6. Macrina, G., Pugliese, L.D.P., Guerriero, F., Laporte, G.: Drone-aided routing: a literature review. Transp. Res. C Emerg. Technol. **120**, 102762 (2020)
7. Moshref-Javadi, M., Winkenbach, M.: Applications and research avenues for drone-based models in logistics: a classification and review. Exp. Syst. Appl. **177**, 114854 (2021)

8. Murray, C.C., Chu, A.G.: The flying sidekick traveling salesman problem: optimization of drone-assisted parcel delivery. Transp. Res. C Emerg. Technol. **54**, 86–109 (2015)
9. Ouest-France, https://www.ouest-france.fr/economie/aeronautique/drones/la-poste-livrera-des-colis-par-drone-dans-une-zone-montagneuse-pres-de-grenoble-6600371
10. Schindl, D., Zufferey, N.: A learning tabu search for a truck allocation problem with linear and nonlinear cost components. Naval Res. Logist. **62**(1), 32–45 (2015)
11. Thevenin, S., Zufferey, N., Widmer, M.: Order acceptance and scheduling with earliness and tardiness penalties. J. Heuristics **22**(6), 849–890 (2016)
12. Thevenin, S., Zufferey, N., Potvin, J.-Y.: Makespan minimisation for a parallel machine scheduling problem with preemption and job incompatibility. Int. J. Product. Res. **55**(6), 1588–1606 (2017)

A Fast Heuristic Approach for the Assignment and Sequencing Storage Location Problem Under a Two Level Storage Policy

Giacomo Lanza, Mauro Passacantando, and Maria Grazia Scutellà

Abstract We consider a storage allocation problem which combines storage location assignment with sequencing decisions about the assigned storage locations, and which originates from a real-world application context. We propose a very efficient successive constrained shortest path method, which outperforms a matheuristic approach recently proposed in the literature in terms of both the computational time required and regarding the quality of the solutions found.

Keywords Storage location assignment · Storage location sequencing · Mixed-integer linear programming · Multicommodity flows · Heuristic

1 Introduction

Storage Location Assignment Problems (SLAPs) are operational problems aiming at defining the exact physical location of a set of items in a storage area, which broadly could be a warehouse, a yard, the bunt of a container ship or even a tram/bus depot. Such decisions are made by considering some long-term storage assignment policies (random, dedicated and class-based are the most popular [1, 2]), that broadly prescribe the rules to follow when stocking is needed, by respecting additional requirements related to the specific application context and, generally, by optimizing criteria such as material handling cost or storage space utilization [3–5].

The problem addressed in this paper has been motivated by a real application involving a production site of an Italian company, in Tuscany, whose large warehouse (more than 10,000 m^2) is the subject of a big modernization project requesting the resolution of a SLAP with operations research techniques. Specifically, a set of

---

G. Lanza (✉) · M. G. Scutellà
Department of Computer Science, University of Pisa, Pisa, Italy
e-mail: giacomo.lanza@di.unipi.it; maria.grazia.scutella@unipi.it

M. Passacantando
Dipartimento di Informatica, University of Pisa, Pisa, Italy
e-mail: mauro.passacantando@unipi.it

L. Amorosi et al. (eds.), *Optimization in Artificial Intelligence and Data Sciences*,
AIRO Springer Series 8, https://doi.org/10.1007/978-3-030-95380-5_14

storage locations (SLs) has to be assigned to a set of different product types, each with its own storage demand expressed in number of items to store in a given time horizon. Each SL has a fixed capacity and can be assigned to at most one product type, i.e., different types of products cannot share the same SL. The majority of the product types can be stored in any available SL, i.e., a random storage policy is considered. However, special product types do exist, which have to be preferably managed according to a dedicated storage policy.

In addition, a suitable sequencing of the assigned SLs must be devised for each product type, i.e., it has to be decided the ordering with which the assigned SLs will be filled up during the storing operations. A motivation is that a FIFO (First-In First-Out) order picking policy based on the time of permanence of the items in the warehouse has to be pursued, separately per product type, when items must be retrieved to fulfill customers' orders. The sequencing established for the assigned SLs will thus allow to easily implement the FIFO policy in the successive order picking steps. Moreover, the selected sequencing also determines the availability of additional extra storage per product type. Specifically, an additional amount of storage can be made available on the top of pairs of consecutive SLs along the sequence, provided that they are fully replenished and physically contiguous, thus allowing a two level stocking policy. The objective is to maximize the storage capacity which remains available after the assignment of the SLs.

The recent paper [6] describes the problem more formally, providing two Mixed Integer Linear Programming (MILP) formulations and the proof of its NP-hardness. Additionally, since the state-of-the-art commercial solver CPLEX is not able to address real-size instances, such those faced daily by our industrial partner, a simple yet effective matheuristic approach to tackle such instances is proposed in [6]. Further models to SLAP have been proposed in [7].

In this paper, we propose a heuristic solution method based on successive constrained shortest paths. The heuristic is able to find solutions to real-size instances in a few seconds, also improving the quality of those found in [6] in terms of both available storage capacity and other crucial features that our industrial partner is interested in.

The paper is organized as follows. We briefly describe the problem statement in Sect. 2. The heuristic method designed to tackle the problem is presented in Sect. 3. Section 4 describes the experimental plan and reports the results of the preliminary computational experiments we performed. Finally, Sect. 5 concludes the paper and identifies some future directions of research.

2 Problem Statement

Let $\mathcal{K}$ be the set of the different product types requiring storage in a given time horizon, and q^k be the number of items of product type k that needs storage, for each $k \in \mathcal{K}$. Let $\mathcal{S}$ be the set of available SLs in the warehouse where the products in $\mathcal{K}$ have to be stocked, each SL $s \in \mathcal{S}$ having a capacity u_s. Two subsets of

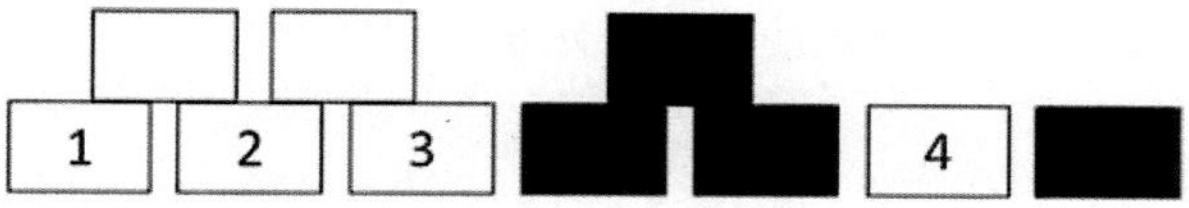

Fig. 1 Available SLs in a storage area

$\mathcal{K}$ of special product types are given: $\mathcal{K}_P \subseteq \mathcal{K}$ denoting perishable products and $\mathcal{K}_{HR} \subseteq \mathcal{K}$ denoting high rotational products. Products in $\mathcal{K}_P$ and $\mathcal{K}_{HR}$ should be preferably stocked in specific SLs, denoted as $\mathcal{S}_P \subseteq \mathcal{S}$ and $\mathcal{S}_{HR} \subseteq \mathcal{S}$, respectively.

The addressed problem consists in assigning a sequence of available SLs to each $k \in \mathcal{K}$, by satisfying the following constraints:

- each SL in $\mathcal{S}$ can be assigned to a unique product type in $\mathcal{K}$;
- the sum of the capacities of the SLs assigned to a product type plus the extra storage made available for it on the top level (in case of pairs of fully replenished and physically contiguous SLs) must be greater than or equal to the storage demand of the product type;
- the special product types in $\mathcal{K}_P$ and $\mathcal{K}_{HR}$ should be preferably assigned to SLs in $\mathcal{S}_P$ and $\mathcal{S}_{HR}$, respectively;

with the aim of maximizing the residual storage capacity, i.e., the one which remains available after the assignment of the SLs.

Figure 1 depicts a storage area where three SLs are occupied by some items in the first level (two are consecutive and one is isolated) and four SLs are available for stocking (three are consecutive and one is isolated), depicted as full black rectangles and as white rectangles, respectively. An example of the two level storage policy is shown for the first two occupied SLs. The residual storage capacity, in this case, is defined as the sum of the capacities of the 4 available SLs at the ground level, plus the capacities exploitable on top of SLs 1 and 2, as well as on top of 2 and 3.

3 A Successive Constrained Shortest Path Method

For each product type $k \in \mathcal{K}$, the proposed heuristic finds a constrained shortest path on a suitable auxiliary graph, whose set of nodes describes the current availability of SLs in the warehouse. This path specifies the SLs assigned to k and the order in which they must be filled up, in such a way as to guarantee that the total capacity of the assigned SLs is enough to store the q^k items required by k, taking into account the possibility of exploiting extra storage on the top of the assigned SLs. After the assignment to k, the auxiliary graph is suitably pruned by removing the assigned nodes and the corresponding incident arcs, to avoid that the corresponding SLs can be assigned to product types other than k.

In the following subsections, we introduce the auxiliary graph, we present the constrained shortest path problem to be solved for each product type and we describe the overall heuristic method.

3.1 The Auxiliary Graph

In the auxiliary graph $\mathcal{G} = (\mathcal{N}, \mathcal{A})$, the set of nodes $\mathcal{N}$ consists of:

- a fictitious source node Σ,
- a fictitious target node Θ,
- a set $\mathcal{L}$ containing one node for each currently available SL in the warehouse.

The set $\mathcal{L}$ is partitioned into two subsets, $\mathcal{L}^I$ and $\mathcal{L} \setminus \mathcal{L}^I$: $\mathcal{L}^I$ is composed of all those nodes of $\mathcal{L}$ corresponding to isolated SLs in the warehouse (that is, no contiguous SL is available for storage, like SL 4 in Fig. 1), while $\mathcal{L} \setminus \mathcal{L}^I$ is in turn partitioned into $\cup_{h=1,\dots,H}\mathcal{F}_h$ subsets, each of them defining a group of nodes associated with physically contiguous SLs. For example, SLs 1, 2 and 3 in Fig. 1 define one of these groups. A generic subset $\mathcal{F}_h$ contains nodes of type $\{j_1^h, j_2^h, \dots, j_{|\mathcal{F}_h|}^h\}$, where j_1^h and $j_{|\mathcal{F}_h|}^h$ respectively denote the nodes associated with the first and the last SL of the physically contiguous group $\mathcal{F}_h$ (like SLs 1 and 3 in Fig. 1), while the remaining nodes j_m^h, with $1 < m < |\mathcal{F}_h|$, are associated with intermediate SLs (like SL 2 in Fig. 1). In particular, the SL associated with node j_m^h is in between the SLs associated with nodes j_{m-1}^h (being the previous one) and j_{m+1}^h (being the next one).

A capacity u_j is associated with each node $j \in \mathcal{L}$. It coincides with the capacity of the SL the node j is associated with, if the SL is isolated or it is the first SL in a group of physically contiguous SLs; otherwise, it coincides with its double, so as to model the two level storage policy. Moreover, a profit δ_j^k is defined for each product type $k \in \mathcal{K}$ and each node $j \in \mathcal{L}$. This profit aims to favour the assignment of k to the preferable subsets $\mathcal{S}_P$ or $\mathcal{S}_{HR}$, if $k \in \mathcal{K}_P$ or $k \in \mathcal{K}_{HR}$. Otherwise, it tends to favour the assignment of k to SLs in $\mathcal{S} \setminus (\mathcal{S}_P \cup \mathcal{S}_{HR})$.

The set of arcs $\mathcal{A}$ is defined in order to model the assignment of a sequence of SLs to each product type in $\mathcal{K}$. The set $\mathcal{A}$ contains:

- arcs (Σ, j), with $j \in \mathcal{L}^I$, and arcs (Σ, j_1^h), with $h = 1, \dots, H$, to model the assignment of the first SL to a product type;
- arcs (j, Θ), with $j \in \mathcal{L}$, to model the assignment of the last SL to a product type;
- arcs (j_m^h, j_{m+1}^h), with $h = 1, \dots, H$ and $m = 1, \dots, |\mathcal{F}_h| - 1$, to model the assignment of the available SL j_{m+1}^h immediately after the available and contiguous SL j_m^h;
- arcs $(j_{|\mathcal{F}_h|}^h, j_1^{h'})$, with $h, h' = 1, \dots, H$, and $h \neq h'$, to model the assignment of the SL $j_1^{h'}$ of group $\mathcal{F}_{h'}$ immediately after the SL $j_{|\mathcal{F}_h|}^h$ of group $\mathcal{F}_h$;
- arcs $(j_{|\mathcal{F}_h|}^h, i)$, with $h = 1, \dots, H$, and $i \in \mathcal{L}^I$, to model the assignment of the isolated SL i immediately after the SL $j_{|\mathcal{F}_h|}^h$ of group $\mathcal{F}_h$;
- arcs (i, j_1^h), with $i \in \mathcal{L}^I$ and $h = 1, \dots, H$, to model the assignment of the SL j_1^h of group $\mathcal{F}_h$ immediately after the isolated SL i;
- arcs (i, j), $i, j \in \mathcal{L}^I$, and $i \neq j$, to model the assignment of the isolated SL j immediately after the isolated SL i.

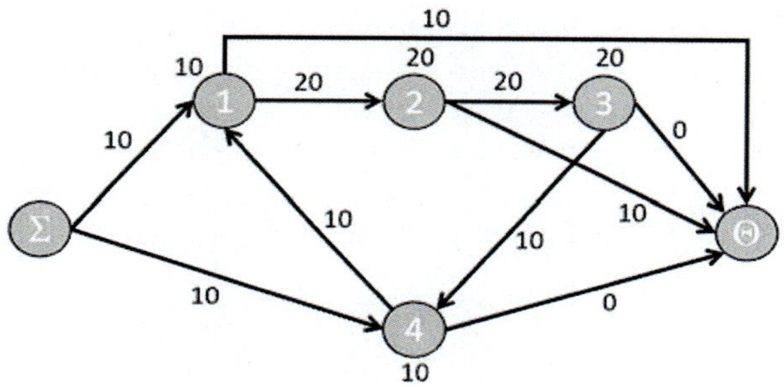

Fig. 2 Graph representation of the available SLs depicted in Fig. 1

Finally, a weight c_{ij} is associated with each arc $(i, j) \in \mathcal{A}$, which indicates the amount of space which becomes unavailable for future assignments due to the joint assignment of i and j to k:

$$c_{ij} = \begin{cases} u_j & \text{if } j \neq \Theta, \\ u_i & \text{if } j = \Theta \text{ and } \exists\, h \text{ such that } i = j_1^h, \\ u_i/2 & \text{if } j = \Theta \text{ and } \exists\, h \text{ such that } i = j_m^h, \text{ with } 1 < m < |\mathcal{F}_h|, \\ 0 & \text{if } j = \Theta \text{ and } \exists\, h \text{ such that } i = j_{|\mathcal{F}_h|}^h \text{ or } i \in \mathcal{L}^I. \end{cases}$$

Figure 2 reports the auxiliary graph associated with the available SLs depicted in Fig. 1. In this example, the capacity of each SL is 10 items. Thus, according to the definition above, the capacity of nodes 1 and 4 is equal to 10, while the capacity of nodes 2 and 3 is equal to 20 to model the two level storage policy. Nodes 1 and 4 are linked with Σ through an entering arc and each node is linked to Θ through an exiting arc. The weight associated with each arc (Σ, j), $j = 1, 4$, is equal to the capacity of node j, i.e., 10. The weight associated with the arcs (j, Θ) is 0 for $j = 3$ and $j = 4$ (since 3 is the last SL of a group of contiguous SLs, and 4 is an isolated SL), while it is 10 for $j = 1$ and $j = 2$ (since 1 represents the first SL and 2 an intermediate SL of a group of contiguous SLs). Finally, the weight of an arc entering node j, with $j \neq \Theta$, is equal to the capacity of j.

3.2 The Constrained Shortest Path Problem

Given the current product type $k \in \mathcal{K}$, the problem of determining a directed path from Σ to Θ in the auxiliary graph $\mathcal{G}$, which represents the sequence of SLs assigned to k, is formulated using the following two families of variables:

- $x_{ij} \in \{0, 1\}$, for any $(i, j) \in \mathcal{A}$, to model the sequence of SLs assigned to k in terms of a directed path in $\mathcal{G}$ from node Σ to node Θ;
- $y_i \in \mathbb{Z}_+$, for any $i \in \mathcal{N}$, to model Miller-Tucker-Zemlin-like constraints, aimed at avoiding subtours.

For each $i \in \mathcal{N}$, we also define $\mathcal{N}^+(i)$ and $\mathcal{N}^-(i)$ as the sets of nodes linked to $i \in \mathcal{N}$ via an exiting and an entering arc, respectively:

$$\mathcal{N}^+(i) = \{j \in \mathcal{N} : \exists\,(i,j) \in \mathcal{A}\}, \qquad \mathcal{N}^-(i) = \{j \in \mathcal{N} : \exists\,(j,i) \in \mathcal{A}\}. \tag{1}$$

The constrained shortest path problem related to k can be formulated as follows:

$$\min \sum_{(i,j)\in\mathcal{A}} c_{ij}x_{ij} - \sum_{j\in\mathcal{N}} \delta_j^k \sum_{i\in\mathcal{N}^-(j)} x_{ij} \tag{2}$$

$$\sum_{j\in\mathcal{N}^-(i)} x_{ji} - \sum_{j\in\mathcal{N}^+(i)} x_{ij} = \begin{cases} -1 & \text{if } i = \Sigma \\ 0 & \text{if } i \in \mathcal{L} \\ 1 & \text{if } i = \Theta \end{cases} \qquad \forall\, i \in \mathcal{N}, \tag{3}$$

$$\sum_{j\in\mathcal{N}} \sum_{i\in\mathcal{N}^-(j)} u_j x_{ij} \geq q^k, \tag{4}$$

$$y_j - y_i - 1 + (1 - x_{ij})|\mathcal{N}| \geq 0 \qquad \forall\, (i,j) \in \mathcal{A}. \tag{5}$$

The objective function (2) consists of two parts: the first summation defines the primary optimization goal to be minimized, i.e., the space no longer available in the warehouse after storing items of type k along the nodes of the path; the second sum is related to the secondary optimization goal, i.e., the request that special product types should be preferably stored in specific SLs. It involves parameters δ_j^k which are set in such a way that it is convenient to assign SL j to the product type k, if j is one of the preferable SLs for k. Constraints (3) define a directed path for k, by means of the binary variables x_{ij}, in terms of a unitary flow sent from the source node Σ to the target node Θ, with the aim of modeling the assignment of a sequence of SLs to k. Constraint (4) imposes that the sum of the capacities of the nodes along the path be greater than or equal to the storage demand of the product type k. Finally, constraints (5) are Miller-Tucker-Zemlin-like constraints, which avoid subtours in the returned solution [8].

Figure 3 shows two feasible solutions referring to the auxiliary graph in Fig. 2, assuming $q^k = 25$. The solution on the left assigns to k the sequence of SLs 4, 1, and 2, whose total capacity, given by the sum of the node capacities, is 40, enough to stock all the items of k. The selected SLs will be filled starting from 4 (10 items), passing then to 1 (other 10 items), and finally considering 2 (the remaining 5 items). The space no longer available in the warehouse for future assignments, i.e., the first sum of (2), is 50. Notice that the assignment of the SL 2 will make unavailable the extra storage on top of 2 and 3 in the future: this is why a weight 10 is associated with the arc (2, Θ). The solution on the right, instead, which is optimal, assigns the

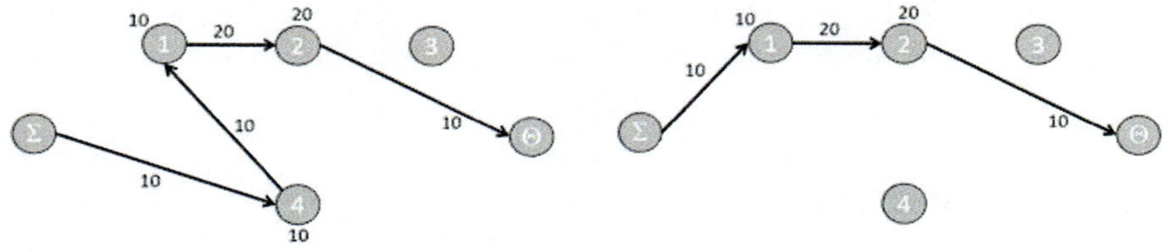

Fig. 3 Two feasible solutions representations

SLs 1 and 2 to k, with total capacity 30. The selected SLs will be filled starting from 1 (10 items) and then passing to 2, thus exploiting the capacity made available on top of 1 and 2 (10 items at the ground level and the remaining 5 on top of 1 and 2). The space no longer available for future assignments is 40, better than in the previously considered solution.

3.3 *The Overall Heuristic Method*

As outlined before, the idea underlying the heuristic approach is to address the product types in cascade, each time solving the constrained shortest path problem described in Sect. 3.2 over an auxiliary graph which is progressively pruned. The steps of the heuristic method are summarized in Algorithm 1.

Algorithm 1: Successive constrained shortest path method

1: Sort the product types in $\mathcal{K}$ in a nonincreasing order with respect to the number of items to stock. Set $\mathcal{K} = \{k_1, \ldots, k_{|\mathcal{K}|}\}$.
2: Define $\mathcal{L}$ as the set of nodes corresponding to all the available SLs in the warehouse.
3: **for** $t = 1, \ldots, |\mathcal{K}|$ **do**
4: Solve the constrained shortest path problem related to k_t on the graph induced by $\mathcal{L}$.
5: Define Φ_t as the set of nodes corresponding to the SLs assigned to k_t.
6: $\mathcal{L} := \mathcal{L} \setminus \Phi_t$.
7: **end for**
8: Unify the subproblem solutions $\Phi_1, \ldots, \Phi_{|\mathcal{K}|}$.

At the beginning, the product types in $\mathcal{K}$ are sorted in a nonincreasing order with respect to the number of items to stock. In the first iteration, the first product type in the resulting ordered set is considered. The constrained shortest path problem in Sect. 3.2 is solved on the graph defined in Sect. 3.1, where the set of nodes $\mathcal{L}$ corresponds to all the available SLs in the warehouse. In any successive iteration, say t, the t-th product type in the considered order is analysed and the corresponding constrained shortest path problem is solved over a graph obtained from the initial one by removing all the nodes corresponding to the SLs assigned till iteration t and their incident arcs. The complete solution is finally given by the set of the paths which have been separately determined for all the product types in $\mathcal{K}$.

4 Numerical Experiments

Two types of experiments have been performed. The first one aims at analysing the performance of the heuristic for different values of the parameters δ_j^k, while in the latter we compare the results of the heuristic here proposed with the ones returned by the matheuristic in [6]. The heuristic has been implemented by using the language OPL and solved via CPLEX 12.6 (IBM ILOG, 2016). All the experiments have been conducted on an Intel Xeon 5120 computer with 2.20 GHz and 32 GB of RAM.

4.1 *The Instances*

We considered the same set of real instances solved in [6], corresponding to 20 randomly selected days. The instances are divided into two classes, called ClassHA and ClassLA, each referring to 10 days where the number of items to stock is higher (ClassHA) or lower (ClassLA) than the average number of items to stock over the 20 selected days. Two kinds of special product types exist, i.e., perishable (P) and high rotational (HR), which should be preferably assigned to specific SLs. More in detail, the instances in ClassHA have to assign 1150 items of 14 different product types on average: 1.2% are items of type P, whereas 21.2% are items of type HR. The instances in ClassLA have to assign 787 items of 11 different product types on average: 1.5% are items of type P, whereas 40.7% are items of type HR.

4.2 *Efficacy and Efficiency of the Heuristic Approach*

As specified, parameters δ_j^k are used to favour the assignment of specific SLs to special product types. In particular, if product k is of type P, then its preferable SLs are those specified in subset $\mathcal{S}_P \subset \mathcal{S}$. In this case, $\delta_j^k > 0$ if $j \in \mathcal{S}_P$, and $\delta_j^k = 0$ otherwise. The same logic applies to a product k of type HR, for which δ_j^k are set in such a way to favour the assignment of SLs belonging to subset $\mathcal{S}_{HR} \subset \mathcal{S}$. On the other hand, for a product k neither of type P nor HR, δ_j^k tend to favour the assignment of SLs in $\mathcal{S} \setminus (\mathcal{S}_P \cup \mathcal{S}_{HR})$. Three settings for positive values of δ_j^k have been investigated, of form $\delta_j^k = p \cdot u_j$, where $p = 0.2, 0.5$ and 0.8.

Table 1, for both the instances in ClassLA and ClassHA, shows the features of the solutions our industrial partner is interested in for the different settings of δ_j^k. Specifically, we report the average solving time (in seconds), the average space

Table 1 Performance of the heuristic for ClassLA and ClassHA

	ClassLA				ClassHA			
p	Solving Time (sec.)	Available space (items)	%P in $\mathcal{S}_P$	%HR in $\mathcal{S}_{HR}$	Solving Time (sec.)	Available space items)	% P in $\mathcal{S}_P$	%HR in $\mathcal{S}_{HR}$
0.8	2.92	2874	100%	75.40%	2.83	2062	96.08%	84.33%
0.5	3.32	2886	100%	75.37%	3.21	2080	84.97%	81.85%
0.2	3.73	2904	56.67%	57.57%	3.29	2096	77.12%	75.46%

available after the assignment (given by the total capacity of the empty SLs at the ground level plus, whenever two empty SLs are contiguous, the capacity of the SL that can be created on top of them), the percentage of items of type P assigned to SLs in $\mathcal{S}_P$, and the percentage of items of type HR assigned to SLs in $\mathcal{S}_{HR}$.

As expected, by increasing p, i.e., the value of δ^k_j, the percentage of items of types in P and HR assigned to their preferable SLs increases, at the expenses of the space available after the assignment. In fact, as already observed in [6], the two objectives are often conflicting, since forcing the assignment of items to specific locations may be in contrast with the maximization of the total space available after the assignment. Interestingly, the higher is value p, the lower is the average solving time.

ClassHA was identified in [6] as the hardest group of instances to solve with the matheuristic, mainly due to the higher number of items to stock. As opposed, the heuristic seems to address such instances more easily. The higher number of items to stock per product type in ClassHA requires longer paths. This implies a wide pruning of the associated graphs at each iteration of the heuristic, and smaller and easier constrained shortest path problems to address during the resolution process.

4.3 Comparison with a Matheuristic Approach

The SLAP here addressed has been formulated in [6] as a multicommodity flow model, where the assignment of SLs is simultaneously addressed for all the product types in $\mathcal{K}$. Specifically, given an auxiliary graph whose set of nodes correspond to the SLs available in the warehouse, $|\mathcal{K}|$ directed paths are sought along which the quantity q^k is sent for each k, by taking into account the two level storage policy. The objective function aims at maximizing the available storage capacity after the assignment and the number of items of types $\mathcal{K}_P$ and $\mathcal{K}_{HR}$ assigned to $\mathcal{S}_P$ and $\mathcal{S}_{HR}$, respectively. A two-phase matheuristic has been proposed in [6], which is based on decomposition. In the first phase, $\mathcal{K}$ is partitioned into Λ subsets, in such a way that each group contains about the same number of items to store. Thus, Λ subproblems are generated and sequentially solved by CPLEX, each time removing those SLs already assigned in the previous solved subproblems. Finally, the Λ

Table 2 Comparison between the proposed heuristic and the matheuristic in [6]

	ClassLA				ClassHA			
	Solving time (sec.)	Available space (items)	%P in $\mathcal{S}_P$	%HR in $\mathcal{S}_{HR}$	Solving time (sec.)	Available space (items)	%P in $\mathcal{S}_P$	%HR in $\mathcal{S}_{HR}$
Heuristic	2.98	2886	100%	75.37%	3.21	2080	84.97%	81.85%
Matheuristic [6]	3330	2897	83.89%	70.60%	3442	2073	85.89%	82.15%

solutions obtained are merged into a unique solution, which is provided as initial feasible solution to the Branch and Bound algorithm of CPLEX. The matheuristic relies on several parameters, whose impact on the performance of the resolution process has been deeply investigated, by identifying a suitable parameter setting.

Table 2 compares the results obtained by the heuristic here proposed with the ones achieved by the matheuristic in [6], on the set of instances described in Sect. 4.1. For the matheuristic, the most suitable parameter setting devised in [6] has been used. For the heuristic, we selected $p = 0.5$ as a good compromise between efficiency and efficacy of results. Table 2 reports, separately for ClassLA and CLassHA, the average solving time (in seconds), the average space available in the warehouse after the assignment, the percentage of items of type P assigned to their preferable SLs, and the percentage of items of type HR assigned to their preferable storage locations.

For ClassLA, the successive constrained shortest path heuristic outperforms the matheuristic for the preferable SLs assigned to the special product types. The available space after the assignment is almost the same, just decreasing of a few units on average. For ClassHA, the results are similar for both the approaches, with the successive constrained shortest path heuristic outperforming the matheuristic regarding the available space after the assignment. The strength of the heuristic is the time required to obtain solutions. In fact, it is of about 3 seconds on average, i.e., about one thousandth of the time required by the matheuristic. The proposed heuristic thus appears to be a very promising tool to solve the addressed SLAP problem, rapidly determining solutions which, for the majority of the tested instances, also improve the quality of the solutions found by the matheuristic approach.

5 Conclusions

A very fast successive constrained shortest path heuristic has been proposed to address the assignment of items to SLs jointly with sequencing decisions about the assigned SLs, as required in a real application context. The heuristic ouperforms the matheuristic in [6] on real-size instances, in terms of both solving time and solution quality. Future research will address a more general problem where storage allocation decisions are considered jointly with the routing of the vehicles in charge of moving items towards the selected locations.

Acknowledgments The research has been supported by Region of Tuscany-Regional Government (POR FESR 2014-2020-Line 1-Research and Development Strategic Projects) through the Project IREAD4.0 under Grant CUP 7165.24052017.112000028.

References

1. Hausman, W.H., Schwarz, L.B., Graves, S.C.: Optimal storage assignment in automatic warehousing systems. Manage. Sci. **22**, 629–638 (1976)
2. Ashayeri, J., Gelders, L.F.: Warehouse design optimization. Eur. J. Oper. Res. **21**, 285–294 (1985)
3. Gu, J., Goetschalckx, M., McGinnis, L.F.: Research on warehouse operation: a comprehensive review. Eur. J. Oper. Res. **177**, 1–21 (2007)
4. Lehnfeld, J., Knust, S.: Loading, unloading and premarshalling of stacks in storage areas: survey and classification. Eur. J. Oper. Res. **239**, 297–312 (2014)
5. Reyes, J., Solano-Charris, E., Montoya-Torres, J.: The storage location assignment problem: a literature review. Int. J. Ind. Eng. Comput. **10**, 199–224 (2019)
6. Lanza, G., Passacantando, M., Scutellà, M.G.: Assigning and sequencing storage locations under a two level storage policy: optimization model and matheuristic approach. Technical Report 2398, University of Pisa (2021)
7. Lanza, G., Passacantando, M., Scutellà, M.G.: Assigning and sequencing storage locations under a two level storage policy: Optimization model and matheuristic approaches. Omega **108**, 102565 (2022)
8. Miller, C.E., Tucker, A.W., Zemlin, R.A.: Integer programming formulation of traveling salesman problems. J. ACM. **7**, 326–329 (1960)

Local Search for Aircraft-Landing Planning

Nicolas Zufferey, Marie-Sklaerder Vié, and Roel Leus

Abstract In collaboration with EUROCONTROL (the *European Organisation for the Safety of Air Navigation*), the considered *Aircraft Landing Planning* (ALP) problem aims at minimizing delays (with respect to the published airline schedules) while satisfying the *separation constraint* (which imposes minimum threshold times between planes, ranging from 90 to 240 s). In this study, the landing sequence of the planes has to be determined first, and subsequently their associated landing times and *Holding-Stack Patterns* (HSPs) needed to meet such landing times. HSPs consist of making a plane wait for its planned landing time by making circular patterns close to the airport. The uncertainty due to winds is taken into account in the simulation procedure (it has an impact on the arrival times). The proposed solution method is a descent local search with restarts. It is quick enough with respect to implementation in real situations as it can be applied within seconds. Furthermore, the obtained results show that the delays can be reduced by approximately 50% on average when compared to a common practice rule.

Keywords Aircraft landing · Scheduling · Descent local search · Uncertainty

1 Introduction

From take-off to landing, planes are subjects to many uncertainties (e.g., wind, traffic). Moreover, the runway capacity of an airport is the bottleneck capacity of the landing process. Therefore, the landing time of each plane arriving at an airport has to be adjusted all along its flight trajectory to use the runway capacity at best despite uncertainty. One way to delay an airplane is to make it perform holding

N. Zufferey (✉) · M.-S. Vié
GSEM, University of Geneva—Uni Mail, Geneva, Switzerland
e-mail: marie-sklaerder.vie@unige.ch; n.zufferey@unige.ch

R. Leus
Faculty of Economics and Business, Leuven, Belgium
e-mail: roel.leus@kuleuven.be

L. Amorosi et al. (eds.), *Optimization in Artificial Intelligence and Data Sciences*,
AIRO Springer Series 8, https://doi.org/10.1007/978-3-030-95380-5_15

stack patterns in the vicinity of the airport (approach area) before its landing, but this creates bigger fuel consumption and noise pollution. Another way is to decrease (sometimes increase) the speed of the plane during its cruise (or make it do a detour using path stretching), but due to larger uncertainties at longer distance from destination, this may create unnecessary additional delays. Also, the fewer number of modifications there are on a landing time, the better it is for the *air traffic controllers*' (ATC) workload, as they have other priorities (mainly ensuring safety and avoiding collisions). The goal of *Aircraft Landing Planning* (ALP) is to optimize the landing times of the planes arriving at an airport runway. For this purpose, the planes to delay have to be selected, as well as how and when to delay them. Safety constraints imposing threshold distances between planes have to be satisfied, obviously.

The paper is organized as follows. A literature review is first conducted in Sect. 2. Second, the considered problem is formally introduced in Sect. 3. Next, the proposed optimization method is proposed in Sect. 4. Results are presented in Sect. 5, followed by conclusions in Sect. 6.

2 Literature Review

A comprehensive review covering the period up to 2011 can be found in [2]. It results that the main objective functions can be divided into four categories, presented below.

- *Safety and efficiency objectives.* Maximize: runway throughput, fairness among the aircrafts. Minimize: approach time before landing, ATCs' workload, aircraft taxi-out time, arrival delays.
- *Airline's objectives.* Maximize: punctuality with respect to the published landing times, adherence to airline priorities within their own flights, connectivity between flights. Minimize: operation costs (mainly fuel costs), total passenger delays.
- *Airport objectives.* Maximize: punctuality according to the operating schedule. Minimize: the need for gate changes due to delays.
- *Government objectives.* Minimize: environmental effects (noise and air pollution).

The review of [2], covering the literature up to 2011, leads to different observations. First, many theoretical studies show a great potential improvement of the runway utilization, but may not be feasible in practice due to ignored operational constraints (e.g., minimum time before landing, precedence constraints) or unreasonable computing times. Indeed, ATCs will always prefer fast (i.e., able to generate a solution within seconds) and satisfying (with respect to the considered objective) solution methods rather than optimal but time-consuming ones. Second, the definition of the objective functions and constraints varies a lot among the articles. Indeed, the involved parties (e.g., airport, companies, customers) have conflicting interests.

Therefore, selecting an appropriate and realistic objective function is a critical issue. Third, and most importantly, uncertainty occurs at different levels (e.g., weather, precision of the equipment, departure time, accuracy of information). However, in contrast with the supply chain management literature (e.g., [18]), very few studies take this aspect into account and actually use simulation, as [6, 15]. Indeed, most of the literature considers the static case rather than a dynamic (but realistic) environment. Interestingly, it is showed in [7] that: deterministic algorithms are sub-optimal in a dynamic environment; a FCFS method is robust (i.e., not too sensitive to variations in problem characteristics or data quality) and it has many advantages over many existing algorithms (e.g., generated sequence understandable by ATCs, stable and easy-to-estimate delays, already in place in many airports).

From 2012, some work on static ALP has still been published [11, 23]. But as highlighted above, from a practical standpoint, only the approaches that are able to take into account random events are relevant. The three main axes that integrate this dimension are queuing theory [16], robust optimization [12], and on-line optimization [3]. The approaches including uncertainty use Monte-Carlo simulation, and some dynamic solution methods (but without uncertainty, except the appearance of new flights) employ a rolling-horizon approach. Other relevant and recent references can be found in [4, 20, 21].

In the light of the weaknesses of the existing literature, various promising research directions are identified below.

(1) Consideration of uncertainty. Obviously, as most variables in ALP are stochastic, a quick, accurate and dedicated simulation tool is mandatory to evaluate the true quality of a solution. The Monte-Carlo simulation that is usually proposed for ALP is probably not the best tool according to these criteria. Generally, the existing literature does not take uncertainty into account. A scarce literature proposes either robust solution within a static approach, or a dynamic solution method but without any robustness guarantee. An interesting approach could integrate robustness in the on-line process (to avoid a prohibitive number of rescheduling actions) [14].

(2) Appropriate solution methods. The existing metaheuristics (generally genetic and ant colony algorithms) do not seem the most appropriate for ALP. Indeed, they need a long learning phase (e.g., Variable Neighborhood Search [5], Tabu Search [22]), and even though they can provide feasible solutions at any time in the computation, there is little chance that the quality of the solutions is good after only a few seconds of computation. Filtering techniques would be an efficient option to definitely discard unpromising decisions from the solution space (e.g., [13]).

(3) Instance calibration. An instance has to be designed in a realistic way. In most of the literature, instances with hundreds of flights are tackled (with metaheuristics), as well as instances with around 50 flights (with exact methods or metaheuristics). However, everyday and for many important European airports, the demand follows patterns with marked peaks (e.g., hub periods of 3 h) and then lower traffic levels [10]. For these reasons, a relevant instance can

be defined by a time horizon of 3 h (i.e., covering a peak) and a single landing runway. Knowing that the time between two consecutive aircraft belongs to interval [90, 240] s, the most relevant instance size ranges from 90 to 120 flights. Surprisingly, this range is usually not considered in the literature.

(4) *Planning horizon.* In a dynamic case, an important optimization would be the size of the considered rolling horizon H, as it can have a crucial impact on the quality of the obtained solution [19]. A tradeoff has to be found between (1) waiting as much as possible before delaying a flight (which minimizes the uncertainties, but increases the noise pollution and fuel costs), and (2) anticipating and delaying a flight as early as possible (more risky approach if too many uncertainties, but more efficient otherwise). The challenge is that the larger H is, the more flexibility we have in the optimization, but the more uncertainty we have too (mainly because of pop-up flights, as presented below).

3 Considered Problem

Each instance covers a 3-h planning horizon, which allows capturing the peak period of most airports. A rolling planning window $H^t = [t, t + w[$ (with $w = 45$ min) is associated with the current time t, with time steps of $\Delta t = 30$ s. At each time t, we only consider the flights that are in cruise and have their landing planned in H^t. The flights that have their landing in H^t but are not yet in cruise are only considered when they take-off. They are called the *pop-up* flights.

Each flight has different stages: (1) take-off, (2) cruise, (3) approach, (4) landing (the last $L = 15$ min, during which no modification is performed). In this paper, we only consider stages (2) and (3). From a practical standpoint, an initial schedule is first built when each flight enters the planning window (i.e., when it has taken off in the case of a pop-up flight, or when its expected landing time is within the next 45 min). Next, we can reschedule it (within the landing sequence) or make it wait to meet its planned arrival time (through HSPs). The popular *First-Come-First-Served* (FCFS) rule is employed to build the initial schedule. FCFS ranks the flights according to their entry times in H^t (i.e., with respect to increasing published arrival times, like the earliest-due-date rule in job scheduling). FCFS used to be the most employed current-practice approach [9], and it is an optimal rule for the single-machine job-scheduling problems when the maximum tardiness has to be minimized [17] (in our case we have to minimize the average tardiness).

We propose the following mathematical model (P^t) for each time t. Among the flights that have already taken off, we only consider the flights with planned landing times up to time $t + w$. Let J^t be the set of (say n) flights considered in H^t. For each flight $j \in J^t$, the following data is given:

- r_j: release date (i.e., take-off time).
- d_j: due date (i.e., published landing time).
- p_j^t: processing time (i.e., remaining time—in seconds—during the cruise phase).

- $s_{j,j'}$: set up time between flights j and j'. More precisely, for each pair (j, j') of flights such that j has to land before j', their landing times must be separated by $s_{j,j'} \in \{90, 120, 150, 180, 210, 240\}$ s, depending on the involved plane types.

We have two types of decision variables:

- determine the vector Π^t of the positions of the flights involved at time t (i.e., improve the current landing sequence by performing an optimization method);
- for each flight j, determine a feasible landing time C_j^t (with respect to the separation constraint) and assign a HSP of duration W_j^t in order to meet C_j^t.

The objective function f to minimize is the sum of all positive delays (i.e., the total tardiness), and it is given in Eq. (1). In contrast with the production-planning literature, negative delays are not penalized (e.g., [24]).

$$f = \sum_{j \in J^t} \max\{C_j^t - d_j, 0\} \tag{1}$$

Below, Constraints (2) impose that two flights are not scheduled in the same position. Constraints (3) capture the separation constraints. Constraints (4) determine the expected landing times. Constraints (5) are the domain constraints.

$$\Pi_j^t \neq \Pi_{j'}^t \qquad \forall j, j' \in J^t \tag{2}$$

$$C_{j'}^t \geq C_j^t + s_{j,j'} \qquad \forall j, j' \in J^t \text{ such that } \Pi_j^t + 1 = \Pi_{j'}^t \tag{3}$$

$$C_j^t = t + p_j^t + W_j^t + L \qquad \forall j \in J^t \tag{4}$$

$$\Pi_j^t \in \{1, \ldots, n\}, C_j^t \geq 0, W_j^t \geq 0 \qquad \forall j \in J^t \tag{5}$$

This problem can be seen as a variant of a single-machine total-tardiness problem with setup times, which is NP-hard even without setup times [8].

4 Optimization Method

Algorithm 1 presents how to roll the planning window H^t over the full 3-h planning horizon.

In Step 2, the landing positions Π^t of the new flights are computed with the following insertion rules used in practice:

- each pop-up flight j that just entered H^t (i.e., $t \geq r_j$ but $t - \Delta t < r_j$, and $t \geq d_j - w$) is added to the landing sequence at a position Π^t such that its due date is respected (i.e., j is placed before all flights j' such that $C_{j'}^t \geq d_j$ but after all the other flights);

Algorithm 1: Optimization for each time step t

Initialization: set $t = 0$, $J^t = J^{t-\Delta t} = \emptyset$ and $\Pi^t = \Pi^{t-\Delta t} = ()$.
While (not all flights have landed), **do:**
1. Update J^t: remove the flights that have started landing (i.e., each flight j for which $t \geq C^t_j - l$), and add the flights that have just entered the updated planning window H^t (i.e., each flight j for which $t \geq r_j$ and $t \geq d_j - w$).
2. Compute the positions of the new flights (i.e., the flights that are in J^t but not in $J^{t-\Delta t}$) to obtain the vector Π^t, based on $\Pi^{t-\Delta t}$ and the insertion rules.
3. Update the remaining cruise time for each flight j: set $p^t_j = p^{t-\Delta t}_j - \Delta t \cdot (1 + u^t_j)$.
4. Update C^t and W^t according to the new flight positions Π^t and the processing times p^t.
5. Improve solution (Π^t, C^t, W^t) with a solution method.
6. Move to the next time step: set $t = t + \Delta t$ and $H^t = [t, t + w[$.

- each flight j that took off a while ago but just entered H^t (i.e., $t \geq r_j$ and $t \geq d_j - w$, but $t - \Delta t < d_j - w$) is put at the end of the landing sequence (FCFS rule).

In Step 3, each remaining processing times p^t_j is updated while considering an uncertainty parameter u_t randomly generated following the EUROCONTROL specifications. u_t generates a deviation (e.g., due to wind) of the cruise speed of around 7% (with an average of 0%, as positive deviations are compensated by negative ones). In Step 4, and after each modification of Π^t, the values of C^t and W^t are updated with the following current-practice rules. First, we re-number all flights of J^t as $j_1, j_2, \ldots, j_n$ such that $\Pi^t_{j_1} < \Pi^t_{j_2} < \ldots < \Pi^t_{j_n}$. Next, for $k = 1$ to n, we perform steps (S1) and (S2).

- *Step (S1).* $C^t_{j_k} = \max\{C^t_{j_{k-1}} + s_{j_{k-1},j_k}, t + p^t_{j_k} + L\}$ (i.e., the arrival time of j_k is as close as possible to the arrival time of the previous flight j_{k-1}, or as soon as j_k can land).
- *Step (S2).* $W^t_{j_k} = C^t_{j_k} - (t + p^t_{j_k} + L)$ (i.e., the flight turns over the airport if it is too early with respect to the planned landing time).

As (1) the considered problem is NP-hard, (2) up to 24 flights are involved in H^t, and (3) the allowed computing-time limit T is very short ($T = \Delta t = 30$ s), quite a number of potential solution methods are not suitable for Step 5. Indeed, exact methods, cumbersome population-based metaheuristics (e.g., genetic algorithms, ant algorithms) or metaheuristics using a somewhat long learning process (e.g., simulated annealing) are too slow. In contrast, a descent local search (DLS) appears as a promising candidate.

DLS takes as input the solution from Step 4. At each iteration, a neighbor solution S' is generated from the current solution $S = (\Pi^t, C^t, W^t)$ by performing the best *Reinsert* move on S. A move *Reinsert* consists of changing the position Π^t_j of a flight $j \in J^t$ within the landing sequence. After each modification of Π^t, the associated variables (C^t, W^t) must be updated to have a feasible solution S' (separation constraint) and to know $f(S')$. The search process stops when no improvement of S is achieved during an iteration. In order to use the full time budget T, DLS is restarted when it encounters a local minimum (which occurs almost every second). The best visited solution is returned at the end.

At each iteration, two mechanisms are used for reducing the computational effort. First, the new position for the investigated flight j must be in $[\Pi^t_j - 5; \Pi^t_j + 5]$. This kind of *Constrained Position Shifting* is standard [1]. Indeed, from a practical standpoint, it seems straightforward to reschedule a flight not too far away from its initial position. Second, only a random proportion ρ (tuned to 50%) of the possible neighbor solutions is generated. These mechanisms allows to perform more iterations during T seconds, which increases the exploration capability of DLS.

5 Results

The algorithms were coded in C++ (under Linux, 3.4 GHz Intel Quad-core i7 processor, 8 GB of DDR3 RAM).

Table 1 compares the proposed DLS approach with FCFS (i.e., a common practice rule, see Algorithm 1 without Step 5). Average results (over all the instances) are indicated in bold face in the last line. For each instance (provided by EUROCONTROL), the following information is provided: the number N of flights, the largest number n_{max} of flights encountered in a planning window, the average delay and the maximum delay (for both DLS and FCFS). The two latter quantities

Table 1 Comparison of FCFS with DLS for 15 instances provided by EUROCONTROL

			FCFS		DLS		% Gain	
Instance	N	n_{max}	Avg. delay	Max delay	Avg. delay	Max delay	Avg. delay	Max delay
1	59	16	91.36	305.00	59.96	450.20	34%	−48%
2	35	10	156.08	528.20	96.98	490.40	38%	7%
3	64	20	154.41	447.60	93.05	456.00	40%	−2%
4	79	24	388.30	782.60	228.31	1672.60	41%	−114%
5	53	14	189.53	545.00	110.36	538.40	42%	1%
6	79	21	328.86	709.20	181.40	1629.20	45%	−130%
7	75	18	208.88	558.80	111.38	475.40	47%	15%
8	75	24	288.57	651.40	143.88	1480.60	50%	−127%
9	62	16	240.26	539.20	118.33	505.20	51%	6%
10	70	22	207.41	503.20	99.57	563.40	52%	−12%
11	72	22	280.79	631.20	131.57	800.20	53%	−27%
12	71	18	170.19	514.80	77.72	475.20	54%	8%
13	97	23	386.69	903.00	174.56	1247.60	55%	−38%
14	61	15	195.54	644.60	86.58	612.80	56%	5%
15	97	20	234.52	569.00	97.04	662.20	59%	−16%
Average results			**234.76**	**588.85**	**120.71**	**803.96**	**48%**	**−31%**

are computed with respect to all flights (in seconds), and averaged over 5 runs (with different uncertainty scenarios). The percentage gains of DLS (compared to FCFS) are given in the two last columns (a negative value indicates a better performance for FCFS).

We can see that DLS can significantly reduce the average delays (almost 50%). Interestingly, the improvement is somewhat increasing with the difficulty of the instance (i.e., with N and n_{max}), but further investigations are required to understand the benefit of DLS with respect to the instance characteristics. FCFS is often better regarding the maximum delay. This makes sense as FCFS guarantees optimality for minimizing the maximum delay (but not the average delay) for single-machine job-scheduling contexts. However, DLS can sometimes do better even for the maximum delay, as it reacts to uncertainties whereas FCFS does not.

6 Conclusion

A lot of work has been done on *Aircraft Landing Planning* (ALP), from exact methods (mainly branch-and-bound) to metaheuristics (e.g., numerous genetic algorithms). These methods, however, entail two main weaknesses. First, they are usually static (i.e., assume that all data is well-known), whereas the problem presents many uncertainties (e.g., weather, traffic, interaction with other flights). Second, the existing approaches generally do not match the quickness of the decision environment in which the decision makers have to work.

ALP is a challenging problem as the runway capacity is the bottleneck of many airports. In collaboration with EUROCONTROL, this study proposes a quick and efficient descent-based solution method for minimizing delays. Indeed, solutions can be obtained within seconds (which is appropriate for real-world implementation) and the average delay is reduced by almost 50%. Possible future works include the joint consideration of various objectives (for instance, in a lexicographic fashion as in [25]), and the development of refined algorithms and other techniques (e.g., speed adjustments, detours) to make the flights meet their landing times in order to reduce the over-the-airport traffic.

Acknowledgments This study was partially financed by EUROCONTROL (SESAR2020 program). The authors would like to thank Mr. Raphaël Christien for his availability and advices.

References

1. Balakrishnan, H., Chandran, B.G.: Algorithms for scheduling runway operations under constrained position shifting. Operations Research **58**(6), 1650–1665 (2010)
2. Bennell, J.A., Mesgarpour, M., Potts, C.N.: Airport runway scheduling. 4OR Q. J. Oper. Res. **9**(2), 115–138 (2011)

3. Bennell, J.A., Mesgarpour, M., Potts, C.N.: Dynamic scheduling of aircraft landings. Eur. J. Oper. Res. **258**(1), 315–327 (2017)
4. Bianco, L., Dell'Olmo, P., Giordani S.: Scheduling models for air traffic control in terminal areas. J. Sched. **9**(3), 180–197 (2006)
5. Bierlaire, M., Thémans, M., Zufferey, N.: A heuristic for nonlinear global optimization. INFORMS J. Comput. **22**(1), 59–70 (2010)
6. Beasley, J.E., Krishnamoorthy, M., Sharaiha, Y.M., Abramson, D.: Displacement problem and dynamically scheduling aircraft landings. J. Oper. Res. Soc. **55**(1), 54–64 (2004)
7. Brentnall, A.R., Cheng, R.C.H.: Some effects of aircraft arrival sequence algorithms. J. Oper. Res. Soc. **60**(7), 962–972 (2009)
8. Du, J., Leung, J.Y.T.: Minimizing total tardiness on one machine is NP-hard. Math. Oper. Res. **15**(3), 483–495 (1990)
9. Erzberger, H.: Design Principles and Algorithms for Automated Air Traffic Management. AGARD Lecture Series No. 200: Knowledge-Based Functions in Aerospace Systems, Madrid, Paris, and San Francisco (1995)
10. EUROCONTROL.: Performance Review Report 2015. Technical report, 2016
11. Faye, A.: Solving the aircraft landing problem with time discretization approach. Eur. J. Oper. Res. **242**(3), 1028–1038 (2015)
12. Heidt, A., Helmke, H., Liers, F., Martin, A.: Robust runway scheduling using a time-indexed model. In: Proceedings of the 4th SESAR Innovation Days, Madrid, pp. 1–8 (2014)
13. Hertz, A., Schindl, D., Zufferey, N.: Lower bounding and tabu search procedures for the frequency assignment problem with polarization constraints. 4OR **3**(2), 139–161 (2005)
14. Krishnan, S., Barton, G.W., Perkins, J.D.: Robust parameter estimation in on-line optimization - part I. Methodology and simulated case study. Comput. Chem. Eng. **16**(6), 545–562 (1992)
15. Moser, I., Hendtlass, T.: Solving dynamic single-runway aircraft landing problems with extremal optimisation. In: Proceedings of the 2007 IEEE Symposium on Computational Intelligence in Scheduling, Honolulu, pp. 206–211 (2007)
16. Nikoleris, T., Hansen, M.: Queueing models for trajectory-based aircraft operations. Transportation Science **46**(4), 501–511 (2012)
17. Pinedo, M.: Scheduling: Theory, Algorithms, and Systems. Springer (2016)
18. Respen, J., Zufferey, N., Wieser, P.: Three-level inventory deployment for a luxury watch company facing various perturbations. J. Oper. Res. Soc. **68**(10), 1195–1210 (2017)
19. Russell, R.A., Urban, T.L.: Horizon extension for rolling production schedules: Length and accuracy requirements. Int. J. Prod. Econ. **29**(1), 111–122 (1993)
20. Samà, M., D'Ariano, A., Pacciarelli, D.: Rolling horizon approach for aircraft scheduling in the terminal control area of busy airports. Transp. Res. E Logist. Transp. Rev. **60**, 140–155 (2013)
21. Samà, M., D'Ariano, A., Pacciarelli, D.: Scheduling models for optimal aircraft traffic control at busy airports: tardiness, priorities, equity and violations considerations. Omega **67**, 81–98 (2017)
22. Schindl, D., Zufferey, N.: A learning tabu search for a truck allocation problem with linear and nonlinear cost components. Naval Res. Logist. **62**(1), 32–45 (2015)
23. Tavakkoli-Moghaddam, R., Yaghoubi-Panah, M., Radmehr, F.: Scheduling the sequence of aircraft landings for a single runway using a fuzzy programming approach. J. Air Transp. Manag. **25**, 15–18 (2012)
24. Thevenin, S., Zufferey, N., Widmer, M.: Order acceptance and scheduling with earliness and tardiness penalties. J. Heurist. **22**(6), 849–890 (2016)
25. Thevenin, S., Zufferey, N., Potvin, J.-Y.: Makespan minimisation for a parallel machine scheduling problem with preemption and job incompatibility. Int. J. Prod. Res. **55**(6), 1588–1606 (2017)

An Emission Pollution Permit System for Time-Dependent Transportation Networks Based on Origin-Destination Pairs

Annamaria Barbagallo

Abstract In the paper an emission pollution permit system for a dynamic traffic equilibrium model based on origin/destination pairs is presented. The time-dependent equilibrium conditions are expressed by an evolutionary variational inequality. Thanks to the variational formulation, existence and continuity results for equilibrium distributions are established.

Keywords Evolutionary variational inequalities · Time-dependent traffic equilibrium problem · Environmental policy · Emission pollution permits

1 Introduction

Scientific and technological progress has reached a remarkable development today. It has brought considerable improvements in human life, but it has also caused an alteration of the natural balance due to the increase in pollution. In last years several goverments adopted policies in order to safeguard the environment. To this aim the activation of emission control systems has a fundamental role to reduce the pollution. For this reason we investigate on a new time-dependent model of transportation networks which takes into account of a permit system. More precisely we consider a new dynamic traffic policy model based on origin-destination licenses. Such a model improves the model presented in [8], not only because the time dependence occurs but also for the presence of capacity constraints on path flows and emissions. It is worth to underline that each phenomenon of the socio-economic and physical world is not stable with respect to the time and that static models of equilibria are a first useful abstract approach. As J. Gwinner rightly mentions in [4]: "*Equilibrium per definition excludes time. On the other hand,*

A. Barbagallo (✉)
Department of Mathematics and Applications "R. Caccioppoli", University of Naples Federico II, Naples, Italy
e-mail: annamaria.barbagallo@unina.it

L. Amorosi et al. (eds.), *Optimization in Artificial Intelligence and Data Sciences*, AIRO Springer Series 8, https://doi.org/10.1007/978-3-030-95380-5_16

time is central in our reality, in the physical-technological world as well as in the socio-economic world. Thus we are led to the study of time-dependent variational inequalities or evolutionary variational inequalities that model various constrained evolution problems."

In [8] and [3], the authors studied models in which vehicular travel and pollution, owing to emissions, is controlled by emission pollution licenses or permits. The equilibrium license price, for these emission permits, allows the reduction in travel and pollution. In [7], the author analyzed two permit system models based on origin/destination pairs and on paths. The travelers try to determine their minimal cost paths of travels from their origins to their destinations. As in the model in [8], the volume of emissions is equal to the product of a composite emission factor times the vehicular activity at the link levels (see [2]).

The goal of the paper is to introduce a more general and time-dependent model starting from the static emission pollution permit system. Furthermore, the characterization by means of a suitable evolutionary variational inequality is established. Such a variational formulation allows us to make use of theoretical results in order to show the existence of time-dependent equilibrium distributions. Since we study the evolution in time of the emission pollution permit system, another aspect to analyze is the continuity of equilibrium solutions with respect to the time variable. Such a result is obtained thanks the Kuratowski set convergence property and under continuity assumptions on data.

The paper is organized as follows. In Sect. 2, we present the time-dependent transportation network model with an emission pollution permit system based on origin-destination pairs. We characterize its equlibrium conditions with an evolutionary variational inequality. Then, in Sect. 3, we prove some existence and continuity results for time-dependent equilibrium distributions.

2 The Dynamic Traffic Equilibrium Problem with the O/D Pair-Based Permit System

Let us consider a traffic network which is represented by a graph $G = [\mathcal{N}, \mathcal{L}]$, where $\mathcal{N}$ is the set of nodes and $\mathcal{L}$ is the set of directed links interlocked the nodes. Let r be a path consisting of a sequence of links which connects an origin-destination (O/D) pair of nodes. In the network there are m paths. Let $\mathcal{W}$ be the set of O/D pairs with typical O/D pair w_j, with $|\mathcal{W}| = l$ such that $m > l$. The set of paths connecting the O/D pair w_j is denoted by $\mathcal{R}_j$ and the entire set of paths in the network by $\mathcal{R}$. The topology of the network is described by the pair-link incidence matrix $\Phi = (\varphi_{jr})$, where φ_{jr} is 1 if path r connects the pair w_j and 0 otherwise. We are interested to a time-dependent transportation network equilibrium problem, for this reason we consider a time interval $[0, T]$, with $T > 0$. Therefore, let us introduce the time-dependent flow vector on route $F(t) = (F_r(t))^T_{r=1,\dots,m}$, a.e. in $[0, T]$, which has to satisfy the time-dependent capacity constraints $\lambda_r(t) \leq F_r(t) \leq$

$\mu_r(t)$, for every $r = 1, \ldots, m$, a.e. in $[0, T]$, and the traffic conservation law states $\Phi F(t) = \rho(t)$, a.e. in $[0, T]$, where $\rho(t) = (\rho_j(t))^T_{j=1,\ldots,l}$ is the total demand vector. We assume that F belongs to $L^2([0, T], \mathbb{R}^m_+)$, as well as, λ and μ, moreover $\rho \in L^2([0, T], \mathbb{R}^l_+)$. Let us consider the user travel cost on paths $C(t, F(t)) = (C_r(t, F(t)))^T_{r=1,\ldots,m}$, a.e. in $[0, T]$. Let $E_r(t)$ be the emission factor on path r, a.e. in $[0, T]$, then $E(t) = (E_r(t))^T_{r=1,\ldots,m}$, a.e. in $[0, T]$. Let $L_j(t)$ be the number of licenses or permits for O/D pair w_j that enables the travelers between O/D pair w_j to release pollutants at a certain index, a.e. in $[0, T]$, so $L(t) = (L_j(t))^T_{j=1,\ldots,l}$, a.e. in $[0, T]$. Moreover, let us denote the initial allocation of licenses for O/D pair w_j by L^0_j, and assume that such a quantity is nonnegative. We consider the price of a license in the O/D pair-based and denote it by $P(t)$, a.e in $[0, T]$. We suppose that a license in the O/D pair-based system is such that $\underline{P}(t) \leq P(t) \leq \overline{P}(t)$, a.e. in $[0, T]$, where $\underline{P}, \overline{P} \in L^2([0, T], \mathbb{R}_+)$. Furthermore let $T_j(t)$ be the marginal cost of emission reduction for the O/D pair w_j, at time $t \in [0, T]$, hence $T(t) = (T_j(t))^T_{j=1,\ldots,l}$, a.e. in $[0, T]$. We suppose that the marginal costs of reduction for the O/D pair w_j satisfies $0 \leq T_j(t) \leq \overline{T}_j(t)$, for every $j = 1, \ldots, l$, a.e. in $[0, T]$. In this model, the network user on a path r not only has to pay the user travel cost but also has to pay the price or cost of his emissions. Hence, the dynamic equilibrium conditions in presence of the O/D pair-based emission pollution permit system are given by:

For each O/D pair $w_j \in W$, for each paths $r, s \in \mathcal{R}_j$ and a.e. in $[0, T]$:

$$C_r(t, H(t)) + E_r(t)T^*_j(t) < C_s(t, H(t)) + E_s(t)T^*_j(t)$$
$$\Rightarrow H_r(t) = \mu_r(t) \text{ or } H_s(t) = \lambda_s(t). \tag{1}$$

Equilibrium conditions (1) assert that the road users choose minimum cost paths, where the cost path is equal to the sum of the user travel cost plus the emission cost for traveling on the path r.

We suppose that the transportation authority is responsible for warning the traffic users of the license price and the corresponding payments requested, and also the availability of the licenses or permits for O/D pairs. The government cashes the payment $E_r(t)T^*_j(t)$ associated with traveling on path $r \in \mathcal{R}_j$, at the time $t \in [0, T]$. Moreover the following equilibrium conditions must hold:

For each O/D pair w_j and a.e. in $[0, T]$, it results:

$$\sum_{r \in \mathcal{R}_j} E_r(t)H_r(t) \begin{cases} \leq L^*_j(t), & \text{if } T^*_j(t) = 0, \\ = L^*_j(t), & \text{if } 0 < T^*_j(t) < \overline{T}_j(t), \\ \geq L^*_j(t), & \text{if } T^*_j(t) = \overline{T}_j(t). \end{cases} \tag{2}$$

The previous conditions state that if the equilibrium marginal cost of emission reduction, $T^*_j(t)$, is zero for an O/D pair w_j, at the time $t \in [0, T]$, then the emissions by that O/D pair do not exceed the pollution permits of that O/D pair, at the same time $t \in [0, T]$; instead if the equilibrium marginal cost of emission reduction is maximum for an O/D pair w_j, at the time $t \in [0, T]$, then the emissions by that O/D pair exceed the pollution permits of that O/D pair, at the same time $t \in [0, T]$; finally if the equilibrium marginal cost of emission reduction satisfies strictly the capacity constraints, at the time $t \in [0, T]$, then the emissions by that O/D pair are exactly equal to the pollution license holdings of that O/D pair, at the same time $t \in [0, T]$. We also assume that:

For each O/D pair $w_j \in \mathcal{W}$ and a.e. in $[0, T]$, one has:

$$T_j^*(t) \begin{cases} = P^*(t), & \text{if } L_j^*(t) > 0, \\ \leq P^*(t), & \text{if } L_j^*(t) = 0. \end{cases} \tag{3}$$

Therefore, a positive equilibrium holding of licenses by an O/D pair, at the time $t \in [0, T]$, implies that the marginal cost of reduction must be equal to the price of the license, at the same time $t \in [0, T]$. Instead, if the price of the license exceeds the marginal cost of reduction, at the time $t \in [0, T]$, then the number of licenses for that O/D pair is zero, at the same time $t \in [0, T]$.

We impose that if the equilibrium price of the license is maximum, at the time $t \in [0, T]$, then we have less of an O/D pair-based demand for licenses, at the same time $t \in [0, T]$; instead if there is an excess supply of licenses, at the time $t \in [0, T]$, then the equilibrium price is minimum, at the same time $t \in [0, T]$; finally we have an O/D pair-based demand for licenses, at the time $t \in [0, T]$, if the equilibrium price of the license satisfies strictly the capacity constraints, at the same time $t \in [0, T]$; namely:

$$\sum_{j=1}^{l} (L_j^0 - L_j^*(t)) \begin{cases} \leq 0, & \text{if } P^*(t) = \overline{P}(t), \\ = 0, & \text{if } \underline{P}(t) < P^*(t) < \overline{P}(t), \\ \geq 0, & \text{if } P^*(t) = \underline{P}(t), \end{cases} \quad \text{a.e. in } [0, T]. \tag{4}$$

For technical reasons, the functional setting for trajectories $\Xi = (F, T, L, P)$ is the reflexive Banach space:

$$\mathscr{L} = L^2([0, T], \mathbb{R}^m) \times L^2([0, T], \mathbb{R}^l) \times L^2([0, T], \mathbb{R}^l) \times L^2([0, T], \mathbb{R}).$$

Hence, the feasible set $\mathbb{K}$ is given by

$$\begin{aligned} \mathbb{K} &= \mathbb{K}_1 \times \mathbb{K}_2 \times \mathbb{K}_3 \times \mathbb{K}_4 \\ &= \Big\{F \in L^2([0, T], \mathbb{R}^m_+) : \lambda(t) \leq F(t) \leq \mu(t), \quad \text{a.e. in } [0, T], \\ &\qquad \Phi F(t) = \rho(t), \quad \text{a.e. in } [0, T]\Big\} \\ &\quad \times \Big\{T \in L^2([0, T], \mathbb{R}^l_+) : 0 \leq T(t) \leq \overline{T}(t), \quad \text{a.e. in } [0, T]\Big\} \end{aligned}$$

$$\times \left\{ L \in L^2([0,T], \mathbb{R}^l_+) : \; L(t) \geq 0, \quad \text{a.e. in } [0,T] \right\}$$

$$\times \left\{ P \in L^2([0,T], \mathbb{R}_+) : \; \underline{P}(t) \leq P(t) \leq \overline{P}(t), \quad \text{a.e. in } [0,T] \right\}.$$

We are able to state the dynamic O/D pair-based permit system equilibrium conditions.

Definition 1 A vector-function $\Xi^* = (H, T^*, L^*, P^*) \in \mathbb{K}$ is an equilibrium distribution of the dynamic traffic equilibrium problem with the O/D pair-based emission permit system if and only if it satisfies conditions (1)–(4).

We establish the time-dependent variational formulation of the equilibrium distribution.

Theorem 1 *A vector-function* $\Xi^* = (H, T^*, L^*, P) \in \mathbb{K}$ *is an equilibrium distribution of the dynamic traffic equilibrium problem with the O/D pair-based emission permit system if and only if it is a solution to:*

$$\int_0^T \sum_{j=1}^n \sum_{r \in \mathcal{R}_j} \left(C_r(t, H(t)) + E_r(t) T_j^*(t) \right) (F_r(t) - H_r(t))\, dt \tag{5}$$

$$+ \int_0^T \sum_{j=1}^n \left(L_j^*(t) - \sum_{r \in \mathcal{R}_j} E_r(t) H_r(t) \right) \left(T_j(t) - T_j^*(t) \right) dt$$

$$+ \int_0^T \sum_{j=1}^n \left(P^*(t) - T_j^*(t) \right) \left(L_j(t) - L_j^*(t) \right) dt$$

$$+ \int_0^T \sum_{j=1}^n \left(L_j^0 - L_j^*(t) \right) \left(P(t) - P^*(t) \right) dt \geq 0, \quad \forall (F, T, L, P) \in \mathbb{K}.$$

Proof Firstly we assume that $(H, T^*, L^*, P^*) \in \mathbb{K}$ satisfies equilibrium conditions (1)–(4) and we prove that it is a solution to evolutionary variational inequality (5). For each O/D pair $w_j \in W$, let $A = \{ q \in \mathcal{R}_j : \; H_q(t) < \mu_q(t), \quad \text{a.e. in } [0,T] \}$, $B = \{ s \in \mathcal{R}_j : \; H_s(t) > \lambda_s(t), \quad \text{a.e. in } [0,T] \}$. Making use of (1), it follows

$$C_q(t, H(t)) + E_q(t) T_j^*(t) > C_s(t, H(t)) + E_s(t) T_j^*(t), \quad \forall q \in A, \forall s \in B, \text{ a.e. in } [0,T].$$

Consequently, there exists a function $\gamma_{w_j} : [0,T] \to \mathbb{R}$ such that, $\inf_{q \in A} \left[C_q(t, H(t)) + E_q(t) T_j^*(t) \right] \geq \gamma_{w_j}(t) \geq \inf_{s \in B} \left[C_s(t, H(t)) + E_s(t) T_j^*(t) \right]$, a.e. in $[0,T]$. Let $F \in \mathbb{K}_1^W$ be arbitrary. Then for every $r \in \mathcal{R}_j$ such that $C_r(t, H(t)) + E_r(t) T_j^*(t) < \gamma_{w_j}(t)$, a.e. in $[0,T]$, it results $r \notin A$; then, $H_r(t) = \mu_r(t)$, a.e. in $[0,T]$, and $F_r(t) - H_r(t) \leq 0$, a.e. in $[0,T]$. Hence we deduce

$\left(C_r(t, H(t)) + E_r(t)T_j^*(t) - \gamma_{w_j}(t)\right)(F_r(t) - H_r(t)) \geq 0$, a.e. in $[0, T]$. Analogously, for every $r \in \mathcal{R}_j$ such that $C_r(t, H(t)) + E_r(t)T_j^*(t) > \gamma_{w_j}(t)$, a.e. in $[0, T]$, we have $r \notin B$ and $\left(C_r(t, H(t)) + E_r(t)T_j^*(t) - \gamma_{w_j}(t)\right)(F_r(t) - H_r(t)) \geq 0$, a.e. in $[0, T]$. As a consequence, it follows

$$\sum_{r\in\mathcal{R}_j} \left(C_r(t, H(t)) + E_r(t)T_j^*(t)\right)(F_r(t) - H_r(t)) \geq \gamma_{w_j}(t) \sum_{r\in\mathcal{R}_j} (F_r(t) - H_r(t)) \tag{6}$$

$$= \gamma_{w_j}(t)\left(\rho_{w_j}(t) - \rho_{w_j}(t)\right) = 0.$$

Summing (6) over all the O/D pairs w_j, $j = 1, \ldots, l$, taking into account (2)–(4) and summing over all the links $j = 1, \ldots, l$, after integrating on the time interval $[0, T]$, and summing all the inequalities, we obtain (5).

Vice versa, we suppose that (5) holds. Assuming in turns $T = T^*$, $L = L^*$ and $P = P^*$, $F = H$, $L = L^*$ and $P = P^*$, $F = H$, $T = T^*$ and $P = P^*$, $F = H$, $T = T^*$ and $L = L^*$, we have:

$$\int_0^T \sum_{j=1}^n \sum_{r\in\mathcal{R}_j} \left(C_r(t, H(t)) + E_r(t)T_j^*(t)\right)(F_r(t) - H_r(t))\, dt \geq 0, \quad \forall F \in \mathbb{K}_1, \tag{7}$$

$$\int_0^T \sum_{j=1}^n \left(L_j^*(t) - \sum_{r\in\mathcal{R}_j} E_r(t)H_r(t)\right)\left(T_j(t) - T_j^*(t)\right) dt \geq 0, \quad \forall T \in \mathbb{K}_2, \tag{8}$$

$$\int_0^T \sum_{j=1}^n \left(P^*(t) - T_j^*(t)\right)\left(L_j(t) - L_j^*(t)\right) dt \geq 0, \quad \forall L \in \mathbb{K}_3, \tag{9}$$

$$\int_0^T \sum_{j=1}^n \left(L_j^0 - L_j^*(t)\right)\left(P(t) - P^*(t)\right) dt \geq 0, \quad \forall P \in \mathbb{K}_4. \tag{10}$$

Proceeding by absurdum we can prove that (1), (2), (3) and (4) follow by (7), (8), (9) and (10), respectively. □

For further considerations, we set $\Sigma(\Xi^*(t)) = \Bigg(\left(\sum_{r\in\mathcal{R}_j} C_r(t, H(t)) + E_r(t)T_j^*(t)\right)_{j=1,\ldots,n}, \left(L_j^*(t) - \sum_{r\in\mathcal{R}_j} E_r(t)H_r(t)\right)_{j=1,\ldots,n}, \left(P^*(t) - T_j^*(t)\right)_{j=1,\ldots,n}, \left(L_j^0 - L_j^*(t)\right)_{j=1,\ldots,n}\Bigg)$, a.e. in $[0, T]$. Hence, we rewrite (5) as:

$$\ll \Sigma(\Xi^*), \Xi - \Xi^* \gg_{\mathcal{L}} \geq 0, \quad \forall \Xi \in \mathbb{K} \tag{11}$$

where $\ll \cdot, \cdot \gg_{\mathscr{L}}$ is the duality pairing of the Hilbert space $\mathscr{L}$. It is easy to show that (11) is equivalent to the following point-to-point variational inequality:

$$\langle \Sigma(t, \Xi^*(t)), \Xi(t) - \Xi^*(t) \rangle \geq 0, \quad \forall \Xi(t) \in \mathbb{K}(t), \text{ a.e in } [0, T], \tag{12}$$

where

$$\begin{aligned}\mathbb{K}(t) &= \mathbb{K}_1(t) \times \mathbb{K}_2(t) \times \mathbb{K}_3(t) \times \mathbb{K}_4(t) \\ &= \left\{ F(t) \in \mathbb{R}^m_+ : \lambda(t) \leq F(t) \leq \mu(t), \ \Phi F(t) = \rho(t) \right\} \\ &\quad \times \left\{ T(t) \in \mathbb{R}^l_+ : 0 \leq T(t) \leq \overline{T}(t) \right\} \\ &\quad \times \left\{ L(t) \in \mathbb{R}^l_+ : L(t) \geq 0 \right\} \\ &\quad \times \left\{ P(t) \in \mathbb{R}_+ : \underline{P}(t) \leq P(t) \leq \overline{P}(t) \right\}.\end{aligned}$$

3 Existence and Continuity Results

Let us start recalling some concepts. Let X be a reflexive Banach space with X^* its dual space and K be a subset of E. An operator $A : K \to X^*$ is said to be:

- *pseudomonotone in the sense of Karamardian* (*K-pseudomonotone*) if $\langle A(x_2), x_1 - x_2 \rangle \geq 0 \Rightarrow \langle A(x_1), x_1 - x_2 \rangle \geq 0$, for every $x_1, x_2 \in K$;
- *strictly pseudomonotone* if $\langle A(x_2), x_1 - x_2 \rangle \geq 0 \Rightarrow \langle A(x_1), x_1 - x_2 \rangle > 0$, for every $x_1, x_2 \in K$, $x_1 \neq x_2$;
- *pseudomonotone in the sense of Brézis* (*B-pseudomonotone*) if
  (a) for every sequence $\{x_n\}$ weakly converging to x (shortly, $x_n \rightharpoonup x$) in K and such that $\limsup_n \ll A(x_n), x_n - x \gg \leq 0$ it results $\liminf_n \ll A(x_n), x_n - y \gg \geq \ll A(x), x - y \gg$, for every $y \in K$;
  (b) for every $x \in K$ the function $y \mapsto \ll A(x), x - y \gg$ is lower bounded on the bounded subsets of K.

Now let K be a convex subset of X, an operator $A : K \to X^*$ is said to be:

- *hemicontinuous in the sense of Fan* (*F-hemicontinuous*) if for every $x \in K$ the function $y \mapsto \langle A(y), y - x \rangle$ is weakly lower semicontinuous on K;
- *lower hemicontinuous along line segments* if for every $x_1, x_2 \in K$ the function $y \mapsto \langle A(y), x_1 - x_2 \rangle$ is lower semicontinuous on the line segment $[x_1, x_2]$.

The following existence result holds (see [6], for results on variational inequalities in reflexive Banach spaces).

Theorem 2 *If there exist A, B two nonempty compact subsets of $\mathbb{K}$, with $B \subset A$ and B with finite dimension such that for each $\Xi \in \mathbb{K} \setminus A$ there exists $\widehat{\Xi} \in B$ such that $\ll \Sigma(\Xi), \Xi - \widehat{\Xi} \gg_{\mathscr{L}} > 0$, each of the following conditions is sufficient to ensure the existence of a solution to (12):*

1. *Σ is B-pseudomonotone,*
2. *Σ is F-hemicontinuous,*
3. *Σ is K-pseudomonotone and lower hemicontinuous along line segments.*

In order to ensure the uniqueness of the solution to (11), it needs to assume that the operator C is strictly pseudomonotone. Moreover, we remark that if Σ verifies the following condition:

$$\exists c > 0 : \ \|C(t, F(t))\| \leq c\|F(t)\|, \quad \forall F \in \mathbb{K}_1, \text{ a.e. in } [0, T],$$

then Σ belongs to the class of the Nemytskii operators (see [6]) and, hence, is lower hemicontinuous along line segments.

In the next, we prove a continuity result for the dynamic traffic equilibrium problem with the path-based emission pollution permit system. For this purpose we make use of the Kuratowski convergence for subsets of a metric space (X, d) (see [5]).

Definition 2 Let (X, d) be a metric space and K be a nonempty, closed and convex subset of X. We say that the sequence $\{K_n\}$ of nonempty, closed and convex subsets of X converges to K in Kuratwoski's sense, if the following conditions hold:

(K1) for any $x \in K$, there exists a sequence $\{x_n\}$ converging to $x \in X$ such that $x_n \in K_n$, for every $n \in \mathbb{N}$,
(K2) for any subsequence $\{x_n\}$ converging to $x \in X$ such that $x_n \in K_n$, for every $n \in \mathbb{N}$, then $x \in K$.

We are able to obtain the next preliminary result.

Lemma 1 *Let $\lambda, \mu \in C^0([0, T], \mathbb{R}^m_+)$, $\rho \in C^0([0, T], \mathbb{R}^l_+)$, $\overline{T} \in C^0([0, T], \mathbb{R}^l_+)$ and $\underline{P}, \overline{P} \in C^0([0, T], \mathbb{R}_+)$. Let $t \in [0, T]$ and $\{t_n\} \subseteq [0, T]$ such that $t_n \to t$, as $n \to +\infty$. Then, the sequence of sets*

$$\begin{aligned}\mathbb{K}(t_n) &= \mathbb{K}_1(t_n) \times \mathbb{K}_2(t_n) \times \mathbb{K}_3(t_n) \times \mathbb{K}_4(t_n)\\ &= \left\{F(t_n) \in \mathbb{R}^m_+ : \lambda(t_n) \leq F(t_n) \leq \mu(t_n), \ \Phi F(t_n) = \rho(t_n)\right\}\\ &\quad \times \left\{T(t_n) \in \mathbb{R}^l_+ : 0 \leq T(t_n) \leq \overline{T}(t_n)\right\}\\ &\quad \times \left\{L(t_n) \in \mathbb{R}^l_+ : L(t_n) \geq 0\right\}\\ &\quad \times \left\{P(t_n) \in \mathbb{R}_+ : \underline{P}(t_n) \leq P(t_n) \leq \overline{P}(t_n)\right\}, \quad \forall n \in \mathbb{N},\end{aligned}$$

converges to $\mathbb{K}(t)$ in Kuratowski's sense.

Proof We start to show that the (K1) is true. We fix $\Xi(t) = (F(t), T(t), L(t), P(t)) \in \mathbb{K}(t)$ and, for each $j = 1, \dots, l$, we consider

$$A_j = \Big\{r \in \{1, \dots, m\} : \varphi_{jr} = 1, \ F_r(t) = \lambda_r(t)\Big\},$$

$$B_j = \Big\{r \in \{1, \dots, m\} : \varphi_{jr} = 1, \ F_r(t) = \mu_r(t)\Big\},$$

$$C_j = \Big\{r \in \{1, \dots, m\} : \varphi_{jr} = 1, \ \lambda_r(t) < F_r(t) < \mu_r(t)\Big\}.$$

Assuming $C_j \neq \emptyset$, we can deduce that there exists an index ν_j such that for $n > \nu_j$ and $r \in C_j$ it results

$$\lambda_r(t) \le F_r(t) + \frac{\rho_j(t_n) - \rho_j(t)}{\sum_{r \in C_j} \varphi_{jr}} - \frac{\sum_{r \in A_j}[\lambda_r(t_n) - \lambda_r(t)]}{\sum_{r \in C_j} \varphi_{jr}} - \frac{\sum_{r \in B_j}[\mu_r(t_n) - \mu_r(t)]}{\sum_{r \in C_j} \varphi_{jr}} \le \mu_r(t).$$

Then we introduce the sequence $\{\Xi(t_n)\} = \{(F(t_n), T(t_n), L(t_n), P(t_n))\}$ such that for $n > \nu_j$ and $\xi_{ji} = 1$, $j = 1, \dots, l$

$$F_r(t_n) = \begin{cases} \lambda_r(t_n), & \text{for } r \in A_j \\ \mu_r(t_n), & \text{for } r \in B_j \\ F_r(t) + \dfrac{\rho_j(t_n) - \rho_j(t)}{\sum_{r \in C_j} \xi_{ji}} - \dfrac{\sum_{r \in A_j}[\lambda_r(t_n) - \lambda_r(t)]}{\sum_{r \in C_j} \varphi_{jr}} & \\ \quad - \dfrac{\sum_{r \in B_j}[\mu_r(t_n) - \mu_r(t)]}{\sum_{r \in C_j} \varphi_{jr}}, & \text{for } r \in C_j \end{cases}$$

and $F_r(t_n) = P_{\mathbb{K}(t_n)} F_r(t)$ for $n \le \nu_j$, $\varphi_{jr} = 1$, $j = 1, \dots, l$, where $P_{\mathbb{K}(t_n)}$ is the Hilbert projection operator on $\mathbb{K}(t_n)$, $T(t_n) = \min\{T(t), \overline{T}(t_n)\}$, $L(t_n) = L(t)$ and $P(t_n) = \underline{P}(t_n) + \min\{P(t) - \underline{P}(t), \overline{P}(t_n) - \underline{P}(t_n)\}$, for every $n \in \mathbb{N}$. It results that $\Xi(t_n) \in \mathbb{K}(t_n)$, for every $n \in \mathbb{N}$, and $\lim_{n \to +\infty} \Xi(t_n) = \Xi(t) = (F(t), T(t), L(t), P(t))$.

Whereas if $C_j = \emptyset$, we can prove that there exists an index $\overline{\nu}_j$ such that for $n > \overline{\nu}_j$

$$\lambda_r(t_n) \le \lambda_r(t_n) + \frac{1}{\sum_{r \in A_j} \varphi_{jr}} \max\left(0, \rho_j(t_n) - \sum_{r \in A_j} \lambda_r(t_n) - \sum_{r \in B_j} \mu_r(t_n)\right) \le \mu_r(t_n),$$

$$\lambda_r(t_n) \le \mu_r(t_n) + \frac{1}{\sum_{r \in B_j} \varphi_{jr}} \min\left(0, \rho_j(t_n) - \sum_{r \in A_j} \lambda_r(t_n) - \sum_{r \in B_j} \mu_r(t_n)\right) \le \mu_r(t_n).$$

Hence, we consider the sequence $\Xi(t_n) = (F(t_n), T(t_n), L(t_n), P(t_n))$ such that for $n > \nu_j, \varphi_{jr} = 1, j = 1, \dots, l$

$$F_r(t_n) = \begin{cases} \lambda_r(t_n) + \dfrac{1}{\sum_{r \in A_j} \varphi_{jr}} \max\left(0, \rho_j(t_n) - \sum_{r \in A_j} \lambda_r(t_n) - \sum_{r \in B_j} \mu_r(t_n)\right), \\ \hfill \text{for } r \in A_j, \\ \mu_r(t_n) + \dfrac{1}{\sum_{r \in B_j} \varphi_{jr}} \min\left(0, \rho_j(t_n) - \sum_{r \in A_j} \lambda_r(t_n) - \sum_{r \in B_j} \mu_r(t_n)\right), \\ \hfill \text{for } r \in B_j, \end{cases}$$

and $F_r(t_n) = P_{\mathbb{K}(t_n)} F_r(t)$, for $n \le \nu_j$, $\varphi_{jr} = 1$, $j = 1, \dots, l$, $\underline{T}(t_n) = \min\{T(t), \overline{T}(t_n)\}$, $L(t_n) = L(t)$ and $P(t_n) = \underline{P}(t_n) + \min\{P(t) - \underline{P}(t), \overline{P}(t_n) - \underline{P}(t_n)\}$, for every $n \in \mathbb{N}$. Also in this case, we can verify that $\Xi(t_n) \in \mathbb{K}(t_n)$, for every $n \in \mathbb{N}$, and $\lim_{n \to +\infty} \Xi(t_n) = \Xi(t) = (F(t), T(t), L(t), P(t))$. Then condition (K1) is achieved. Furthermore condition (K2) can be easily verified. Hence the claim is completely achieved. □

We aim to establish a continuity result for our model. Preliminary it is easy to verify that the following statement holds.

Proposition 1 *If C is strictly monotone then Σ is also strictly monotone.*

Making use of Theorem 4.2 in [1] and Proposition 1, we get the following result.

Theorem 3 *Let $\lambda, \mu \in C^0([0, T], \mathbb{R}^m_+)$, $\rho \in C^0([0, T], \mathbb{R}^l_+)$, $E \in C^0([0, T], \mathbb{R}^m_+)$, $\overline{T} \in C^0([0, T], \mathbb{R}^l_+)$ and $\underline{P}, \overline{P} \in C^0([0, T], \mathbb{R}_+)$. Let us assume that $C \in C^0([0, T], \mathbb{R}^m_+)$ is a strictly monotone operator such that*

$$\exists c \ge 0 : \|C(t, F(t))\| \le c\|F(t)\|, \quad \forall F \in \mathbb{K}_1, \ \forall t \in [0, T].$$

Then (11) *admits a unique continuous solution.*

References

1. Barbagallo, A.: On the regularity of retarded equilibria in time-dependent traffic equilibrium problems. Nonlinear Anal. **71**, e2406–e2417 (2009)
2. Decorla-Souza, P., Everett, J., Cosby, J., Lim, P.: Trip based approach to estimate emissions with environmental protection Agency's MOBILE model. Transp. Res. Record **1444**, 118–125 (1995)
3. Dhanda, K.K., Nagurney, A., Ramanujam, P.: A Framework for Economic Decision-Making and Policy Analysis. Edward Elgar, Cheltenham (1999)
4. Gwinner, J.: Time dependent variational inequalities – some recent trends. In: Daniele, P., Giannessi, F., Maugeri, A. (eds.) Equilibrium Problems and Variational Models, pp. 225–264. Kluwer Academic Publishers, Dordrecht (2003)
5. Kuratowski, K.: Topology. Academic Press, New York (1996)

6. Maugeri, A., Raciti, F.: On existence theorems for monotone and nonmonotone variational inequalities. J. Convex Anal. **16**, 899–911 (2009)
7. Nagurney, A.: Alternative pollution permit systems for transportation networks based on origin/destination pairs and paths. Transp. Res. D **5**, 37–58 (2000)
8. Nagurney, A., Ramanujam, P., Dhanda, K.K.: A multimodal traffic network equilibrium model with emission pollution permits: compliance versus noncompliance. Transp. Res. E **3**, 349–374 (1998)

Power Network Design with Line Activity

Daniel Bienstock, Martina Cerulli, Mauro Escobar, and Leo Liberti

Abstract We discuss the problem of optimally designing a power transportation network with respect to line activity. We model this problem as an alternating current optimal power flow with on/off variables on lines. We formulate this problem as a nonconvex MINLP in complex numbers, then we propose two convex MINLP relaxations. We test our formulations on some small-scale standard instances.

Keywords Optimal power flow · MINLP · Relaxation

1 Introduction

Every network routing problem naturally yields a design counterpart which optimally decides some part of the network topology. Network routing problems based on multicommodity flows yield design problems where arcs, nodes and/or other features are installed/removed according to flow cost and demand. Such optimization problems often arise in telecommunication networks [1], supply chain [2], logistics, and more. They are usually solved using a mixture of Mathematical Programming (MP) formulations of the mixed-integer sort, decomposition strategies, combinatorial algorithms, and heuristics. On the other hand, the first

This paper has received funding from the European Union's Horizon 2020 research and innovation programme under the Marie Sklodowska-Curie grant agreement n. 764759.

D. Bienstock
IEOR, Columbia University, New York, NY, USA
e-mail: dano@columbia.edu

M. Cerulli · M. Escobar (✉) · L. Liberti
LIX CNRS Ecole Polytechnique, Institut Polytechnique de Paris, Palaiseau, France
e-mail: mcerulli@lix.polytechnique.fr; escobar@lix.polytechnique.fr; liberti@lix.polytechnique.fr

L. Amorosi et al. (eds.), *Optimization in Artificial Intelligence and Data Sciences*, AIRO Springer Series 8, https://doi.org/10.1007/978-3-030-95380-5_17

approach—which we follow in this paper—is always the deployment of off-the-shelf MP solvers on such problems.

Power networks are used to transport and distribute electricity. The transportation occurs at very high voltage levels (hundreds of thousands of volts), while distribution occurs at lower voltage levels (hundreds of volts). Typically, such networks are reasonably sparse, but have some cycles for redundancy-based protection. Moreover, these networks route Alternating Current (AC) rather than Direct Current (DC) [3]. The typical network routing problem for current is known as Optimal Power Flow (OPF). It is well known that the OPF for DC can be well approximated by a Linear Program (LP) (see [4, §1.2.4], [5, Eq. (5.48)]). On the other hand, the OPF for AC, commonly known as ACOPF, is the object of intense research [4, 5] because of its difficulty and importance.

We shall see in the following that the ACOPF can be naturally formulated in MP in many ways, e.g. Quadratically Constrained Quadratic Programming (QCQP), Polynomial Programming (PP), and general Nonlinear Programming (NLP), all of which involve nonconvexities [4]. Common relaxations are LP, Second-Order Cone Programs (SOCP), Semidefinite Programs (SDP) [5]. The variables (voltage, current, power) are naturally defined on continuous domains. A very interesting feature of the ACOPF is that its variables range in complex numbers. While a separation in real and imaginary parts is always possible, matrix formulations and relaxations generally take up twice the amount of storage w.r.t. working directly in complex numbers [6].

Network design problems defined on the OPF in DC can be readily formulated as Mixed-Integer Linear Programs (MILP) [7, 8]. This is also done for problems arising in grid robustness analysis [9, 10], where binary variables model attacks and vulnerabilities [4, Ch. 3]. Binary variables in the ACOPF have also been used to discretize continuous variables arising in nonconvex constraints, so as to obtain an approximate reformulation turning nonconvexities into a finite set of binary choices [5, §4.3.5–4.3.6], which can be dealt with using standard Mixed-Integer Programming (MIP) solvers.

To the best of our knowledge, the first paper exhibiting computational results for the ACOPF with binary variables used for design (rather than approximation) purposes is [11], where binary variables are used to switch generators and shunts on and off: a local NLP solver is deployed on a well-known continuous NLP reformulation of the corresponding nonconvex MINLP. A perspective cut based relaxation of an ACOPF formulation with binary variables for switching generators on and off was proposed in [12]. Another possible approach for working with ACOPF involving binary variables is to apply network design modeling techniques involving binary variables to an LP or SOCP relaxation of the ACOPF. This was done in [13], which proposed inner and outer mixed-integer Diagonally Dominant Programming (DDP) formulations. DDP [14] is a MP technique to approximate the Semidefinite (PSD) cone using LP. The ACOPF is **NP**-hard [15], and remains hard even when the goal is to minimize the number of active generators [12].

In this paper we move a step towards solving a "network design ACOPF" by integrating binary variables that control whether a line is active or not. Our objective is to decrease the number of active lines while still satisfying demand. While this is similar to the optimal switching problem [16], here we start from the ACOPF formulation rather than its DC counterpart. We shall present a (nonconvex) MINLP formulation of the network design problem derived from the ACOPF, and two convex MINLP relaxations. While there is little hope of solving even the tiniest ACOPF instances with the nonconvex MINLP, we show that some results for small ACOPF instances can be obtained using convex MINLPs.

The rest of this paper is organized as follows. We present the ACOPF formulation and a nonconvex MINLP formulation for the corresponding network design problem in Sect. 2. We then propose some new mixed-integer SOCP (MISOCP) relaxations in Sect. 3. We test our formulations with some standard instances in Sect. 4.

2 The ACOPF Formulation

Modeling the ACOPF can be daunting. Most of the literature refers to Matlab-style modeling: painstakingly filling the correct components of a huge constraint matrix with the correct values. This is the low-level kind of interface to MP solvers which produces "flat" formulations that can be read directly by solvers: extremely fast in execution, but a debugging nightmare. See [4, Eg. 1.2.1] and [17, 18] for some introductory material.

Today, most MP formulations are presented in "structured" form: index sets first, then parameters, decision variables, objective function(s), and constraints, all parametrized by and quantified over the aforementioned indices. Each MP entity (parameter, variable, objective, constraint) is stored in a multi-dimensional jagged array, possibly not completely defined. Structured formulations convey the problem definition much more clearly than flat ones, at least to MP-versed readers. Detailed formulations can be found in [19]. Modeling tools such as AMPL [20] allow for fast(er) debugging.

We model the electrical network as a loopless multi-digraph $G = (B, L)$ where B is the set of nodes and L the set of arcs. In power engineering terminology a node is called a *bus* and an arc is called a *line* or *branch*. We assume $|B| = n$ and $|L| = m$. Parallel arcs occur whenever parallel cables are deployed on connections that must transport excessive amounts of power for a single cable. The h-th line ℓ_{bah} joining two buses b and a is represented by a pair of anti-parallel arcs $\ell_{bah} = \{(b, a, h), (a, b, h)\}$. We assume that L is partitioned in two sets L_0, L_1 with $|L_0| = |L_1|$: for each pair of antiparallel arcs, one is in L_0 and the other in L_1, according to the asymmetry of the branch admittance matrix $\mathbf{Y}_{bah}$ matrix below.

Ohm's law expresses the current I_{bah} injected on a line ℓ_{bah} in function of the voltages V_b, V_a at the endpoints b and a, and of the physical properties of the line. The fundamental difference with Ohm's law in DC is that AC yields an asymmetry. While in DC we have $I_{bah} = -I_{abh}$, in AC we instead have:

$$\forall (b, a, h) \in L_0 \quad I_{bah} = Y^{\mathsf{ff}}_{bah} V_b + Y^{\mathsf{ft}}_{bah} V_a \quad \wedge \quad I_{abh} = Y^{\mathsf{tf}}_{bah} V_b + Y^{\mathsf{tt}}_{bah} V_a. \qquad (1)$$

The Y constants in the above equations are defined as follows [4, 19]:

$$\mathbf{Y}_{bah} = \begin{pmatrix} Y^{\mathsf{ff}}_{bah} & Y^{\mathsf{ft}}_{bah} \\ Y^{\mathsf{tf}}_{bah} & Y^{\mathsf{tt}}_{bah} \end{pmatrix} = \begin{pmatrix} (\frac{1}{r_{bah}+ix_{bah}} + i\frac{\mathfrak{b}_{bah}}{2})/\tau^2_{bah} & -\frac{1}{(r_{bah}+ix_{bah})\tau_{bah}e^{-i\nu_{bah}}} \\ -\frac{1}{(r_{bah}+ix_{bah})\tau_{bah}e^{i\nu_{bah}}} & \frac{1}{r_{bah}+ix_{bah}} + i\frac{\mathfrak{b}_{bah}}{2} \end{pmatrix}, \tag{2}$$

where $r, x, \mathfrak{b}, \tau, \nu$ measure some physical properties of the line, and are given as part of the instance. The suffixes ff, ft, tf, tt to Y stand for "from-from", "from-to", "to-from", and "to-to": they are a reminder of the direction of the routed quantities w.r.t. the line ℓ_{bah}.

We can now introduce sets, parameters and decision variables of the ACOPF.

- *Sets:* B, L and a set $\mathscr{G}$ of generators partitioned as $\{\mathscr{G}_b \mid b \in B\}$, where $\mathscr{G}_b$ contains the generators attached to bus b.
- *Parameters:* power demand (or *load*) $\tilde{S}_b$, *shunt admittance* A_b; voltage magnitude bounds $\underline{V}_b, \overline{V}_b$ at each bus $b \in B$; *admittance matrix* $\mathbf{Y}_{bah}$; upper bound $\bar{S}_{bah}$ to injected power magnitude; lower/upper bounds $\underline{\eta}_{bah}, \overline{\eta}_{bah}$ to phase difference at each line $(b, a, h) \in L$; cost coefficients C_{g2}, C_{g1}, C_{g0}; lower/upper bounds $\underline{\mathscr{S}}_g, \overline{\mathscr{S}}_g$ to power generated at $g \in \mathscr{G}$; a *reference bus* $r \in B$.
- *Decision variables*: voltage V_b at bus $b \in B$, injected current I_{bah}, injected power S_{bah} at each line $(b, a, h) \in L$, and generated power $\mathscr{S}_g$ at each generator $g \in \mathscr{G}$.

All variables range in $\mathbb{C}$. Among the parameters, the power magnitude, voltage magnitude, phase difference bounds, cost coefficients are in $\mathbb{R}$; r ranges in the bus set; the generated power bounds are in $\mathbb{C}$. Limited to this paper we assume that, for two complex numbers $\alpha = \alpha^{\mathsf{r}} + i\alpha^{\mathsf{c}}$ and $\beta = \beta^{\mathsf{r}} + i\beta^{\mathsf{c}}$, $\alpha \le \beta$ means $\alpha^{\mathsf{r}} \le \beta^{\mathsf{r}}$ and $\alpha^{\mathsf{c}} \le \beta^{\mathsf{c}}$. We also recall that $|\alpha| = \sqrt{(\alpha^{\mathsf{r}})^2 + (\alpha^{\mathsf{c}})^2}$ is the magnitude of α, that $\alpha^* = \alpha^{\mathsf{r}} - i\alpha^{\mathsf{c}}$ is the conjugate of α, and that $|\alpha|^2 = \alpha\,\alpha^*$.

We present now objective function and constraints of what we call the (S, I, V)-formulation of the ACOPF.

- *Objective function:* $\min \sum_{g \in \mathscr{G}} (C_{g2}(\mathscr{S}^{\mathsf{r}}_g)^2 + C_{g1}\mathscr{S}^{\mathsf{r}}_g + C_{g0})$, which is quadratic and separable in generated power.
- *Bound constraints:* on voltage magnitude $\underline{V}^2_b \le |V_b|^2 \le \overline{V}^2_b$ for each $b \in B$; on power magnitude $|S_{bah}|^2 \le \bar{S}^2_{bah}$ for each $(b, a, h) \in L$; on phase difference $\tan(\underline{\eta}_{bah})(V_b\, V_a{}^*)^{\mathsf{r}} \le (V_b\, V_a{}^*)^{\mathsf{c}} \le \tan(\overline{\eta}_{bah})(V_b\, V_a{}^*)^{\mathsf{r}}$ together with $(V_b\, V_a{}^*)^{\mathsf{r}} \ge 0$ for $(b, a, h) \in L_0$; on generated power $\underline{\mathscr{S}}_g \le \mathscr{S}_g \le \overline{\mathscr{S}}_g$ for each $g \in \mathscr{G}$. Moreover, we have $V^{\mathsf{c}}_r = 0$ and $V^{\mathsf{r}}_r \ge 0$ on the reference bus.
- *Functional constraints:*
  - Power flow equations:

$$\forall b \in B \qquad \sum_{(b,a,h)\in L} S_{bah} + \tilde{S}_b = -\, A_b{}^*|V_b|^2 + \sum_{g\in\mathscr{G}_b} \mathscr{S}_g. \tag{3}$$

– The relationship between S, V, I:

$$\forall (b,a,h) \in L \quad S_{bah} = V_b \, I_{bah}{}^*. \tag{4}$$

– Ohm's laws Eq. (1), which we write equivalently as:

$$\forall (b,a,h) \in L_0 \quad I_{bah} = Y^{\mathsf{ff}}_{bah} V_b + Y^{\mathsf{ft}}_{bah} V_a \tag{5}$$

$$\forall (b,a,h) \in L_1 \quad I_{bah} = Y^{\mathsf{tf}}_{abh} V_a + Y^{\mathsf{tt}}_{abh} V_b. \tag{6}$$

2.1 The Network Design ACOPF

We now introduce a binary variable y_{bah} for each (b, a, h) in L. We have $y_{bah} = 1$ iff the corresponding line is active, and we must ensure that both antiparallel arcs are active/inactive at the same time by $y_{bah} = y_{abh}$. We control activation/deactivation of a line by limiting the injected power magnitude bound:

$$\forall (b,a,h) \in L \quad |S_{bah}|^2 \le \bar{S}^2_{bah} y_{bah}. \tag{7}$$

In order to ensure that Eq. (7) does not impose constraints on V_b and V_a when the line (b, a, h) is not active, we introduce a new complex variable z_{bah} in Eq. (4), such that:

$$\forall (b,a,h) \in L \quad S_{bah} = V_b \, I_{bah}{}^* + z_{bah}, \tag{8}$$

and

$$\forall (b,a,h) \in L \quad |z_{bah}|^2 \le M^2_{bah}(1 - y_{bah}), \tag{9}$$

where M_{bah} is a large enough constant. Note that Eqs. (7)–(9) do not cut the global optima of the ACOPF: it suffices to set $y_{bah} = 1$ for each $(b, a, h) \in L$ to see this. Instead, we add an objective function $\min \sum_{(b,a,h) \in L_0} y_{bah}$. We can tackle this bi-objective MINLP either by scalarization or by adding a constraint $\sum_{(b,a,h) \in L_0} y_{bah} \le \xi$ and letting ξ vary in $\{1, \ldots, m/2\}$. In this paper we consider scalarization approach, so that the objective function becomes:

$$\min \sum_{g \in \mathscr{G}} (C_{g2}(\mathscr{S}^{\mathsf{r}}_g)^2 + C_{g1} \mathscr{S}^{\mathsf{r}}_g + C_{g0}) + \rho \sum_{(b,a,h) \in L_0} y_{bah}, \tag{10}$$

where $\rho > 0$ is a scalar weight which we set to 1 for testing purposes. We denote the network design ACOPF problem with binary variables on lines by $\mathrm{ACOPF}_{\mathsf{L}}$.

3 Linearizing Relaxations

The material in this section is motivated by the solution difficulty posed by the nonconvex MINLP formulation of the ACOPF_L. First of all we propose some valid relaxation for ACOPF problem.

The decision variables I for current can be eliminated from the (S, I, V)-formulation by replacing them in Eq. (4) with their expressions in Eqs. (5)–(6). This yields the (S, V)-formulation, which is still a nonconvex NLP. In turn, using Eqs. (7)–(10), this NLP yields a nonconvex MINLP formulation for the ACOPF_L.

3.1 (S, V, X)-Relaxation

The only nonlinear terms appearing in the nonconvex constraints of the ACOPF (S, V)-formulation are quadratic in voltage: they are products $V_b V_a{}^*$ for some $b, a \in B$. Every such product term can be linearized, i.e. replaced by a new (complex) variable X_{ba} for $b, a \in B$ (we do not include the corresponding defining constraint $X_{ba} = V_b V_a{}^*$). Let us call this the (S, V, X)-relaxation. This turns out to be a convex QCQP (more specifically a SOCP). The quadratic terms are: $\mathscr{S}_g^2$ in the minimizing objective and $|S_{bah}|^2$ in the LHS of the power magnitude bound constraints.

3.2 (S, V, X)-SDP

Note that the (S, V, X)-relaxation is an exact reformulation if we enforce $X = VV^\mathsf{H}$, where the apex stands for "hermitian transpose", i.e. the transpose of the componentwise complex conjugate. Accordingly, since X is a PSD rank-one matrix, we get a stronger relaxation w.r.t. the (S, V, X)-relaxation presented in Sec. 3.1, if we replace $X = VV^\mathsf{H}$ by $X \succeq 0$, which yields a complex SDP relaxation called (S, V, X)-SDP.

3.3 (S, V, X)-$\frac{1}{2}$DDP

Given the scarcity of off-the-shelf mixed-integer SDP solvers, we consider a DDP approximation of the PSD cone [14]: since every Diagonally Dominant (DD) matrix is also PSD [21], the constraint "X is DD" yields an inner approximation (i.e. a restriction) of the complex SDP.

Writing the DDP constraints corresponding to $X \succeq 0$ requires splitting X into real and imaginary parts, which yields $\bar{X} = \begin{pmatrix} X^{\mathsf{rr}} & X^{\mathsf{rc}} \\ X^{\mathsf{cr}} & X^{\mathsf{cc}} \end{pmatrix} \in \mathbb{R}^{2n\times 2n}$, where $X^{\mathsf{rr}} = (X^{\mathsf{r}}_{ba})$, $X^{\mathsf{cc}} = (X^{\mathsf{c}}_{ba})$, X^{rc} linearizes the matrix $(V^{\mathsf{r}}_b V^{\mathsf{c}}_a)$, and X^{cr} linearizes the matrix $(V^{\mathsf{c}}_b V^{\mathsf{r}}_a)$. We remark that X^{rr}, X^{cc} are symmetric matrices, while X^{rc}, X^{cr} are not; on the other hand, $X^{\mathsf{rc}}_{ba} = X^{\mathsf{cr}}_{ab}$ for each $b, a \in B$.

Now the DDP inner approximation of $\bar{X} \succeq 0$ states that any diagonal component of $\bar{X}$ is greater than or equal to the sum of the absolute values of the components in the same row. This corresponds to:

$$\forall b \in B \quad X^{\mathsf{rr}}_{bb} \geq \sum_{\substack{a\in B\\ a\neq b}} T^{\mathsf{rr}}_{ba} + \sum_{a\in B} T^{\mathsf{rc}}_{ba} \tag{11}$$

$$\forall b \in B \quad X^{\mathsf{cc}}_{bb} \geq \sum_{\substack{a\in B\\ a\neq b}} T^{\mathsf{cc}}_{ba} + \sum_{a\in B} T^{\mathsf{cr}}_{ba}, \tag{12}$$

where $\bar{T} = \begin{pmatrix} T^{\mathsf{rr}} & T^{\mathsf{rc}} \\ T^{\mathsf{cr}} & T^{\mathsf{cc}} \end{pmatrix}$ is a real variable matrix such that $-\bar{T} \leq \bar{X} \leq \bar{T}$ [14].

The issue with inner DDP approximations is that they may be infeasible even if the corresponding SDP is feasible. Experimentally, this was verified to be the case in every ACOPF instance we tested. This issue can be addressed algorithmically [14], but this would require solving a sequence of DDPs, which would in turn take excessive time. Instead, we chose to only impose Eq. (11), which yielded feasible "half-DDP" relaxations (which we refer to as (S, V, X)-$\frac{1}{2}$DDP relaxation) of the tested ACOPF instances. Note that we do not have a general feasibility proof for $\frac{1}{2}$DDP relaxations. So far, we have not found any counterexamples yet, either.

3.4 *Jabr Relaxation*

Another SOCP relaxation of the ACOPF, called "Jabr relaxation", was proposed in [22]. It can be constructed from the (S, V)-formulation as follows:

1. transform cartesian coordinates V^{r}, V^{c} to polar coordinates v, θ by replacing $V^{\mathsf{r}} = v\cos\theta$ and $V^{\mathsf{c}} = v\sin\theta$: this will result with nonlinear terms in $v_b v_a \cos(\theta_b - \theta_a)$ and $v_b v_a \sin(\theta_b - \theta_a)$;
2. define an index set $R = \{(b, b) \mid b \in B\} \cup \{(b, a) \mid (b, a, 1) \in L\}$;
3. linearize (replace) the nonlinear terms with new variables $c_{ba} = v_b v_a \cos(\theta_b - \theta_a)$ and $s_{ba} = v_b v_a \sin(\theta_b - \theta_a)$ for all $(b, a) \in R$: this also yields $c_{ba} = c_{ab}$, $s_{ba} = -s_{ab}$, $c^2_{ba} + s^2_{ba} = v^2_b v^2_a$ $(\star)$ for all $(b, a, 1) \in L_0$, as well as $s_{bb} = 0$ and $c_{bb} = v^2_b$ $(\dagger)$ for each $b \in B$;

4. replace v_b^2, v_a^2 in ($\star$) with c_{bb}, c_{aa} by means of (†), and relax ($\star$) to a convex (conic) constraint $c_{ba}^2 + s_{ba}^2 \le c_{bb}c_{aa}$;
5. replace $|V_b|^2$ in the voltage magnitude bounds with c_{bb};
6. remark that $V_b V_a{}^* = c_{ba} + i s_{ba}$, and infer the phase difference bounds as $c_{ba} \ge 0$ and $\tan(\underline{\eta}_{bah})c_{ba} \le s_{ba} \le \tan(\overline{\eta}_{bah})c_{ba}$ for each $(b, a, h) \in L_0$;
7. the injected power variables S_{bah} satisfy the linear equations:

$$\forall (b,a,h) \in L_0 \quad (S_{bah})^{\mathsf{r}} = (Y^{\mathsf{ff}}_{bah})^{\mathsf{r}} c_{bb} + (Y^{\mathsf{ft}}_{bah})^{\mathsf{r}} c_{ba} + (Y^{\mathsf{ft}}_{bah})^{\mathsf{c}} s_{ba}$$

$$\forall (b,a,h) \in L_0 \quad (S_{bah})^{\mathsf{c}} = -(Y^{\mathsf{ff}}_{bah})^{\mathsf{c}} c_{bb} + (Y^{\mathsf{ft}}_{bah})^{\mathsf{r}} s_{ba} - (Y^{\mathsf{ft}}_{bah})^{\mathsf{c}} c_{ba}$$

$$\forall (b,a,h) \in L_1 \quad (S_{bah})^{\mathsf{r}} = (Y^{\mathsf{tt}}_{abh})^{\mathsf{r}} c_{bb} + (Y^{\mathsf{tf}}_{abh})^{\mathsf{r}} c_{ba} + (Y^{\mathsf{tf}}_{abh})^{\mathsf{c}} s_{ba}$$

$$\forall (b,a,h) \in L_1 \quad (S_{bah})^{\mathsf{c}} = -(Y^{\mathsf{tt}}_{abh})^{\mathsf{c}} c_{bb} + (Y^{\mathsf{tf}}_{abh})^{\mathsf{r}} s_{ba} - (Y^{\mathsf{tf}}_{abh})^{\mathsf{c}} c_{ba}.$$

3.5 ACOPF$_{\mathsf{L}}$ Relaxations

We derive ACOPF$_{\mathsf{L}}$ relaxations from the (S, V, X)-relaxation, the (S, V, X)-$\frac{1}{2}$DDP and Jabr relaxations of the ACOPF, by employing the binary variables y as in Sect. 2.1, i.e. by imposing Eqs. (7)–(9) and minimizing Eq. (10). A few preliminary results showed that the active lines do not form a connected set at the optimum. In order to enforce connectivity, we therefore also added a set of multicommodity flow constraints on added variables f^{ba}_{deh}, defined for each distinct pair $b, a \in B$ and line $(d, e, h) \in L$:

$$\forall b < a \in B \qquad \sum_{(b,d,h)\in L} f^{ba}_{bdh} - \sum_{(d,b,h)\in L} f^{ba}_{dbh} = 1$$

$$\forall b < a \in B \qquad \sum_{(d,a,h)\in L} f^{ba}_{dah} - \sum_{(a,d,h)\in L} f^{ba}_{adh} = 1$$

$$\forall b < a \in B, d \in B \smallsetminus \{b,a\} \qquad \sum_{(e,d,h)\in L} f^{ba}_{edh} - \sum_{(d,e,h)\in L} f^{ba}_{deh} = 0,$$

as well as the linking constraints: $\forall b < a \in B, (d, e, h) \in L \quad f^{ba}_{deh} \le y_{deh}$.

In Table 1, we shall refer to the ACOPF$_{\mathsf{L}}$ relaxations from (S, V, X)-$\frac{1}{2}$DDP, and Jabr as "ddp", and "Jabr" respectively.

Table 1 Numerical results limited to 300s using a single CPU processor. We report instance name, number of lines, known optimal value; then, for each relaxation type (mod1,mod2), we report obtained optimal value, number of active lines, solver status, CPU time. Best results are in boldface, invalid results are grayed (the solver could not find any feasible solution in the allotted time)

Name	Lines	Known	mod1	mod2	opt1	opt2	act1	act2	stat1	stat2	cpu1	cpu2
case5	6	17551.89	ddp	Jabr	0.00	**15169.08**	4	4	Solved	Limit	**2.03**	300
case9	9	5296.67	ddp	Jabr	2244.81	**5296.67**	9	9	Solved	Solved	**3.58**	3.98
case14	20	8081.52	ddp	Jabr	0.00	**1786.93**	13	14	Limit	Limit	300	300
case18	17	11.85	ddp	Jabr	−0.00	**11.85**	17	17	Solved	Solved	**0.36**	0.49
case22	21	0.068	ddp	Jabr	0.00	**0.068**	21	21	Solved	Solved	**0.56**	6.01
case24	38	63352.20	ddp	Jabr	47320.2	**63345.20**	38	38	Limit	Limit	300	300
case30	41	576.89	ddp	Jabr	0.00	**568.86**	29	30	Limit	Limit	300	300

4 Computational Experiments

The standard reference testbed for computational assessments in ACOPF is the PGLib library [23], which also includes "case files" from MATPOWER [18]. We compare performances of the two convex MINLP relaxations of the ACOPF_L (ddp and Jabr) on the small case instances `case`i for $i \in \{5, 9, 14, 18, 22, 24, 30\}$. Our implementation is carried out in AMPL [20]. We solve both formulations, which are of the Mixed-Integer SOCP sort, with CPLEX 12.9 [24], which is given 300s as maximum CPU time. Only instance "`case5`" is solved using Baron, because AMPL failed to successfully pass it to Cplex.

The results in Table 1 are obtained on a a 2.53GHz Intel(R) Xeon(R) CPU with 49.4 GB RAM. They show that 300s are only sufficient to obtain meaningful results for small instances.

An encouraging feature of the results in Table 1 is that the slacker ddp relaxation takes less time to solve than Jabr, provides a worst bound, but still identifies a valid connectivity for active lines for all the tested instances. In Fig. 1, e.g., we report two solutions found by solving the ddp relaxation, which appear to be the same found by Jabr relaxation, as well as the two different solutions obtained by solving the same instance "`case30`" with ddp, and Jabr.

In Table 2 we report results from the (S, V, X)-relaxation of ACOPF_L on slightly larger instances, solved using CPLEX limited to 7200s. When solutions are found atypically quickly (e.g. `case69`, `case85`), it is because the networks have no cycles.

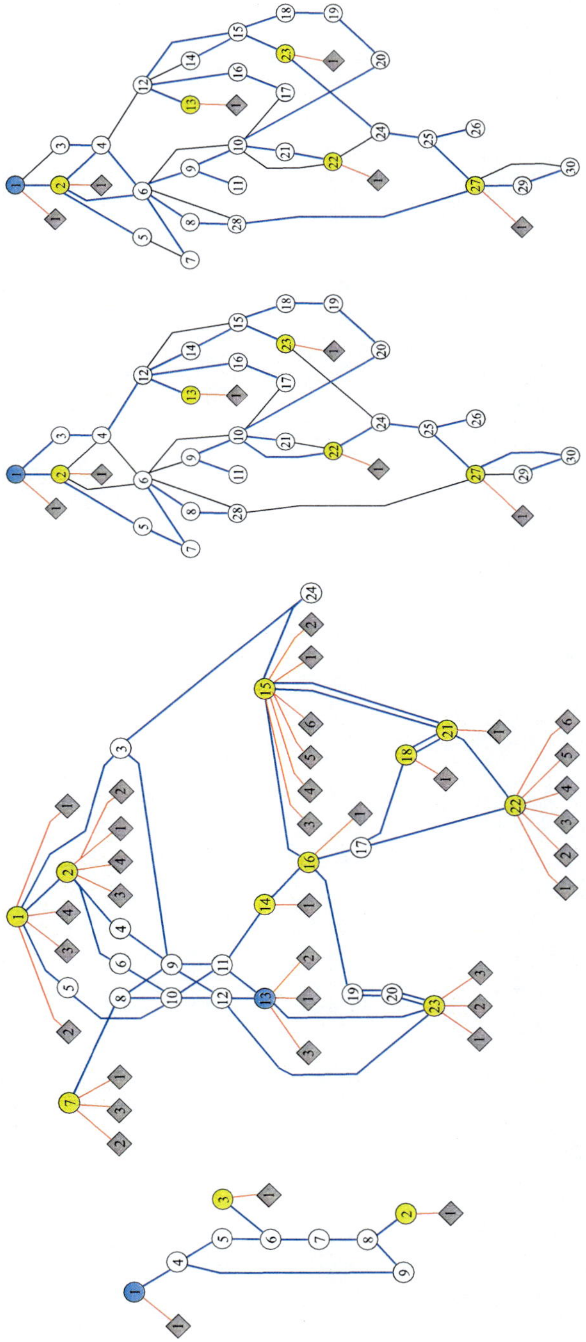

Fig. 1 The solutions of three instances from results in Table 1. Buses in circles, generators in parallelograms (buses with generators are colored, reference bus is colored differently); active lines are thick and colored

Table 2 Computational results on the (S, V, X)-relaxation limited to 7200s (left). Nontrivial solution for `case30` (right)

Name	Lines	Known	Opt	Act	Stat	Cpu
case24	38	63352.20	47320.20	38	Limit	7200
case30	41	576.89	0.00	29	Limit	7200
ieee30	41	9974.99	0.00	29	Limit	7200
case39	46	41864.17	27417.26	46	Limit	7200
case69	68	0.39	0.00	68	Solved	9.74
case85	84	0.00	0.00	84	Solved	21.50

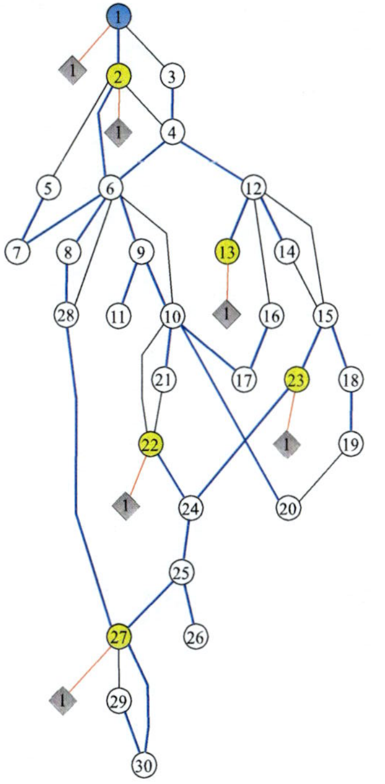

References

1. Gendron, B., Crainic, T., Frangioni, A.: Multicommodity Capacitated Network Design. In: Sansò, B., Soriano, P. (eds.) Telecommunications Network Planning, pp. 1–32. Springer, New York (1999). https://doi.org/10.1007/978-1-4615-5087-7_1
2. Bruglieri, M., Liberti, L.: Optimal running and planning of a biomass-based energy production process. Energy Policy **36**(7), 2430–2438 (2008). https://doi.org/10.1016/j.enpol.2008.01.009
3. Glover, J., Sarma, M., Overbye, T.: Power System Analysis and Design. Cengage Learning, Stamford CT (2010)
4. Bienstock, D.: Electrical Transmission System Cascades and Vulnerability: an Operations Research Viewpoint. No. 22 in MOS-SIAM Optimization. SIAM, Philadelphia (2016). https://doi.org/10.1137/1.9781611974164
5. Molzahn, D., Hiskens, I.: A Survey of Relaxations and Approximations of the Power Flow Equations. Foundations and Trends in Electric Energy Systems. Now Publishers, Hanover, MA (2019). https://doi.org/10.1561/3100000012
6. Gilbert, J.C., Josz, C.: Plea for a semidefinite optimization solver in complex numbers. Tech. rep., HAL Archives-Ouvertes: hal-01422932 (2017)
7. Knight, U.: The logical design of electrical networks using linear programming methods. Proc. IEE **107**(33), 306–314 (1960). https://doi.org/10.1049/pi-a.1960.0063
8. Adams, R., Laughton, M.: Optimal planning of power networks using mixed-integer programming. Part 1: Static and time-phased network synthesis. Proc. IEE **121**(2), 139–147 (1974). https://doi.org/10.1049/piee.1974.0024
9. Bienstock, D., Mattia, S.: Using mixed-integer programming to solve power grid blackout problems. Discrete Optimization **4**(1), 115–141 (2007). https://doi.org/10.1016/j.disopt.2006.10.007
10. Bienstock, D., Verma, A.: The N-k problem in power grids: New models, formulations, and numerical experiments. SIAM J. Optim. **20**(5), 2352–2380 (2010). https://doi.org/10.1137/08073562X
11. Ruiz, M., Maeght, J., Marié, A., Panciatici, P., Renaud, A.: A progressive method to solve large-scale AC optimal power flow with discrete variables and control of the feasibility. In: 2014 Power Systems Computation Conference (2014). https://doi.org/10.1109/PSCC.2014.7038395
12. Salgado, E., Gentile, C., Liberti, L.: Perspective cuts for the ACOPF with generators. In: Daniele, P., Scrimali, L. (eds.) New Trends in Emerging Complex Real Life Problems, AIRO Series, vol. 1, pp. 451–461. Springer, Cham (2018). https://doi.org/10.1007/978-3-030-00473-6_48
13. Salgado, E., Scozzari, A., Tardella, F., Liberti, L.: Alternating current optimal power flow with generator selection. In: Lee, J., Rinaldi, G., Mahjoub, R. (eds.) Combinatorial Optimization ISCO 2018, LNCS, vol. 10856, pp. 364–375 (2018). https://doi.org/10.1007/978-3-319-96151-4_31
14. Ahmadi, A., Majumdar, A.: DSOS and SDSOS optimization: More tractable alternatives to sum of squares and semidefinite optimization. SIAM J. Appl. Algebra Geometry **3**(2), 193–230 (2019). https://doi.org/10.1137/18M118935X
15. Bienstock, D., Verma, A.: Strong NP-hardness of AC power flows feasibility. Oper. Res. Lett. **47**(6), 494–501 (2019). https://doi.org/10.1016/j.orl.2019.08.009
16. Fisher, E., O'Neill, R., Ferris, M.: Optimal transmission switching. IEEE Trans. Power Syst. **23**(3), 1346–1355 (2008). https://doi.org/10.1109/TPWRS.2008.922256
17. Molzahn, D., Holzer, J., Lesieutre, B., DeMarco, C.: Implementation of a large-scale optimal power flow solver based on semidefinite programming. IEEE Trans. Power Syst. **28**(4), 3987–3998 (2013). https://doi.org/10.1109/TPWRS.2013.2258044
18. Zimmermann, R., Murillo-Sánchez, C.: MatPower 7.0b1 User's Manual. Power Systems Engineering Research Center (2018)

19. Bienstock, D., Escobar, M., Gentile, C., Liberti, L.: Mathematical programming formulations for the alternating current optimal power flow problem. 4OR **18**(3), 249–292 (2020). https://doi.org/10.1007/s10288-020-00455-w
20. Fourer, R., Gay, D.: The AMPL Book. Duxbury Press, Pacific Grove (2002)
21. Gerschgorin, S.: Über die abgrenzung der eigenwerte einer matrix. Izvestia Akademii Nauk USSR **6**, 749–754 (1931)
22. Jabr, R.: Radial distribution load flow using conic programming. IEEE Trans. Power Syst. **21**(3), 1458–1459 (2006). https://doi.org/10.1109/TPWRS.2006.879234
23. Babaeinejadsarookolaee, S., Birchfield, A., Christie, R., Coffrin, C., DeMarco, C., Diao, R., Ferris, M., Fliscounakis, S., Greene, S., Huang, R., Josz, C., Korab, R., Lesieutre, B., Maeght, J., Mak, T., Molzahn, D., Overbye, T., Panciatici, P., Park, B., Snodgrass, J., Tbaileh, A., Van Hentenryck, P., Zimmerman, R.: The power grid library for benchmarking AC optimal power flow algorithms. Tech. rep., (2019). arXiv:1908.02788
24. IBM: ILOG CPLEX 12.9 User's Manual. IBM (2019)

Risk Assessment in Transactions Under Threat as Partially Observable Markov Decision Process

Vassil Vassilev, Doncho Donchev, and Demir Tonchev

Abstract This paper presents a theoretical model and algorithms for calculating the security risks for planning active counteractions in transaction processing under security threats. It is a part of an integrated cybersecurity framework, which combines AI-based planning of active counteractions with Machine Learning for the detection of security threats during transaction processing. The risk assessment is based on the optimal strategy for decision making which minimizes the security risks in controlled transactions modeled as Partially Observable Markov Decision Process (POMDP). By statistical reduction, this model is converted into a Markov Decision Process (MDP) with full information so that the algorithm for calculating the risks can use the standard dynamic programming. Although developed primarily for applications in fintech industry, this framework can be adapted to a wide range of business process workflows that incorporate both synchronous operations and asynchronous events caused by human errors, technical faults, or external interventions.

Keywords Transaction processing · Security threats · Risk assessment · Partially-observable Markov decision process · Statistical reduction

1 Introduction

Cybersecurity becomes critical for successful digital transformation of the businesses in many areas of human activity—fintech industry, e-commerce, business process management, healthcare, public services, etc. Over the last three years we have been working on a hybrid AI-based framework which combines the power of

V. Vassilev (✉)
Cyber Security Research Centre, London Metropolitan University, London, UK
e-mail: v.vassilev@londonmet.ac.uk

D. Donchev · D. Tonchev
GATE Institute, Sofia University "St. Kliment Ohridski", Sofia, Bulgaria
e-mail: doncho.donchev@gate-ai.eu; demir.tonchev@gate-ai.eu

L. Amorosi et al. (eds.), *Optimization in Artificial Intelligence and Data Sciences*, AIRO Springer Series 8, https://doi.org/10.1007/978-3-030-95380-5_18

logical analysis of security policies, from one side, with machine learning for data analytics, on the other. Such a framework must secure the transaction processing by accounting for both the threat intelligence, obtained in advance from security experts, and the security risks assessed in real time. Our approach to planning is rooted in the traditional AI planning introduced by McCarthy in the situation calculus, but follows different approach from both conceptual and theoretical point of view, which allows to avoid some of the problems encountered in the original deterministic planning such as the qualification and frame problems. Instead of combining the information from the real world with the planning heuristics in a single representational language, like in the original situation calculus, we have adopted multi-level problem formalization which separates the *domain ontology* from the *security policies* and adds two more levels: *analytical level* of decision making for selecting appropriate actions and applying potential counteractions to the security threats, and *implementation level* for executing ML algorithms for security analytics to detect potential security threats at the different steps of the transactions [8]. For the first three levels of the framework we have adopted the standard languages of the Semantic Web—OWL, SWRL and RDF, which have direct logical interpretation [7], while the implementation level utilizes a variety of ML algorithms for detection [9]. In this article we will present an approach for assessment of the security risks at each step of the transactions in the presence of security threats, which is necessary for planning of suitable counteractions during execution of the transactions and progressing towards completion of the transaction.

The paper is organised as follows. First we will briefly review the research in risk assessment from cybersecurity perspective and will set the problem in the context of a hybrid AI-based cybersecurity framework which accounts both the security policies and the threat intelligence to execute active counteractions against the threats detected during transaction execution. After this preliminaries we will introduce the POMDP model, will preform statistical reduction to MDP model and will describe the algorithms for risk assessment, based on the optimal strategy for controlling the transaction under threats. We will illustrate the use of the algorithm by analyzing the decision threshold, which guides the choice of counteraction along the transaction based on the optimal strategy. After brief information about the current state of implementation of the framework we will finish with a discussion and our plans about the future research in this direction.

2 Brief Review of the Relevant Research

The advances in heuristic planning for intelligent control of the transactions and the need to account stochastic factors which interfere with the execution of the transactions, such as errors, faults and intrusions, focused the attention on continuous planning and re-planning. Unfortunately, the heuristic planning faces the need to account the security risks which does not fit within the deterministic models used as a base for the planning algorithms. An adequate formalization of the stochastic planning problem requires working with POMDP model which is significantly more complex than the two popular deterministic models—the classical state-space search and the MDP [1]. An excellent overview of the different models and algorithms for non-deterministic planning from AI perspective is provided in [2]. Despite some recent adoptions of POMDP for the purpose of risk assessment [3–6], the adoption of POMDP remains valuable for mostly offline analytics due to the need to the need to solve multi-step optimization problem of large complexity.

The major contribution of our research is in the integration of the purely deterministic method for controlling the transactions under threat with the stochastic method for decision making using the risk assessment as a heuristic function, which is based on the original POMDP model but reduced to a tractable MDP problem. This reduction makes possible to use more efficient recurrent algorithms for optimization, based on the standard dynamic programming methodology which for realistic transaction lengths can be executed in real time.

3 Controlling Transactions and Decision Making

Contemporary transaction processing requires planning and controlling the execution of a sequence of operations to reach the goal state, namely the commit point of the transaction. In our security framework [8] each step of the transaction is modeled as a separate *situation*. Along the multi-step transition from situation to situation the transactions face multiple challenges due to the unpredictability of the factors which may influence the process—security threats which may require neutralization, safety threats which may need mitigation or logical non-determinism for choosing alternative options. In accordance with our theoretical framework we are considering both synchronous activities (in our framework they are called *actions*) and asynchronous activities (it events). While the actions change the situations in a deterministic way, the event are the main stochastic factors since they may or may not trigger actions, and also can happen at any time. This way the analytical level can be modelled naturally as a directed AND-OR graph. Choosing suitable operation based on risk assessment when the transactions execute under security threats would allow to implement control algorithms with guaranteed chances for successful commit of the transaction.

As an illustration, Fig. 1 presents one such graph which models a typical transaction for reading the emails in the presence of potential security threats on analytical level. The graph nodes represent situations and are painted in white, green or red; the solid arrows represent the deterministic transitions from situation to situation, while the events and threats are associated with the situations in a non-deterministic way and painted in blue and black, respectively. Some of the actions are normal actions which progress the transaction towards its commit situation, which can be prescribed using suitable heuristics, while other actions are outside of the control since they are triggered by asynchronous events or caused by

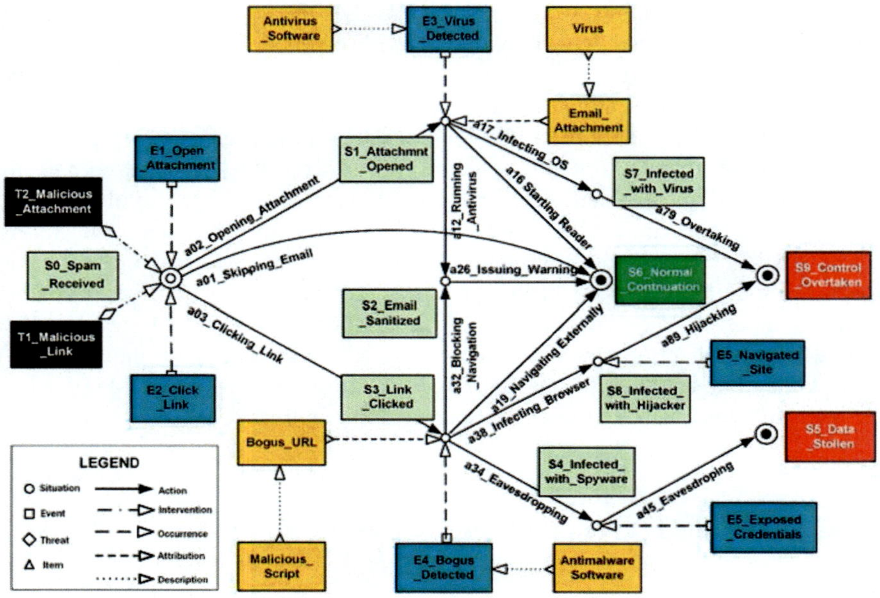

Fig. 1 Threat in the email

security threats. The situations, events and threats can be described qualitatively and quantitatively using various *items*, colored in yellow.

This graph can be created entirely automatically using the domain ontology and the security policies of the framework. However, in order to implement the control strategy, we need to deal with the non-determinism. The graph contains multiple decision points which cannot be resolved without additional information and this is where we need to make informed decision choice of an action to be executed.

4 Transactions Under Threat as POMDP

Before we specify the model we will make some assumptions which can be lifted at a later stage:

1. We will omit the descriptions, which use concepts of the type `Item` and will consider only `Situation`, `Action` and `Event` taxonomies.
2. There will be no distinction between situations, free of any threats, between situations, which are result of malicious actions and between transient situations. So the model will consider only the top classes of the taxonomy—`SafeSituation`, `DangerousSituation` and `TransientSituation`.

3. There will be no distinction between different malicious actions and between different counteractions—we will consider only the top roles in the action taxonomies `MaliciousAction` and `CounterAction`.
4. We assume that the counteractions always bring the system back to a safe situation in a single step. This also means that all transient situations are safe.
5. Only events relevant to the threats will be considered (class `Threat`). The non-threatening events will be addressed by the security policies on logical level.

The above assumptions makes it possible to apply a reduction of the original POMDP problem with partial information to an MDP problem with full information and to use the recurrent algorithm of dynamic programming for solving it. This way, we can have a quantitative evaluation of the risk in each situation incrementally.

Due to the presence of asynchronous events, which can be either unpredictable, but anticipated—like many malicious interventions, or unexpected, but predictable—such as human or technical errors, we must model the transactions under threat as POMDP, rather than as MDP which requires full information. Our model has the following elements:

1. **State space** $S = \{safe, danger, deadend\}$—corresponds to the different top-level types of situations from risk viewpoint
   a. $safe$ Situations along the normal transactions in absence of any threats
   b. $danger$ Situations in which the system is under the influence of security threats but is still able to recover
   c. $deadend$ Situations in which the system experiences severity and crashes completely under the security threats
2. **Control space** $C = \{noact, respond\}$—corresponds to the different top-level types of counteractions for risk mitigation
   a. $noact$—no control intervention, the system goes straight to the next situation according to the planned action in order to continue its normal track of execution of the current transaction
   b. $respond$—counteraction, which brings the system back to a safe situation after malicious action
3. **Observation space** $Z = \{nothreat, threat, crash\}$—corresponds to the different top-level types of events from security viewpoint
   a. $nothreat$—asynchronous event, which is non-threatening and does not require counteraction
   b. $threat$—detection of malicious intervention which requires counteraction
   c. $crash$—losing control of the system without chance for recovery
4. **Transition kernel** $q(s_{n+1}|s_n, c_{n+1})$—probability of the transition from situation s_n to situation s_{n+1} under control c_{n+1}, calculated as follows
   - $q(safe|safe) = p$, p is the probability for absence of threats after transition from a safe situation

- $q(danger|safe) = 1 - p$, $1 - p$ is the probability for presence of threats after transition from a safe situation
- $q(safe|danger, respond) = 1$ because the counteraction in a dangerous situation eliminates the threat
- $q(deadend|danger, noact) = 1$ because the absence of counteraction in dangerous situation leads to an inevitable deadend of the system
- $q(deadend|deadend) = 1$ since there is no way out of the deadend

5. **Occurrence kernel** $t(z_n|s_n)$—probability of occurrence of event z_n in state s_n, calculated as follows

    - $t(nothreat|safe) = p_{11}$—probability of not observing threat in a safe state
    - $t(nothreat|danger) = p_{12}$—probability of not observing threat in a dangerous stage (false negative)
    - $t(threat|safe) = p_{21}$—probability of observing threat in a safe state (false positive)
    - $t(threat|danger) = p_{22}$—probability of observing threat in a dangerous state
    - $t(crash|deadend) = 1$—probability of observing the system crash under threat

    Let's denote the matrix with entries $p_{ij}, i, j = 1, 2$ by P. Its transpose P^T is a stochastic matrix since

$$p_{11} + p_{21} = p_{12} + p_{22} = 1.$$

6. **Rewards**—quantitative measures of the costs of the actions taken as follows

    a. Current reward $r(c)$ calculated as follows: $r(noact) = 0$, $r(respond) = -c$ where $c > 0$ is the cost for using $respond$
    b. Final reward $R(s)$ calculated as follows: $R(safe) = R(danger) = 1$ if either the transaction terminates normally or the threat occurs after finalizing it, and $R(deadend) = 0$ if the crash occurs during the transaction.

7. **Horizon** N—length of the transaction, calculated as the number of safe situations in it.

5 Optimal Strategy for Counteracting and Its Cost

The solution of the risk assessment task can be obtained as a byproduct of the calculation of the optimal strategy for control of the transactions.

Definition *Security decision* $\phi(s)$ *is a function which on each step of the transaction* s chooses either $noact$ or $respond$.

The security decisions may modify the original transactions by enforcing *respond* actions in some of the situations. Therefore, they can extend the transaction path. If the security decisions are wrong it might be even possible to end the transaction in a *deadend* situation. In order to maximize the chances to make the right decisions we will account all information available at the time of decision making, which will turn the security decision into a stochastic function of the parameters of the POMDP model.

Definition *Decision policy* $\pi = (\phi(1), \phi(2), \dots, \phi(N))$ is a collection of security decision functions such that on each step n of the transaction, $\phi(n)$ depends only on the past history till time n, and the prior probabilities of the states at time 0, that is before the transaction has begun.

We assume that the prior probability of state *deadend* is 0, since otherwise any policy makes no sense. Therefore, the sum of the prior probabilities of the other two states is equal to 1, and the prior distribution of the states at time 0 is determined by the prior probability x of state *safe*. So, we are now looking for a decision policy π which maximizes the total reward $v^{\pi}(x) = E_x^{\pi}(R(state_N) - cK)$, where E_x^{π} is the expectation, corresponding to the policy π and the prior probability x, and K is the number of times when we apply the action *respond*. In the above expression $R(state_N)$ is the final income which we get in the last step of the transaction.

Definition *Value function* of the POMDP model is the function

$$v(x) = \max_{\pi} v^{\pi}(x)$$

Definition The policy π such that $v(x) = v^{\pi}(x)$ is an *optimal policy.*

The optimal policy π of the POMDP maximizes the chances to avoid a crash during the transaction, taking into account the total price of counteractions. It solves the following optimization problem:

$$v^{\pi}(x) = E_x^{\pi}(R(s_N) - cK), \tag{1}$$

where x the prior probability of the state *safe* in the moment $n = 0$, s_N is the final state of the controlled process, and K is the total number of times when the counteraction has been used. Here, E_x^{π} is the mathematical expectation corresponding to π and x.

To calculate the optimal strategy we will follow the standard procedure for reducing the POMDP model with partially observable states to a MDP model with fully observable states which would allow to apply the standard algorithm of dynamic programming [7]. The reduction can be done by following the steps bellow:

1. Constructing sufficient statistics for the POMDP model by solving the filtration equations

2. Building a model with fully observable states using the sufficient statistics from step one
3. Solving Bellman's equation for the MDP model built in step two making use of the dynamic programming algorithm

This gives us the optimal strategy in both POMDP and MDP models. Based on it we can now estimate the risks.

Definition The risk corresponding to the prior probability x of the state $safe$ of the POMDP model is equal to $1 - v(x)$, where

$$v(x) = \sup_{\pi} v^{\pi}(x), \tag{2}$$

is the value function of the model.

So the risk in each state can be assessed if the optimal strategy is known. In the general case this is a difficult problem, but fortunately, for the special case of our POMDP there is an elegant solution based on statistical reduction of the POMDP model to deterministic MDP model.

Let f_n (resp. g_n), $n = 0, 1, \ldots, N-1$, be 3x1-vectors with elements equal to the prior (resp. posterior) probabilities of the states $safe$, $danger$ and $deadend$ during the transaction. We assume that $f_n(1)$ and $g_n(1)$ correspond to state $safe$, $f_n(2)$ and $g_n(2)$—to state $danger$, and $f_n(3)$ and $g_n(3)$—to state $deadend$.

We can think of these vectors as points in the two-dimensional simplex in $\mathbf{R}^3$ (Fig. 2). In order to exclude the trivial case of a system's breakdown before any transaction has begun, we assume that $f_0(3) = 0$. Thus, we have $f_0(1) = x$, $f_0(2) = 1 - x$, where x is the same as in formulas (1) and (2). The other vectors f_n and g_n satisfy the following relations:

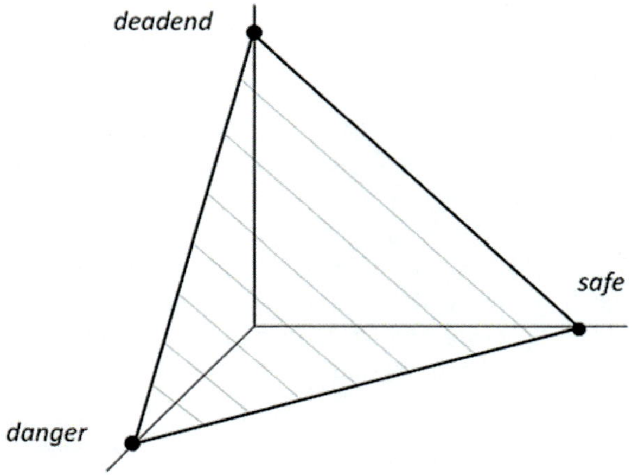

Fig. 2 Situation simplex

- Since the state *deadend* is absorbing, $g_n(3) = 0$ or 1. If $g_n(3) = 1$, then $f_m(3) = g_m(3) = 1$ for all $m > n$.
- Making use of the Bayes formula, the coordinates of the vector g_n can be calculated as follows:

$$g_n(1) = \frac{f_n(1)p_{21}}{f_n(1)p_{21} + f_n(2)p_{22}} := \Gamma^1(f_n(1), f_n(2)), \tag{3}$$

$$g_n(2) = \frac{f_n(2)p_{22}}{f_n(1)p_{21} + f_n(2)p_{22}}, \; g_n(3) = 0, \tag{4}$$

if $z_n = threat$;

$$g_n(1) = \frac{f_n(1)p_{11}}{f_n(1)p_{11} + f_n(2)p_{12}} := \Gamma^2(f_n(1), f_n(2)), \tag{5}$$

$$g_n(2) = \frac{f_n(2)p_{12}}{f_n(1)p_{11} + f_n(2)p_{12}}, \; g_n(3) = 0, \tag{6}$$

if $z_n = nothreat$;

$$g_n(1) = 0, \, g_n(2) = 0, \, g_n(3) = 1, \tag{7}$$

if $z_n = crash$.

On the other hand, if $g_n(3) = 0$ then the coordinates of f_{n+1} are

$$f_{n+1}(1) = pg_n(1), \, f_{n+1}(2) = (1 - p)g_n(1), \, f_{n+1}(3) = g_n(2), \tag{8}$$

whenever $c_n = noact$;

$$f_{n+1}(1) = p, \, f_{n+1}(2) = 1 - p, \, f_{n+1}(3) = 0, \tag{9}$$

if $c_n = respond$ where $\Gamma^1(x, 1-x)$ and $\Gamma^2(x, 1-x)$ are the posterior probabilities to remain safe after detecting absent or present threats, respectively.

The last equations show that if we consider the vectors $g_n, n = 0, 1, \ldots, N - 1$, as points in the two-dimensional simplex, they are located either in the vertex of the simplex, corresponding to state *deadend*, or on the edge, connecting the vertices corresponding to states *danger* and *safe*. On the other hand, the location of the points on this edge is entirely determined by a single coordinate, say that which is equal to the posterior probability of the state *safe*. This observation plays an important role for reducing the POMDP model to a MDP model with fully observable states.

According to the general theory of POMDP (see [1]), sufficient statistics allow to reduce the initial POMDP problem to a fully observable MDP problem on the base of posterior probabilities $g_n, n = 0, 1, \ldots N - 1$.

We consider the following fully observable MDP model. Its state space is the set $S = (0, 1) \cup \{*\}$, where $*$ is an isolated point.

Definition The controlled process in the model with complete information is defined as

$$x_n = \begin{cases} *, & \text{if } g_n(3) = 1 \\ g_n(1), & \text{if } g_n(3) = 0 \end{cases}, \quad n = 0, 1, \ldots N-1.$$

Let us note, that in view of (3), (5), and the total probability formula, the initial distribution of x_0 is the following:

$$x_0 = \begin{cases} \Gamma^1(x, 1-x) \text{ with probability } p_{21}x + p_{22}(1-x) \\ \Gamma^2(x, 1-x) \text{ with probability } p_{11}x + p_{12}(1-x) \end{cases}.$$

The fact that P^T is a stochastic matrix implies that the distribution of x_0 is a proper probability distribution. The same holds for all distributions that appear in the definition of the transition kernel $t(\{y\}|x, c)$ of the model with fully observable states. The filtration equations (3)–(9), and the total probability formula motivate us to define it as follows:

$$t(\{y\}|x, c) = \begin{cases} pxp_{21} + (1-p)xp_{22}, & y = \Gamma^1(px, (1-p)x) \\ pxp_{11} + (1-p)xp_{12}, & y = \Gamma^2(px, (1-p)x) \\ 1-x, & y = * \end{cases}$$

provided that $c = noact$,

$$t(\{y\}|x, c) = \begin{cases} pp_{21} + (1-p)p_{22}, & y = \Gamma^1(p, 1-p) \\ pp_{11} + (1-p)p_{12}, & y = \Gamma^2(p, 1-p) \end{cases},$$

provided that $c = respond$,

$$t(*|*, \cdot) = 1.$$

In all other cases we set $t(\{y\}|x, \cdot) = 0$. The final reward is

$$R(x) = 1, x \in (0, 1), R(*) = 0.$$

The other elements of the model—state space C, running reward r and horizon N remain unchanged after the reduction.

Consider the functions

$$V_n(x) = \max_{\pi} E_x^{\pi}(\Sigma_{k=n}^{N-1} r(c_{k+1}) + R(x_N)), n = 0, 1, \ldots N-1. \tag{10}$$

They satisfy the Bellman's equation

$$V_n(x) = \max(V_n^{noact}(x), V_n^{respond}(x)), \tag{11}$$

and the final condition

$$V_N(x) = R(x). \tag{12}$$

In (11), $V_n^{noact}(x)$ and $V_n^{respond}(x)$ are one-step ahead estimates of both actions $noact$ and $respond$:

$$V_n^{noact}(x) = (pxp_{21} + (1-p)xp_{22})V_{n+1}(\Gamma^1(px, (1-p)x)) \\ + (pxp_{11} + (1-p)xp_{12})V_{n+1}(\Gamma^2(px, (1-p)x)),$$

$$V_n^{respond}(x) = -c + (pp_{21} + (1-p)p_{22})V_{n+1}(\Gamma^1(p, 1-p)) \\ + (pp_{11} + (1-p)p_{12})V_{n+1}(\Gamma^2(p, 1-p)).$$

Let us note that since after action $respond$ the system instantly falls into a $safe$ state ($x = 1$), the right-hand side of the last formula does not depend on x, but still depends on n.

The optimal strategy φ_{n+1} at any moment of time $n = 0, 1, \ldots, N-1$ is the following:

$$\varphi_{n+1}(x) = \begin{cases} noact, & \text{if } V_n(x) = V_n^{noact}(x) \\ respond, & \text{if } V_n(x) = V_n^{noact}(x) \end{cases}.$$

These equations can be solved backwards, starting with the state of successful completion of the transaction. For example, for $n = N-1$ we get:

$$V_{N-1}(x) = \max(1-c, x),$$

$$\varphi_N(x) = \begin{cases} noact, & \text{if } x \geq 1-c \text{ (above the threshold)} \\ respond, & \text{if } x < 1-c \text{ (bellow the threshold)} \end{cases}$$

The remaining iterations until reaching the beginning of the transaction can be performed recursively, taking the previously calculated solution as terminal.

Finally, the connection between the value functions in both models is given by the formula

$$v(x) = (xp_{21} + (1-x)p_{22})V_0(\Gamma^1(x, 1-x)) \\ +(xp_{11} + (1-x)p_{12})V_0(\Gamma^2(x, 1-x)).$$

6 Analysis of the Results

In this section we will analyse the results of applying the optimal strategy to the problem for risk assessment. The model parameters used in the calculations are as follows:

- $p_{11} = 0.9$ is the probability for not detecting attack in a safe situation $t(nothreat|safe)$
- $p_{22} = 0.9$ is the probability for detecting an attack in a dangerous situation $t(threat|danger)$
- $p = 0.9$ is the probability of not having an attack after transition from a safe situation $q(safe|safe)$
- $r(respond) = -0.1$ is the cost of responding to a threat
- $N \in \{1..50\}$ is the horizon of the transaction.

Table 1 presents the optimal policy threshold for making decision to counteract which can be done by comparing it to the posterior probability to remain safe at different steps of the transaction. The risk has been calculated for four different prior probabilities. Their choice reflects the most typical cases of potential distribution of the threats as follows:

- 1.0: No threats are expected in the beginning of the transaction. This is the case when we are operating clean computer, browser or ATM machine.
- 0.25: Low probability to start a transaction in a safe state. This is the case when it is very likely for threats to occur immediately after starting the transaction (for example infected computer, spyware in the browser or tampered ATM machine).
- 0.50: Equal probabilities for presence and absence of attacks at the beginning of the transaction. This is a case of maximum uncertainty about the threats, i.e., we have a weak threat intelligence.

Table 1 Risk thresholds of the optimal strategy for different prior probabilities of having threat

Remaining steps	Probability threshold	0.25	0.5	0.75	1.0
1	0.9	0.1	0.1	0.0552	0.0018064
2	0.87134	0.12866	0.127227	0.083137768	0.030414629
3	0.842985516	0.157014484	0.15416376	0.110777719	0.058717893
5	0.787180101	0.212819899	0.207178904	0.165176837	0.114422501
7	0.732558117	0.267441883	0.259069789	0.218422348	0.168945816
10	0.652789319	0.347210681	0.334850147	0.296180972	0.248570520
15	0.550343126	0.449656874	0.43217403	0.39604552	0.350831653
20	0.404677698	0.595322302	0.570556187	0.53804018	0.496233951
30	0.181781435	0.809352173	0.782307637	0.755319458	0.718727575
50	4.04E-05	0.98990967	0.979799118	0.969688567	0.959630849

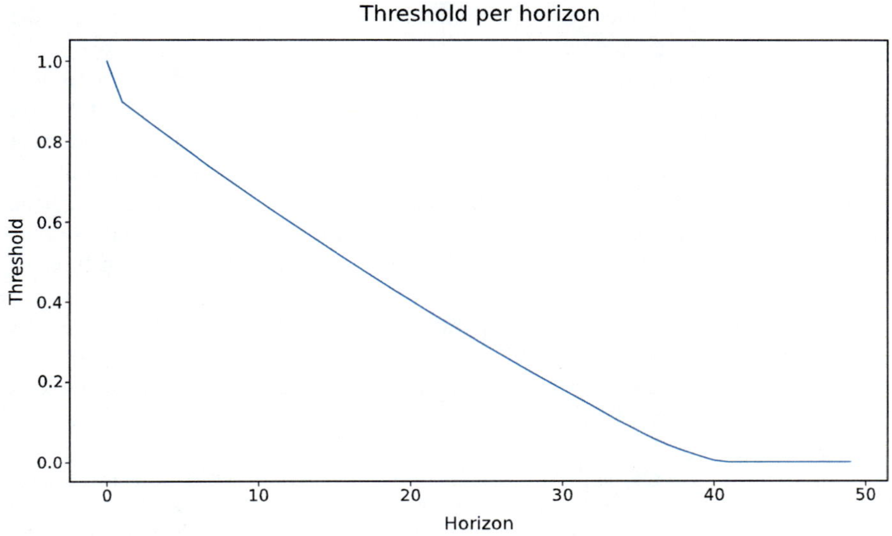

Fig. 3 Decision threshold

- 0.75: More likely to start in a safe state at the beginning but possible intrusion at a later step. This is statistically safe prediction when using clean computer, browser or ATM machine.

The first column of the table contains the number of remaining steps till completion of the transaction, while the second—the threshold of the optimal strategy. Figure 3 illustrates the evolution of the threshold in function of the remaining steps of the transaction. It shows that it is higher when there are fewer remaining steps of the transaction, because counteracting towards the end of the transactions is more efficient due to lower costs. The remaining columns of the table contain the estimations of the risk for fixed priory probabilities. They show that when increasing the prior probability the risks decrease, which matches the intuition and is confirmed for other fixed values of the probability parameter as well. At the same time, the results also show that the risks increase with the length of transactions, which also matches the intuition. These results give enough evidence that estimating the risks on the base of the optimal strategy can be an adequate heuristic to compare alternative paths through the graph for planning countermeasures.

7 Conclusion and Future Plans

Our hybrid cybersecurity framework employs a number of enabling technologies. The risk assessment component presented here adds a decision-making heuristic for choosing an optimal counteraction to neutralize the security threats and commit the transaction, despite the threats.

The method of assessing the security risks based on POMDP model presented here can be used for further analysis of the risk-related problems. Particularly interesting would be to investigate the impact of false negatives p_{12} and false positives p_{21} of the data analytics engines on the security risks and the possibility to account more information about the transactions for further tunning of the control strategy. We are also planning to add reinforcement learning capabilities to the framework for further tuning of the model and improving the algorithms for assessment.

Although developed primarily for applications in fintech industry, this framework can be adapted to a wide range of business process workflows—production line fault management, critical infrastructure protection, public safety management, autonomous agent control, etc.

Acknowledgments The research presented in this paper is a result of collaboration between the Cyber Security Research Centre (CSRC) of London Metropolitan University and GATE Institute of Sofia University (GATE). GATE has received funding from the EU H2020 WIDESPREAD-2018–2020 TEAMING Phase 2 programme under Grant Agreement No. 857155 and BG Operational Programme "Science and Education for Smart Growth" under Grant Agreement No. BG05M2OP001-1.003-0002-C01. CSRC has received funding from the university Rescaling Fund. The opinions presented are, however, of the authors only and do not reflect the official policies of these organisations.

References

1. Dynkin, E., Yushkevich, A.: Controlled Markov Processes. Springer, New York (1979)
2. Masuam, Kolobov, A.: Planning with Markov Decision Processes: An AI Perspective. Morgan & Claypool, San Rafael (2012)
3. Mundt, A.: Dynamic Risk Management with Markov Decision Processes. KIT Sc. Publ., Karlsruhe (2008)
4. Liang, Y.: Risk Management by Markov Decision Processes. PhD Thesis, University of Manitoba, Winipeg (2015)
5. Kreidl, O.: Analysis of a Markov Decision Process Model For Intrusion Tolerance. In: Int. Conf. Dependable Syst. and Networks (DSN2010), IEEE Xplore, 2010, pp. 156–161
6. Jean-Baptiste, E., Rotshtein, P., Russell, M.: POMDP Based Action Planning and Human Error. In: 11th IFIP Int. Conf. on AI Applications and Innovations (AIAI2015), pp. 250–265
7. Bataityte, K., Vassilev, V., Gill, O.: Ontological Foundations of Modelling Security Policies for Logical Analysis. IFIP Advances in Inf. and Comm. Technology, vol. 583, pp. 368–380. Springer (2000)
8. Vassilev, V., Sowinski-Mydlarz, V., Gasiorowski, P., et al.: Intelligence Graphs for Threat Intelligence and Security Policy Validation of Cyber Systems. Advances in Int. Syst. and Computing, vol. 1164, pp.125–140. Springer (2000)
9. Sowinsky-Mydlarz, V., Li, J., Ouazzane, K., Vassilev, V.: In: Proc. 20th International Conference on Security and Management (SAM'21), Las Vegas, USA, 26–29 July 2021. Springer (2021)

A Decomposition Approach to the Clinical Pathway Deployment for Chronic Outpatients with Comorbidities

Paola Cappanera, Marco Gavanelli, Maddalena Nonato, and Marco Roma

Abstract Most chronic patients with comorbidities are cared for at home. Still, they must get treatments, consultancy, and tests at specialized medical units in a hospital setting, according to a given frequency set by their clinical pathways. As such demand is known in advance, it could be scheduled to ensure ideal frequency, avoid potential visit repetitions that arise in case of comorbidities, and minimize hospital access by pursuing decision coordination. Booking involves setting a date for each health service contained in the pathway, and fixing a time on that day, i.e. building a master plan that spans the planning horizon, and a specific daily agenda for each day. The master plan handles time constraints on the dates, while each daily agenda must comply with the staffing level at each care unit for that day and allow transfer time for patients receiving care at different units. Tackling the master plan together with the daily agendas is rather complex. We present a logic-based Benders decomposition approach where the Master Problem solves the master plan with respect to a relaxation of the units resource constraints, and the subproblems return *no-good* cuts to the master when their daily agenda problem is not feasible. We present an Answer Set Programming based approach for the Master Problem, as part of a broader project aimed to tackle the whole problem for the first time.

Keywords Chronic patients with comorbidities · Clinical pathways · Outpatient appointment scheduling · Answer Set Programming

---

P. Cappanera
DINFO, University of Florence, Florence, Italy
e-mail: paola.cappanera@unifi.it

M. Gavanelli · M. Nonato (✉) · M. Roma
DE, University of Ferrara, Ferrara, Italy
e-mail: marco.gavanelli@unife.it; nntmdl@unife.it; rmomrc@unife.it

L. Amorosi et al. (eds.), *Optimization in Artificial Intelligence and Data Sciences*, AIRO Springer Series 8, https://doi.org/10.1007/978-3-030-95380-5_19

1 Introduction

Human life expectancy is increasing world-wide, mostly due to progresses in medical science which have raised survival rates to life threatening diseases. At the same time, though, unhealthy lifestyles such as sedentary habits and high-calorie diets are becoming more common. As a consequence, the number of people who are affected by several chronic diseases at a time, so called comorbidities, is steadily raising. A recent study estimates that more than half of the population over 65 in Europe suffers from two or more non transmissible chronic diseases (NCDs) [3]. It adds to that number the share of patients having survived the acute phase of COVID-19 illness but still experiencing the symptoms of long-COVID syndrome. To increase life quality and keep public spending at bay, such patients are not hospitalized but live at home, often receiving health care at their own domicile. Nevertheless, they need to frequently access a hospital setting to receive treatments and consultancy at specialized medical units, and take periodic tests to monitor their health condition.

Well-assessed medical guidelines are available for the care of chronic patients affected by a specific morbidity, such as diabetes, hypertension, chirrosis, obesity, kidney and heart failure, among many other NCDs. Medical protocols based on such guidelines include recommendations concerning drug based therapies as well as specific health services a patient should receive at given frequencies, which must necessarily be delivered within a hospital setting. Such services include illness-specific treatments (such as dialysis for patients with renal failure), consultancy, and tests, along with more general check up activities (such as blood tests or X rays) that are common to several different illness-specific protocols. Once enrolled in a monitoring and control program, an NCD patient will be given a personalized care plan, built along the above mentioned protocols, to be followed for good and subject to periodic revaluation based on the patient's health status.

In case of comorbidities, different protocols get merged into a single care plan. Service frequencies and drug dosages are then adjusted and tailored to the individual patient, yielding a personalized clinical pathway [2] (PDTA in Italian). Even though an increased level of coordination among different specialists has been highly advocated, along with patient-centered team working, the use of the so-called family nurse—who acts as a single point of contact between the patient and the care facilities—is not yet common practice, at least in Italy. When lacking, no one is in charge of i) ensuring the timely realization of the clinical pathway, and ii) exploiting the advantages of the best synchronization of the different activities the pathway is made of. Actually, because of the periodic health service components of a clinical pathway, an NCD patient may ask for a hospital appointment quite frequently, according to some regularity set in the PDTA. However, the patient is often the one in charge of the booking task. In particular, current practice is what we call *independent incremental booking*, i.e., as soon as a health service has been administered, the appointment for the next occurrence is tentatively booked, potentially along with any correlated secondary activity—think of check up exams whose results are to be brought along when seeing the specialist. When booking, the patient asks for an ideal date, but this decision is rather constrained: beside abiding by service frequency, additional time constraints might have to be taken into account because of necessity as well as opportunity. The former concerns precedence and/or minimum and maximum time gaps in between the date of the primary activity (i.e., the visit) and the secondary ones (i.e., the exams), as well as services that have some kind of interaction. For example, in case of a cardiopathic oncological patient, a chemotherapy session may alter the results of EKG if taken too soon after the treatment. Regarding the latter (i.e., opportunity), when the different medical protocols that got merged into a single care plan share the same health service, redundancies may arise, yielding a waste of time and money. For example, follow-up visits of different diseases may require the same tests to be taken right prior to the visit, while not too soon as results may need time. A single exam session could serve the purpose for both visits, had their dates been synchronized. Finally, frail patients, such as those with comorbidities, should keep the number of hospital access to the minimum, particularly during pandemic situations. Therefore, it is advisable to concentrate different health services on the same date, if possible. As service capacity in public hospitals is limited (in terms of daily working hours and number of operators) while the number of NCDs patients steadily increases, scheduling complex clinical pathways while delivering services in a timely manner to all patients is increasingly challenging and calls for optimization-based decision support tools. In particular, much could be gained from a centralized management able to exploit complete knowledge of service requests and resource availability to yield a feasible and effective medium term plan. However, the resulting scheduling problem is rather complex, which motivated this study.

In this paper, we tackle the scheduling of clinical pathways of NCD patients with comorbidities (*NCDs Agenda* in the following) by an iterative two level approach, exploiting decomposition: for a given planning horizon, at each iteration, at the higher level the Master Problem (MP) sets up a tentative *master plan* which assigns a date to each health service that complies with the time constraints of each pathway, by solving a relaxed problem with respect to resource availability. At the inner level, a Sub Problem SP(h) is solved for each day h. SP(h) builds the daily agenda, i.e., the fine grained schedule of day h, setting the timing of each service MP has assigned to h, while taking into account a full representation of resource availability on day h as well as patient transfer between medical units. Each SP returns a *no good*-like cut to the MP in case a patient cannot be served and the process iterates until all SPs provide a feasible schedule for the day. Then, the daily agendas are consistent and all together provide a feasible NCDs Agenda for the planning period.

We believe this is the first time that the NCDs Agenda problem is formally introduced. This study is part of a larger project aimed at selecting the best solution technology to tackle this challenging planning problem. In particular, here we discuss and present the details of an Answer Set Programming (ASP) based deployment of the Master Problem. The paper is organized as follows: Sect. 2 formalizes the NCDs-Agenda problem and outlines the decomposition based solution approach. Related papers are mentioned in Sect. 3. Fundamentals of ASP technology are recalled in Sect. 4 where the ASP based model for the

MP is presented. Preliminary computational results are discussed in Sect. 5, where conclusions are drawn and on going works are sketched.

2 A Decomposition Approach for the NCDs Agenda Problem

Let us formally introduce the problem components and its constraints, ranging from those concerning individual patients—regarding the frequency of the same service as well as the timing of related pairs of services—up to those affecting different patients who share the same care unit on the same day. Consider the input data described in Table 1. In the following, $s \in S(\pi)$ will be written as $s \in \pi$ to simplify the notation.

Each pathway must be scheduled complying with three types of time-constraints, i.e., *Frequency*, *Interdiction*, and *Necessity*, so called FIN constraints in the following. *Frequency* is ensured once the date of the oth occurrence of $S(\pi)$ belongs to $[h^*_{o\pi} - \rho, h^*_{o\pi} + \rho]$, $\forall o \in [1, .., n_\pi]$, $\forall \pi \in cp(p)$, $\forall p \in P$.

Given two health services s_i, s_j in the same pathway, and letting $\tau(s)$ be the appointment date of s, the following constraints may be given:

- *forward (backward) interdiction*: $\tau(s_i) = h \Rightarrow \tau(s_j) \notin [h, .., h + \delta]$ ($[h - \delta, .., h]$).
- *forward (backward) necessity*: $\tau(s_i) = h \Rightarrow \tau(s_j) \in [h + \delta + 1, .., h + \delta + \Delta] \wedge \tau(s_j) \notin [h, .., h + \delta]$ ($\tau(s_j) \in [h - \delta - \Delta, .., h - \delta - 1] \wedge \tau(s_j) \notin [h - \delta, .., h]$).

Interdiction constraints are satisfied even if s_j is never scheduled, while necessity constraints require s_j to be given a date within a time window next but not too close to s_i. Examples regarding interdiction include the case of one service affecting the

Table 1 Input data: notation

Symbol	Description
$H = \{h\}$	A planning horizon (indexed by h)
S	A set of health services
$P = \{p\}$	A set of patients (indexed by p)
$cp(p) = \{\pi\}$	Clinical pathway of $p \in P$, defined as its set of packets π
$\pi = \langle S(\pi), H(\pi), n(\pi)\rangle$	A packet, as a triplet
$S(\pi) \subseteq S$	The services of packet π, which come as a whole even if delivered at different care units
$H(\pi) \subseteq H$	Time horizon in which $S(\pi)$ are provided with the same frequency
$n(\pi)$	Number of repetitions of $S(\pi)$ during $H(\pi)$ (so called *occurrences*)
$h^*_{o\pi} \in H(\pi)$	The ideal date of π, centered in the middle of the oth of the $n(\pi)$ equally large sub-intervals of $H(\pi)$
ρ	Tolerance with respect to the ideal date (time windows of consecutive occurrences of the same packet are disjoint)

results or the effectiveness of another service when too close in time, i.e. at least δ days must elapse between the two dates.

Necessity typically concerns the timing of a primary activity and a secondary one which is functional to the first one only if sufficiently but not too close in time.

Any time constraint on $s \in \pi$ propagates on all the services in the same packet. Therefore, the available options for feasible dates may become rather limited, despite of tolerance. For a given staffing level, an incremental booking process—which to our knowledge is common practice in most cases—may fail to find time-feasible dates with enough residual capacity to accommodate all requests, since previous booking has been done without knowledge of incoming demand. Furthermore, incremental booking may not be able to synchronize and aggregate different services on the same day, thus spreading appointments for the same patient on several dates. On the contrary, frail patients should minimize the number of trips to the hospital. At the same time, though, strong aggregation challenges the scheduling of daily agendas, that set the timing of the planned health services. In fact, such timing must allow patients to carry out all their activities in sequence, even when they are provided at different hospital care units and time for transfer is not negligible. Furthermore, interaction among patients must be considered when they receive service from the same care unit, as they compete for operators' time and other limited resources.

Summing up, the NCDs Agenda problem consists of scheduling the clinical pathways of all patients in P for the current planning horizon H by taking decisions at two levels: (1) for each $p \in P$ and each $\pi \in cp(p)$, assign a time feasible date $\tau(s)$ to each $s \in \pi$ for each occurrence $1, .., n_\pi$, so that FIN constraints are satisfied; (2) for each day h and for each patient p such that $\exists s \in \pi, \pi \in cp(p)$, and $\tau(s) = h$, set a starting and an ending time for each s so that: (i) the services of each patient can be feasibly sequenced; (ii) an operator can deliver each s with no preemption within her/his working shift, serving at most one patient at a time. At this stage of the project, operators in the same unit have identical skills but potentially different shifts.

As the whole problem, encompassing both levels of decisions, is quite complex, a decomposition approach inspired by Hooker and Ottosson [9] is investigated.

Let the day of each service be given (i.e., $\tau(s)$ is known). Then, the remaining problem can be decomposed in one scheduling problem $SP(h)$ for each day. In particular, $SP(h)$ can be restated in the machine scheduling framework as a multistage open shop with identical parallel machines with set up times and no preemption, by modeling each patient p as a job whose tasks are the services in $S(p, h)$, where $S(p, h)$ denotes the set of services s of patient p with $\tau(s) = h$; each care unit is seen as a pool of identical machines, one per operator, which are active during time periods that map the working shifts. Sequence dependant set up times refer to the walking distance between units within the hospital, or model the case of some rest to be taken in between two services, such as after a treatment before a test. As said, the problem can be modeled as a classical Open Shop, whose constraints ensure there is no overlapping between services of the same patient, as well as between those provided by the same operator. The primary objective is to serve as many requests as possible with no overtime.

At the higher level, the Master Problem (*MP*) builds the master plan by solving the date assignment problem taking into account FIN constraints plus a relaxation of the daily problem. *MP* can be stated as a constrained multidimensional [12] multiple-choice [16] binary knapsack problem, since:

- Each occurrence of $S(\pi)$ is seen as an item in a $|U|$-dimension space, where $|U|$ denotes the number of care units. If $S(\pi)$ contains no services provided by unit u its size on that dimension is 0, otherwise it is the sum of the duration of such services. The item profit is the number of services in $S(\pi)$.
- Each day h is seen as a multidimensional knapsack with $|U|$ dimensions, and each dimension u has capacity equal to C_h^u, i.e., the sum of the duration of the working shifts of the operators on duty at unit u on day h.

MP aims at selecting the maximum profit subset of items by placing each item in at most one knapsack so that, for each knapsack, capacities are not violated, and FIN constraints are met.

As MP disregards the details of the daily agendas, it returns an upper bound to the number of requests that can be timely scheduled at the hospital. If $SP(h)$ fails to accommodate some requests, this information is returned to MP as a no-good cut which is added to the MP constraints. Indeed, no-good cuts carry the information that if (an occurrence of) some packets of some patients have been scheduled on day h then (an occurrence of) the packet of another patient cannot be served on the same day. Rather than explicitly modeling the constraints of each subproblem into the *MP*, the infeasible master plans *MP* would choose are forbidden. Thus, subproblems certify whether or not the current master plan can be mapped to a feasible schedule. In addition, by putting together the schedules computed by each $SP(h)$, a feasible solution can be built by dropping what MP has left out of the master plan as well as what each $SP(h)$ has scheduled overtime. At each iteration, then, a lower bound on the number of unscheduled requests as well as an upper bound is computed. Therefore, once each $SP(h)$ has returned a feasible solution, the problem is solved to optimality. Clearly, when demand exceeds capacity, some services will not be scheduled even at optimality. In those cases the objective function could weight patients according to their status, as in [15]. Anyway, dropped requests can be rerouted to other health structures, such as private clinics integrated into the health care network.

Figure 1 depicts the interaction between the *MP* and the subproblems; it shows an example with 2 patients and 2 units, red and blue, with equal starting time. However, the red one is operative for 4 time slots on both days, while the blue one has 4 slots available on day 1 and 5 on day 2. Patient *p1* asks for a packet made of a 2-slot service from the red unit and a 3-slot service from the blue one, while *p2* asks for a 2-slot red service. According to the MP, both patients can be feasibly assigned to day 1 (bold left arrows, outcoming from each patient). On the contrary, SP(1) detects infeasibility since *p1* services cannot be delivered in sequence within the working hours of day 1. Thus SP(1) returns a no-good cut to the MP. At the next iteration $p1$ is expected to be assigned to day 2 (dotted right arrow outcoming from *p1*).

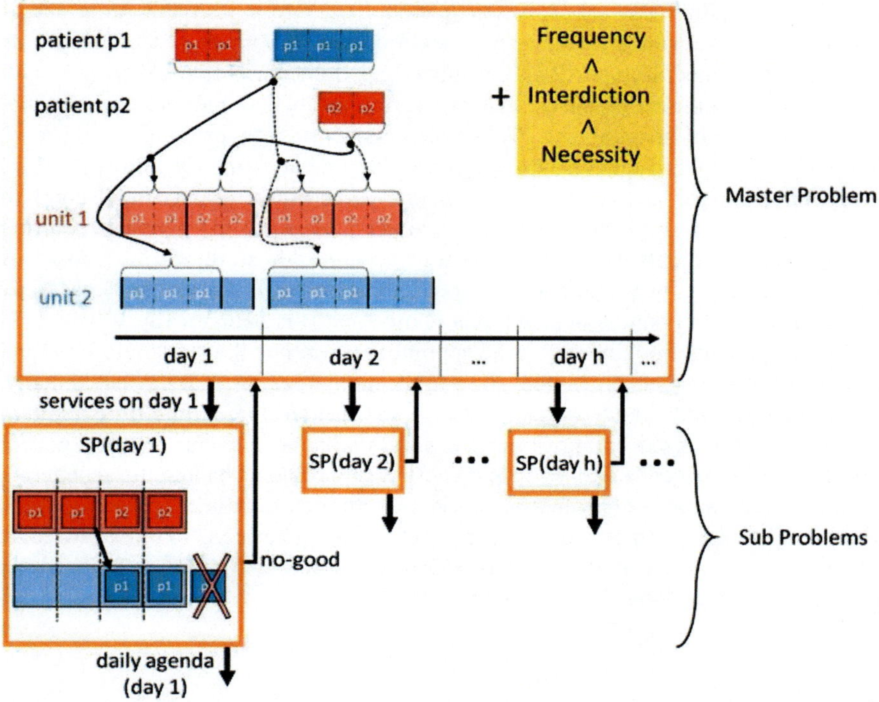

Fig. 1 MP and SP(h) interaction

Finally, to mimic progressive planning in the long term, a rolling horizon mechanism was implemented. As patient history in terms of recent past matters, the planning horizon of two successive iterations should overlap. In particular, the services whose ideal date falls within the far left of H inherit, as a tentative date, the one computed in the last iteration. Only slight variations are admitted and penalized by the second term of the MP objective function. As ASP naturally allows hierarchical objective functions, *MP* aims, in this order, to minimize (1) the number of services not present in the master plan; (2) the displacement in time regarding the services in the far left end of H; (3) the number of patients' trips to the hospital (see Sect. 4).

3 Related Papers

The scientific literature on outpatient appointment scheduling is quite rich and still growing—see [1] for a general review and [14] specifically for the multi-appointment case, covering the case of multiple appointments on the same day, as well as appointment series, where patients need to revisit the same set of resources more than once. Here, we just mention a few. Among those tackling offline deterministic scheduling, ASP is the modelling framework adopted in [8], with the purpose of minimizing unscheduled prescriptions, patient indirect waiting time, and the number of hospital trips. To handle larger instances, patients are clustered based on priority and then groups are scheduled one at a time. While [8] tackles a multidisciplinary service for chronic patients, the planning horizon (from 1 to 4 weeks) spans a single packet occurrence per patient. In [5] patients requiring multiple appointments or series of treatments are addressed, with at most one medical interaction per visit, and the scheduling problem is modelled as a Markov decision process to handle demand uncertainty, and it is heuristically solved. [17] addresses patients booking series of appointments rather than a single one, and proposes a Markov decision model and a much more tractable Index Policy. Finally, [15] schedules multi-mode outpatients of a gastroenterology clinic by exploiting logic-based Benders decomposition in a very similar way to our, i.e., the master problem sets the agenda and the 2-level nested subproblems handle the scheduling part. However, there is no synchronization nor interdiction or necessity constraints, since the patient is seen only once during each planning horizon. To our knowledge, despite of many similarities, none has addressed the NCDs Agenda problem so far, that revolves around the notion of clinical pathway.

4 An ASP Based Approach

Logic programming is one of the four programming paradigms [13]; a logic program consists of a set of clauses, in the form of implications $Head \leftarrow Body$, where $Head$ is an atom (or a disjunction) and $Body$ is a conjunction of literals (atoms, possibly negated). Atoms can have parameters that can be variables or constants. An atom (or a clause) is ground if it does not contain variables. A Declarative Semantics provides a formal meaning to a logic program by assigning a truth value to each ground atom in such a way to satisfy all the clauses. ASP is a logic programming language relying on the Stable Models semantics [7]; a logic program can have zero, one, or more than one stable models. In ASP, each solution to a combinatorial problem is associated with a stable model. There exist several solvers, based on different technologies; however, the best-performing ones are based on a two-phase solution scheme: a grounding phase, that generates a ground program equivalent to the original one, followed by a solving phase, in which a stable model of the ground program is computed. The ground program can be built by substituting to the variables in each clause all the possible constants appearing in the program (although modern grounders may avoid generating useless clauses), and, in the worst case, it is exponentially larger than the original one. One of the best ASP solvers is Clasp [6], which finds stable models by using technologies developed in SAT solvers such as conflict graphs, conflict-directed clause learning, and restarts. These technologies are very efficient in proving satisfiability/unsatisfiability of a problem; optimization problems are usually transformed into a sequence of satisfiability problems. ASP solvers can solve problems up to Σ_2^P. As a rule of thumb, to get an efficient solving algorithm one should produce programs whose grounding is not too large; for example, if some parameter of an atom can take a large variety of different values, the ground program can be very large.

Beside the basic syntax for clauses, recent ASP solvers accept several extensions [4]. A clause of the format $L \le \{Head\} \le U \leftarrow Body$ states that, in case the $Body$ is true, then the solver can choose the truth value of the atom in $Head$; this is the usual syntax for defining decision variables in optimization problems. The bounds L and U are optional; if they are present, then the number of true atoms matching with $Head$ in the stable model must be between L and U. If a clause is without head, $\leftarrow Body$, then the head is intended as the literal *false*, it is called an Integrity Constraint (IC) and in all stable models the $Body$ must be false.

The ASP formulation takes as input the description of the instance by means of a set of predicates. Predicate $\texttt{patient}(p)$ is true for each patient p; the set of services is provided by predicate $\texttt{service}(s, u, rc)$ where s identifies the service, u is the resource type (e.g., the care unit) and rc the resource consumption (service duration). Predicate $\texttt{resource}(h, u, C_h^u)$ provides the capacity C_h^u of each care unit u per each day $h \in H$. An interdiction of δ days between two services s_i and s_j is declared through the predicate $\texttt{interdiction}(s_i, s_j, \delta)$, while $\texttt{necessity}(s_i, s_j, (\delta, \delta + \Delta))$ states that service s_i requires s_j to be scheduled between $[\delta, \delta + \Delta]$ days in advance.

The main decision *MP* takes is the schedule of each packet π for each patient; the next clause declares that if o is an occurrence of packet $\pi \in cp(p)$ and its ideal date is $h^*_{o\pi}$, then π should be scheduled within a tolerance $\tau_{o\pi}$ from the ideal date.

$$0 \le \{\texttt{schedule}(p, cp, \pi, o, h) : horizon(h), h \in [h^*_{o\pi} - \tau_{o\pi}, .., h^*_{o\pi} + \tau_{o\pi}]\} \le 1$$
$$\leftarrow \quad \texttt{occurrence}(p, cp, \pi, o),\ \texttt{ideal\_date}(p, cp, \pi, o, h^*_{o\pi}).$$

Actually, the number of scheduled days for such packet occurrence is required to lie in {0, 1}: in fact, a packet could be not scheduled, and the number of scheduled packet occurrences should be maximized (see the primary objective function).

In ASP, problem constraints are usually imposed as ICs, that are denials, or implications with a *false* conclusion (in the syntax, the *false* conclusion is omitted). The body of the implication must always be false.

Interdiction between services is implemented by the following IC:

$$\leftarrow \texttt{schedule}(p, cp_1, \pi_1, o_1, h_1),\ \texttt{schedule}(p, cp_2, \pi_2, o_2, h_2),$$
$$s_1 \in \pi_1,\ s_2 \in \pi_2,\ \texttt{interdiction}(s_1, s_2, \delta),\ h_1 < h_2 \le h_1 + \delta.$$

stating that if service s_1 interdicts s_2 for δ days, the two services are scheduled for the same patient, and the day h_2 when s_2 is scheduled is between h_1 and $h_1 + \delta$, then such schedule is inconsistent.

Necessity is dealt with by the following IC; if a necessity is not satisfied within the planning horizon nor it can be postponed beyond it, then the schedule is inconsistent

$$\leftarrow \texttt{schedule}(p, cp_1, \pi_1, o_1, h_1),\ s_1 \in cp_1,\ \texttt{necessity}(s_1, s_2, \_),$$

$\texttt{not satisfies\_necessity}(p, cp_1, \pi_1, o_1, s_1, s_2)$,
$\texttt{not necessity\_beyond\_horizon}(p, cp_1, \pi_1, o_1, s_1, s_2)$.

where the two predicates used in the IC are defined as

$\texttt{satisfies\_necessity}(p, cp_A, \pi_A, o_A, s_A, s_B) \leftarrow s_A \in \pi_A$
$\texttt{necessity}(s_A, s_B, (\delta, \delta + \Delta))$, $\texttt{schedule}(p, cp_A, \pi_A, o_A, h_A)$,
$\texttt{schedule}(p, cp_B, \pi_B, o_B, h_B)$, $s_b \in \pi_B$, $h_A + \delta \leq h_B \leq h_A + \delta + \Delta$.
$\texttt{necessity\_beyond\_horizon}(p, cp_A, \pi_A, o_A, s_A, s_B) \leftarrow h_A + \delta + \Delta > \max H$,
$\texttt{necessity}(s_A, s_B, (\delta, \delta + \Delta))$, $\texttt{schedule}(p, cp_A, \pi_A, o_A, h_A)$, $s_A \in \pi_A$.

while the following IC imposes the above mentioned condition $\tau(s_j) \notin [h, .., h+\delta]$:

$\leftarrow \texttt{schedule}(p, cp_1, \pi_1, o_1, h_1), \texttt{schedule}(p, cp_2, \pi_2, o_2, h_2)$, $s_1 \in \pi_1, s_2 \in \pi_2$,
$\texttt{necessity}(s_1, s_2, (\delta, \_))$, $h_1 < h_2 \leq h_1 + \delta$.

Given a limit C_h^u of available resources in day h, if the sum of the consumption of resources of type u exceeds C_h^u, then the schedule is inconsistent:

$\leftarrow h \in H$, $\sum_{S_h} c > C_h^u$, $\texttt{resource}(h, u, C_h^u)$.

where $S_h = \{(c, p, s) | \texttt{schedule}(p, cp, \pi, o, h), s \in \pi, \texttt{service}(s, rt, c)\}$.

The objective is hierarchical, and defined through the so-called *weak constraints*, i.e., integrity constraints that can be relaxed, with the objective to maximize the number of the satisfied ones. Weak constraints are identified syntactically by using $\leftharpoonup$ as implication symbol. Each weak constraint can have a priority and a weight: the ASP solver searches the solution maximizing the weighted sum of the highest-priority weak constraints; among those solutions, the weight of satisfied second-highest-priority weak constraints is maximized, and so on.

The highest-priority weak constraint in our ASP program maximizes the number of scheduled packets whose ideal date lays within the horizon:

$\leftharpoonup \texttt{occurrence}(p, cp, \pi, o)$, $\texttt{not schedule}(p, cp, \pi, o, \_)$,
$\texttt{ideal\_date}(p, cp, \pi, o, D)$, $\texttt{horizon}(D)$.

The second-highest priority weak constraints minimize the number of day changes with respect to the schedule computed in the previous period. The schedule of the previous period is taken as input as a predicate `prev_schedule`.

$\leftharpoonup \texttt{prev\_schedule}(p, cp, \pi, o, h_1)$, $\texttt{schedule}(p, cp, \pi, o, h_2)$, $h_1 \neq h_2$.

The objective function with least priority minimizes the number of hospital trips.

$\texttt{trip}(p, h) \leftarrow \texttt{patient}(p)$, $\texttt{schedule}(p, cp, \pi, o, h)$.
$\leftharpoonup \texttt{trip}(p, h)$.

At the end of the paper, a list of the main predicates has been provided.

5 Computational Results and Conclusions

Preliminary computational results have been obtained on randomly generated pathways, inspired by well assessed and publicly available medical guidelines for the most common NDCs (such as [10] and [11] for diabetes), with service frequencies correlated to the severity of the patient. With such paradigm in mind, first, a set of services has been created, each one with a duration randomly picked within a given interval, according to uniform distribution. Then, time constraints between services have been generated, and protocols have been built as clusters of (packets of) services. For each patient, a pathway is build by picking one or more protocols and setting the value of its parameters (horizon and frequency) according to a uniform distribution within each range of values. For each combination of number of patients in {40, 80, 160} and length of the planning horizon in {60, 120, 150}, 5 instances are generated. The number of resources is 20 for all instances. With regard to efficiency and efficacy, Fig. 2a and b show the average (computed across the 5 instances) ratio between the number of not scheduled occurrences and the total number of occurrences in the instance, in dependence of (1) number of patients for a fixed planning horizon length (2a) and in dependence of (2) both number of patients and planning horizon length (2b). In Fig. 2a colored bars correspond to different time limits, while in Fig. 2b the colors correspond to different number of patients and the time limit is fixed to 600 seconds. Percentage values upon the bars represent the percentage of instances solved to optimality with respect to the first component of the hierarchical objective function. Experiments have been run on a Windows 10 OS, i7-4510U CPU machine. With regard to solution quality, measured in terms of satisfaction of service requests, Fig. 3 shows, separately for each instance (x-axis), boxplots reporting the distribution, among patients, of the ratio between number of satisfied occurrences and number of total occurrences requested by a patient. Very similar results have been obtained when the ratio is computed taking into

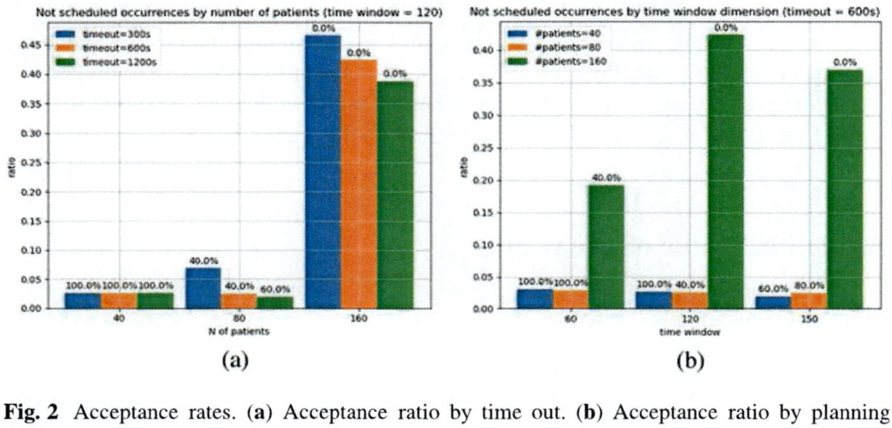

(a) (b)

Fig. 2 Acceptance rates. (**a**) Acceptance ratio by time out. (**b**) Acceptance ratio by planning horizon

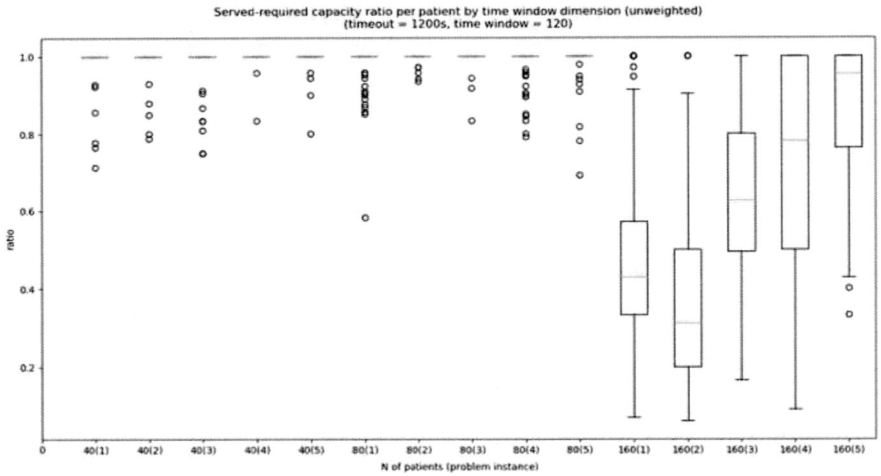

Fig. 3 Satisfaction of service requests

account the time requested by service occurrences instead of their number. Time limit is fixed to 1200 seconds and planning horizon is fixed to 120 working days. We can observe that up to 80 patients the request acceptance ratio is very high for (almost) all the patients, while for 160 patients the ratio can be quite low even when the time limit is increased. On the bigger instances, the dissatisfaction related to not accepted requests is not equally distributed among patients. Further analysis is needed to understand if this depends on instance structure and to evaluate the impact of dynamically generated cuts and bounds on the performance of the proposed methodology.

In a historical moment such as the one we are living, limiting the number of accesses by frail patients to health facilities is of paramount importance both to reduce costs and especially to limit the risk of infections. It is equally important to ensure uniformity of care across the country and among patients. It therefore becomes essential to deploy their clinical paths in order to pursue these objectives. Preliminary results show that the proposed approach is well able to respond to these needs in a short computational time up to 80 patients.

Appendix: Table of Predicates

Input predicates:

`horizon`(h): h is a day in the horizon
`interdiction`(s_i, s_j, δ): declares an interdiction of δ days between two services s_i and s_j

`necessity`$(s_i, s_j, (\delta, \delta + \Delta))$: service s_i requires s_j to be scheduled between $[\delta, \delta + \Delta]$ days in advance.
`patient(p)`: true for each p that is a patient (provides the list of patients)
`prev_schedule`(p, cp, π, o, h_1): In the schedule computed in the previous period, occurrence o of packet π in pathway cp for patient p was scheduled on day h_1.
`resource`(h, u, C_h^u): the capacity of care unit u in day h is C_h^u.
`service`(s, u, rc) : s is a service, u is the corresponding resource type and rc is its resource consumption

Predicates corresponding to decision variables of the optimization problem:

`schedule`(p, cp, π, o, h): occurrence o of packet π in clinical pathway $cp(p)$ for patient p is scheduled in day h

Computed predicates:

`occurrence`(p, cp, π, o): o is an occurrence of packet $\pi \in cp(p)$ for patient p
`ideal_date`$(p, cp, \pi, o, h^*_{o\pi})$: the ideal date of occurrence o of packet π in the clinical pathway $cp(\pi)$ for patient p is $h^*_{o\pi}$
`satisfies_necessity`$(p, cp_1, \pi_1, o_1, s_1, s_2)$: the necessity of occurrence o_1 of packet π_1 in pathway cp_1 of patient p is satisfied by service s_2
`necessity_beyond_horizon`$(p, cp_1, \pi_1, o_1, s_1, s_2)$: the necessity of occurrence o_1 of packet π_1 in pathway cp_1 of patient p is satisfied by service s_2, which can be postponed beyond the horizon.
`trip`(p, h) patient p has a trip to the hospital on day h.

References

1. Ahmadi-Javid, A., Jalali, Z., Klassen, K.J.: Outpatient appointment systems in healthcare: a review of optimization studies. Eur. J. Oper. Res. **258**(1), 3–34 (2017)
2. Aspland, E., Gartner, D., Harper, P.: Clinical pathway modelling: a literature review. Health Syst. **10**(1), 1–23 (2021)
3. Bennett, J.E., et al.: NCD Countdown 2030: worldwide trends in NCDs mortality and progress towards sustainable development goal target 3.4. Lancet **392**, 1072–1088 (2018)
4. Calimeri, F., Faber, W., Gebser, M., Ianni, G., Kaminski, R., Krennwallner, T., Leone, N., Maratea, M., Ricca, F., Schaub, T.: ASP-Core-2 input language format. Theory Pract. Log. Program. **20**(2), 294–309 (2020)
5. Diamant, A.: Dynamic multistage scheduling for patient centered care plans. Health Care Manag. Sci. **4**(4), 827–844 (2021)
6. Gebser, M., Kaminski, R., Kaufmann, B., Schaub, T.: Multi-shot ASP solving with clingo. Theory Pract. Log. Program. **19**(1), 27–82 (2019)
7. Gelfond, M., Lifschitz, V.: The stable model semantics for logic programming. In: International Conference on Logic Programming, pp. 1070–1080 (1988)
8. Guido, R., Ielpa, G., Conforti, D.: Scheduling outpatient day service operations for rheumatology diseases. Flex. Serv. Manuf. J. **32**, 102–128 (2020)
9. Hooker, J.N., Ottosson, G.: Logic-based Benders decomposition. Math. Program. **96**(1), 33–60 (2003)

10. https://salute.regione.emilia-romagna.it/cure-primarie/diabete/gestione-integrata-del-diabete-mellito-di-tipo-2-2017. Last Accessed 7 July 2021
11. https://www.regione.toscana.it/documents/10180/23793180/ALL+A+23-2019+PDTA-Diabete.pdf/f1e8ea87-145f-08c4-6c3d-16b69f5f43c2?t=1578658143393. Last Accessed 7 July 2021
12. Kellerer, H., Pferschy, U., Pisinger, D.: Multidimensional knapsack problems. In: Knapsack Problems, pp. 235–283. Springer, Berlin (2004)
13. Lloyd, J.W.: Foundations of Logic Programming. Springer, Berlin (1987)
14. Marynissen, J., Demeulemeester, E.: Literature review on multi-appointment scheduling problems in hospitals. Eur. J. Oper. Res. **272**(2), 407–419 (2019)
15. Riise, A., Mannino, C., Lamorgese, L.: Recursive logic-based Benders'^{TM} decomposition for multi-mode outpatient scheduling. Eur. J. Oper. Res. **255**(3), 719–728 (2016)
16. Sinha, P., Zoltners, A.A.: The multiple-choice knapsack problem. Oper. Res. **27**(3), 503–515 (1979)
17. Yu, S., Kulkarni, V.G., Deshpande, V.: Appointment scheduling for a health care facility with series patients. Prod. Oper. Manag. **29**(2), 388–409 (2020)

Optimal Sample Size for Evidence and Consensus in Phase III Clinical Trials

Fulvio De Santis and Stefania Gubbiotti

Abstract Design and analysis of clinical trials imply decisions that often involve multiple parties. We focus here on one of the main design issues in phase III trials, that is the choice of the sample size, that influences the final probability of success of the experiment, i.e. showing evidence of superiority of a new treatment over the standard one. Bayesian Statistics allows one to exploit pre-experimental information and uncertainty that can be translated into probability distributions for the effects-difference parameter. Sometimes sources of prior knowledge can be in striking contrast (skeptimism vs optimism), possibly leading to divergent final post-experimental conclusions. We propose a sample size criterion that controls not only the achievement of *minimal evidence* of superiority but also posterior *consensus*. The method is illustrated for trials involving binary outcomes with normal approximation for the log odds ratio with application to a comparative study of two interventions for diabetic patients with coronary artery disease.

Keywords Bayesian inference · Clinical trials · Hellinger distance · Power · Sample size

1 Introduction

Phase III clinical trials are randomized experiments designed to establish the superior efficacy of a novel treatment with respect to a placebo or the standard-of-care. Criteria for sample size determination are typically based on the idea of making the chance of success sufficiently large. The definition of success is related to the ability of the experiment to provide evidence in favor of superiority, i.e. in favor of the hypothesis that the effects-difference parameter θ exceeds a minimally significant clinical threshold δ. Bayesian Statistics allows one to exploit

F. De Santis · S. Gubbiotti (✉)
Department of Statistical Sciences, Sapienza University of Rome, Rome, Italy
e-mail: fulvio.desantis@uniroma1.it; stefania.gubbiotti@uniroma1.it

L. Amorosi et al. (eds.), *Optimization in Artificial Intelligence and Data Sciences*, AIRO Springer Series 8, https://doi.org/10.1007/978-3-030-95380-5_20

pre-experimental information and uncertainty on the unknown parameter that is translated into a prior probability distribution. Prior information is combined with experimental evidence via Bayes theorem, yielding a posterior distribution for the parameter of interest that is the basis to evaluate efficacy of the novel treatment. More specifically we declare success of the experiment if the posterior probability that θ is above δ is sufficiently large.

Some recent articles (see for instance [7]) deal with the problem of sample size determination for clinical trials in the presence of multiple (two, for simplicity) and contrasting sources of prior information. The proposed methods are based on the concept of consensus between specific summaries of the posterior distributions induced by the two alternative priors. Criteria that control *posterior consensus* in general do not guarantee *minimal evidence* provided by the posteriors, whereas generation of statistical evidence is typically the ultimate purpose of a clinical trial. In [4] a two-fold criterion based on intervals estimates is proposed, that takes into account both consensus and evidence. The idea is to select the minimal sample size that simultaneously guarantees a pre-posterior large chance to have final consensus and minimal evidence of superiority. In this paper we propose a further consensus&evidence criterion, that quantifies posterior agreement using a relative measure of global distance between the posterior distributions, that is the Hellinger distance.

This article is related to the literature on the following fields: (i) Bayesian experimental design and sample size determination (see [2, 3, 9, 10]); (ii) measurement of conflict/consensus of Bayesian procedures in the presence of multiple sources of prior information (see [4, 7, 11]); (iii) use of Hellinger distance in Bayesian analysis of clinical trials (see [5, 8]).

The paper is organized as follows. In Sect. 2 we formalize how to quantify consensus and evidence in the presence of two alternative priors. In Sect. 3 the proposed sample size determination criterion is implemented for the log odds ratio using the normal approximation. An example is provided in Sect. 4. Finally, Sect. 5 reports a discussion.

2 Methodology

Let θ be a real one-dimensional parameter denoting the unknown effects-difference. Also, let $\mathbf{X}_n$ be the data based on n observations with density or probability mass function $f_n(\cdot|\theta)$. We assume to be interested in assessing whether θ (on a suitable scale) is larger than a threshold δ, that is a clinically significant difference between two treatments. Adopting a Bayesian perspective, let $\pi(\theta)$ be the prior density of θ, $\mathbf{x}_n$ the observed data and $\pi(\theta|\mathbf{x}_n)$ the corresponding posterior density. We say that the experiment yielding $\mathbf{x}_n$ is *successful* at a prespecified level $\epsilon_s \in [0, 1]$ if $\mathbb{P}(\theta > \delta|\mathbf{x}_n) > \epsilon_s$. Suppose that two alternative sources of prior information are available and formalized by two prior densities $\pi_a(\theta)$ and $\pi_b(\theta)$. Given the observed sample $\mathbf{x}_n$, the corresponding posteriors $\pi_a(\theta|\mathbf{x}_n)$ and $\pi_b(\theta|\mathbf{x}_n)$ yield two different posterior probabilities of success $\mathbb{P}_j(\theta > \delta|\mathbf{x}_n)$, $j = a, b$, and the experiment provides the *minimal posterior evidence of success* if:

$$S_n(\mathbf{x}_n) = \min\{\mathbb{P}_a(\theta > \delta|\mathbf{x}_n), \mathbb{P}_b(\theta > \delta|\mathbf{x}_n)\} > \epsilon_s. \tag{1}$$

In addition to S_n, used to quantify minimal evidence in favor of superiority, we can also take into account the degree of agreement of the two posterior distributions $\pi_a(\theta|\mathbf{x}_n)$ and $\pi_b(\theta|\mathbf{x}_n)$. A natural choice is to consider a measure based on a relative distance, such as for instance the Hellinger distance between the two posterior densities:

$$d[\pi_a(\cdot|\mathbf{x}_n), \pi_b(\cdot|\mathbf{x}_n)] = \left(1 - \int_{\mathbb{R}} \sqrt{\pi_a(\theta|\mathbf{x}_n) \cdot \pi_b(\theta|\mathbf{x}_n)} d\theta\right)^{\frac{1}{2}}. \tag{2}$$

Therefore, we have *posterior consensus* at degree $\epsilon_c \in [0, 1]$ if

$$C_n(\mathbf{x}_n) = 1 - d[\pi_a(\cdot|\mathbf{x}_n), \pi_b(\cdot|\mathbf{x}_n)] > \epsilon_c. \tag{3}$$

Note that, since $d[\cdot, \cdot]$ ranges in $[0, 1]$, $C_n(\mathbf{x}_n)$ provides a relative measure of consensus. Before the experiment is carried out, $\mathbf{X}_n$ and the posterior probabilities $\mathbb{P}_j(\theta > \delta|\mathbf{X}_n)$, $j = a, b$, as well as $S_n(\mathbf{X}_n)$ and $C_n(\mathbf{X}_n)$, are all random objects. Following the standard approach, sample size determination can be based on the average behavior of $S_n(\mathbf{X}_n)$ and $C_n(\mathbf{X}_n)$, resulting in the following *expectation criterion:*

$$n^* = \max\{n : n_s, n_c\}, \tag{4}$$

where $n_s = \min\{n : e_n^s > \epsilon_s\}$, $n_c = \min\{n : e_n^c > \epsilon_c\}$ and $e_n^s = \mathbb{E}[S_n(\mathbf{X}_n)]$, $e_n^c = \mathbb{E}[C_n(\mathbf{X}_n)]$ and $\mathbb{E}[\cdot]$ is the expected value taken with respect to the sampling distribution of the data $f(\mathbf{x}_n|\theta_d)$, where θ_d is a design value, that is a prefixed guess value for θ that describes the design scenario. Since θ_d represents the target effects-difference to be detected, in superiority trials it is convenient to set $\theta_d > \delta$. In this paper we follow this *conditional approach*. As an alternative, one could resort to the so-called *predictive approach* that relies on the specification of an additional prior density $\pi_d(\theta)$ instead of the single guess value θ_d and models uncertainty on the parameter (see [10] and [2] for details). By averaging the sampling distribution $f(\mathbf{x}_n|\theta)$ with respect to $\pi_d(\theta)$ we obtain the predictive distribution $m_d(\mathbf{x}_n) = \int_\Theta f(\mathbf{x}_n|\theta)\pi_d(\theta)d\theta$. It is straightforward to check that $f(\mathbf{x}_n|\theta_d)$ arises as a special case of $m_d(\mathbf{x}_n)$ when a point-mass design prior distribution on the single value θ_d is chosen. In this sense the conditional approach is a special case of the predictive one.

3 Bayesian Analysis with Normal Distributions

Assume now that data relevant to θ are summarized by a statistic Y_n with (at least approximately) normal distribution of parameters $(\theta, \sigma^2/n)$. This model can be used in phase III two-arms clinical trials in one of the following setups (see [9]):

- θ is the true difference in mean responses and y_n is the difference of the sample means in the two arms (normal data);
- θ is the log odds ratio (log OR) and y_n is the sample log OR (binary data);
- θ is the log hazard ratio and y_n is a function of the log-rank test statistics (survival data).

Specifically we assume that σ^2 is known. If we set $\sigma = 2$, n can be interpreted as the effective number of observations (for instance in the second and third cases n is the total number of events occuring in the two arms). We assume that $\pi_j(\theta)$ is a normal density of mean μ_j and variance σ^2/n_j, where n_j is the prior sample size, $j = a, b$. From standard conjugate analysis the posterior density of θ is normal with parameters $\bar{\mu}_j = (ny_n + n_j\mu_j)/(n + n_j)$, $\bar{\sigma}_j = \sigma/(n + n_j)$, $j = a, b$. In order to define S_n according to Eq. (1) note that $\mathbb{P}_j(\theta > \delta | y_n) = 1 - \Phi\left[(\delta - \bar{\mu}_j)/\bar{\sigma}_j\right]$, $j = a, b$, where $\Phi(\cdot)$ is the standard normal cdf. It is easy to check that the Hellinger distance between the two posteriors is

$$d[\pi_a(\cdot|\mathbf{x}_n), \pi_b(\cdot|\mathbf{x}_n)] = \left(1 - \sqrt{\frac{2\bar{\sigma}_a\bar{\sigma}_b}{\bar{\sigma}_a^2 + \bar{\sigma}_b^2}} \cdot \exp\left\{-\frac{1}{4}\frac{(\bar{\mu}_a - \bar{\mu}_b)^2}{\bar{\sigma}_a^2 + \bar{\sigma}_b^2}\right\}\right)^{\frac{1}{2}},$$

which yields the expression of C_n given by Eq. (3).

Remark The values of d and C_n depend crucially on the difference $|\bar{\mu}_a - \bar{\mu}_b|$ as well as on n_a and n_b. To have a better insight consider the special case $n_a = n_b = n_o$, which occurs when the precision of the two sources of prior information is the same. In this case, since $\bar{\sigma}_j^2 = \bar{\sigma}_o^2 = \sigma^2/(n + n_o)$, then $d = \sqrt{1 - \exp\left\{-\frac{1}{2}\left(\frac{\bar{\mu}_a - \bar{\mu}_b}{2\bar{\sigma}_o}\right)^2\right\}}$ is a monotone function of the effect size $\frac{|\bar{\mu}_a - \bar{\mu}_b|}{\bar{\sigma}_o}$ ranging in [0, 1], that is a relative measure of the discrepancy between the two posterior means. More specifically, $d = \sqrt{1 - \exp\left\{-\frac{1}{8}\left(\frac{\mu_a - \mu_b}{\sigma}\right)^2 \cdot \frac{n_o^2}{n + n_o}\right\}}$ which tends to 0 as $n \to \infty$ as fast as $e^{-\frac{c}{n}}$. However, the convergence is influenced by the "prior" effect size $\frac{|\mu_a - \mu_b|}{\sigma}$ and the prior sample size n_o. For a given fixed non negligible effect size, the larger n_o with respect to n the slower the convergence to 0.

Following the conditional approach, determination of the optimal sample size according to (4) requires the computation of e_n^s and e_n^c with respect to the sampling distribution of $Y_n|\theta_d$ that is $N(\theta_d, 4/n)$. For each n the values of e_n^s and e_n^c are obtained as Monte Carlo approximations.

4 Example

In this section we consider an example inspired by the real data application of [1] in which coronary artery bypass graft (CABG) is compared with percutaneous coronary intervention (PCI) in diabetic patients with multivessel coronary artery disease in terms of survival (here we assume that log OR > 0 favours CABG with respect to PCI). In the original paper a Bayesian analysis is performed: the prior distribution for the log OR, based on a metanalysis of 8 historical trials, is combined with the likelihood of the data of the FREEDOM trial. The posterior distribution of the log OR shows evidence in favour of CABG (e.g. posterior mean 0.545 with 95% credible interval $[0.342, 0.734]$ corresponding to a reduction in mortality risk between 29%–52%). Here we assume this posterior as one of the two prior distributions, say π_a, e.g. a normal density of parameters $\mu_a = 0.545$, $n_a = 376$, to be used in the design of a new hypothetical trial. As an alternative in [1] the Authors also consider a skeptical prior yielding a second posterior distribution for the log OR representing a milder preference towards CABG with respect to PCI (e.g. posterior mean 0.198 with 95% credible interval $[0, 0.4]$, corresponding to a reduction in mortality risk between 0%–23%). We take this posterior as the prior distribution π_b, e.g. a normal density of parameters $\mu_b = 0.198$, $n_b = 390$. In addition, in order to define superiority in terms of the posterior probability that $\theta > \delta$, we set $\delta = 0.36$ corresponding to a 30% reduction in mortality in the experimental treatment arm. Finally, as described in Sect. 2 our proposed method requires the specification of a design scenario. Here we consider $\theta_d = 0.69$, which corresponds to a 50% reduction in mortality due to CABG.

Figure 1 panel (a) represents the plots of e_n^s and e_n^c as functions of n with the prior distributions defined above. The sample size required to reach the success threshold $\epsilon_s = 0.9$ is $n_s = 476$, whereas for a sample size as large as 1000 the average consensus e_n^c is at most slightly larger than 0.4. The small values of e_n^c depend on the large values of the prior sample sizes n_a and n_b as discussed in the final remark of Sect. 3. In order to have a better insight on the crucial effect of n_a and n_b on both consensus and evidence measures, we consider the same prior means μ_a and μ_b but we assume the prior sample sizes to be discounted by a factor $\gamma \in (0, 1)$. As an example in panel (b) we set $\gamma = 0.1$ and we obtain $\gamma n_a = 37.6$ and $\gamma n_b = 39$. In this case the values of e_n^c are remarkably larger than those of panel (a), yielding $n_c = 546$ for a threshold $\epsilon_c = 0.9$. With respect to panel (a), the values of e_n^s are larger as well, yielding a much smaller $n_s = 171$ for the same $\epsilon_s = 0.9$. This depends on the minor impact of the more skeptical prior which is decisive in determining the numerical values of S_n in (1) since $\mathbb{P}_a(\theta > \delta | y_n) > \mathbb{P}_b(\theta > \delta | y_n)$. For comparison, in Fig. 1 we also consider a non-informative prior for π_b ($n_b = 0$) in contrast with the same π_a (panel c) and its discounted version (panel d). Note that the non-informative prior is less skeptical towards the new treatment than the original π_b. As a consequence both the values of e_n^s and e_n^c in panel (c) are larger than those observed in panel (a). The optimal value n_c of panel (d) is smaller than that of panel (b), due to the minor distance between the posteriors induced by the

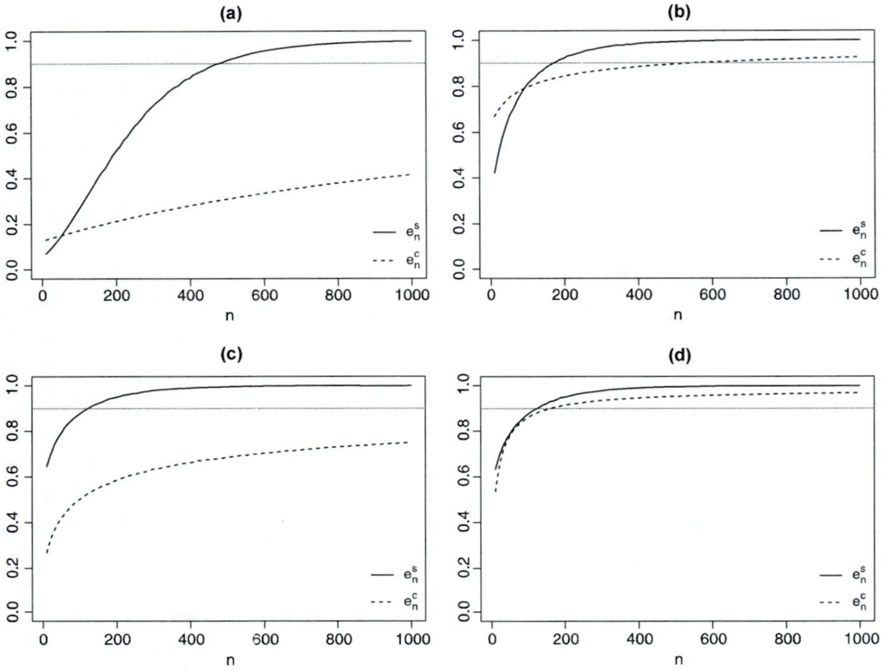

Fig. 1 Plots of e_n^s and e_n^c as functions of n for two alternative choices of the two priors π_a and π_b: (**a**) $\mu_a = 0.545$, $n_a = 376$ $\mu_b = 0.198$, $n_b = 390$; (**b**) $\mu_a = 0.545$, $n_a = 37.6$ $\mu_b = 0.198$, $n_b = 39.0$. (**c**) $\mu_a = 0.545$, $n_a = 376$, $nb = 0$; (**d**) $\mu_a = 0.545$, $na = 37.6$, $nb = 0$. The optimal sample size thresholds are $\epsilon_s = \epsilon_c = 0.9$

non-informative prior and the discounted π_a. Note that the values of e_n^s in panels (c) and (d) are the same since $\mathbb{P}_a(\theta > \delta|y_n) > \mathbb{P}_b(\theta > \delta|y_n)$ and therefore they only depend on the non-informative prior in both cases. Table 1 summarizes the optimal sample sizes in the different four prior setups considered for several choices of the thresholds ϵ_s and ϵ_c (for simplicity we set $\epsilon_s = \epsilon_c$). Due to the behavior of e_n^s and e_n^c increasing with n, the larger the thresholds the larger the optimal sample sizes. For illustrative purposes, we fix here a maximum sample size as large as 1000. When the prior sample sizes are not discounted (see setups (a) and (c)) it is very hard to reach posterior consensus, unless we consider a much smaller threshold (e.g. $\epsilon_c = 0.7$ in setup (c)). As a consequence, the optimal sample size based on criterion (4) is also unfeasible (i.e. n^* is larger than 1000). Conversely, when prior information is downweighted (see setups (b) and (d)) the number of observations required to simultaneously achieve consensus and evidence seems to be reasonably moderate. Notice that it may also happen that $n_c < n_s$ (e.g. setup (b) with $\epsilon_s = \epsilon_c = 0.7$) and therefore $n^* = n_s$.

Table 1 Optimal sample sizes for several choices of the thresholds $\epsilon_s = \epsilon_c$, given the following alternative choices of the priors π_a and π_b: (a) $\mu_a = 0.545$, $n_a = 376$ $\mu_b = 0.198$, $n_b = 390$; (b) $\mu_a = 0.545$, $n_a = 37.6$ $\mu_b = 0.198$, $n_b = 39.0$. (c) $\mu_a = 0.545$, $n_a = 376$, $n_b = 0$; (d) $\mu_a = 0.545$, $n_a = 37.6$, $n_b = 0$

$\epsilon_c = \epsilon_s$	0.9	0.8	0.7
(a)			
n_s	476	363	292
n_c	> 1000	> 1000	> 1000
$n^* = \max\{n_s, n_c\}$	> 1000	> 1000	> 1000
(b)			
n_s	171	98	60
n_c	546	104	22
$n^* = \max\{n_s, n_c\}$	546	104	60
(c)			
n_s	120	53	21
n_c	> 1000	> 1000	574
$n^* = \max\{n_s, n_c\}$	> 1000	> 1000	574
(d)			
n_s	119	55	21
n_c	159	57	28
$n^* = \max\{n_s, n_c\}$	159	57	28

5 Conclusions

The present paper deals with a sample size determination criterion defined to achieve two main goals: evidence (with respect to the aim of the trial, i.e. superiority) and consensus (in the presence of more than one source of prior information). Some comments are in order.

1. The main motivation of the proposed approach is that the optimal sample size based on a single goal might be inadequate to guarantee the other one as well. For instance, as shown by numerical examples (see Sect. 4), the sample size based on evidence only might be too small to provide consensus as well, i.e. $n_s < n_c$. In other examples the opposite situation can be observed.
2. The measure of consensus defined in Eq. (3) is based on the overall discrepancy between the posterior distributions induced by the competing priors. This releases us from selecting an arbitrary posterior summaries as in [7] and [4].
3. The choice of the design value θ_d and of thresholds involved in the criterion in general strongly influences the resulting sample sizes, but it is problem-specific.
4. We investigated the crucial impact of n_a and n_b on the consensus measure and therefore on n_c and n^*. Specifically if the prior sample sizes attached to the two competing priors are excessively large, then the number of observations required to reach consensus can be unaffordable.

5. In the normal model considered in this paper all the posterior quantities are available in closed-form formulae that allow a straightforward interpretation. In more general setups lack of available analytical results can be supplied by standard simulations, that are used here for the predictive analysis only.

Finally, here is a list of potential further developments.

1. In order to take into account the joint preposterior variability of (S_n, C_n) we can replace the expectation criterion with the probability criterion (see [2]).
2. Implementation of the sample size criteria based on the joint control of consensus and evidence can be extended to alternative models relevant to applications in clinical trials, such as models for binary, count and survival data.
3. Success of the experiment is formalized for superiority trials. In the future we plan to extend the methodology to experiments with different goals, such as non-inferiority or equivalence studies.
4. We would like to generalize this approach to experiments with multiple sources of prior information.
5. The need of discounting historical information mentioned in point 4 of the previous list of comments, might be addressed resorting to the power prior methodology (see [6]).

References

1. Bittl, J.A., He, Y.: Bayesian analysis. A practical approach to interpret clinical trials and create clinical practice guidelines. Circ. Cardiovasc. Qual. Outcomes **10**(8), e003563 (2017)
2. Brutti, P., De Santis, F., Gubbiotti, S.: Bayesian frequentist sample size determination: A game of two priors. Metron **72**(2), 133–151 (2014)
3. CDHR/FDA: Guidance for the Use of Bayesian Statistics in Medical Device Clinical Trials. Guidance for industry and FDA staff (2010)
4. De Santis, F., Gubbiotti, S.: Joint control of consensus and evidence in Bayesian design of clinical trials. Biom. J. (2021) https://doi.org/10.1002/bimj.202100035
5. De Santis, F., Gubbiotti, S.: A dynamic power prior for Bayesian non-inferiority trials. Submitted (2021)
6. Ibrahim, J.G., Chen, M.H., Gwonb, Y., Chenc, F.: The power prior: theory and applications. Stat. Med. **34**, 3724–3749 (2015)
7. Joseph, L., Belisle, P.: Bayesian consensus-based sample size criteria for binomial proportions. Stat. Med. **38**(23), 4566–4573 (2019)
8. Ollier, A., Morita, S., Ursino, M., Zohar, S.: An adaptive power prior for sequential clinical trials - Application to bridging studies. Stat. Methods Med. Res. **29**(8), 2282–2294 (2020)
9. Spiegelhalter, D.J., Abrams, K.R., Myles, J.P.: Bayesian Approaches to Clinical Trials and Health-Care Evaluation. Wiley, New York (2004)
10. Wang, F., Gelfand, A.E.: A simulation-based approach to Bayesian sample size determination for performance under a given model and for separating models. Stat. Sci. **17**(2), 193–208 (2002)
11. Weerahandi, S., Zidek, J.V.: Multi-Bayesian statistical decision theory. J. R. Stat. Soc. A 144(1), 85–93 (1981)

Novel Applications of the Team Orienteering Problem in Health Care Logistics

Roberto Aringhieri, Sara Bigharaz, Davide Duma, and Alberto Guastalla

Abstract The team orienteering problem is a routing problem belonging to the class of the vehicle routing problems with profits. We present two problems arising in the health care logistics that are modelled as team orienteering problem. To the best of our knowledge, these are the first applications to health care logistics problems. The former is a problem arising in the digital contact tracing system as a measure for the containment of the Covid-19 pandemic. The latter is a problem arising in post-disaster management to transport the injured to hospitals. We discuss the novelty of some of their features with respect to the current literature. We present and discuss the mathematical formulation of a new variant of the team orienteering problem that includes such new features.

Keywords Team orienteering problem · Health care logistics · Fairness

1 Introduction

The Team Orienteering Problem (TOP) [6, 7] is a routing problem belonging to the class of the Vehicle Routing Problems with Profits. The TOP is characterised by the fact that not all customers can be served. This implies the need to consider two different decisions [2], that is (i) which customers to serve, and (ii) how to cluster

R. Aringhieri (✉)
Department of Computer Engineering, University of Torino, Torino, Italy
e-mail: roberto.aringhieri@unito.it

A. Guastalla
Computer Science Department, University of Turin, Turin, Italy

S. Bigharaz
Department of Industrial Economics and Technology Management, Faculty of Economics and Management, Trondheim, Norway

D. Duma
Department of Mathematics "Felice Casorati", University of Pavia, Pavia, Italy

L. Amorosi et al. (eds.), *Optimization in Artificial Intelligence and Data Sciences*, AIRO Springer Series 8, https://doi.org/10.1007/978-3-030-95380-5_21

the customers to be served in different routes (if more than one) and order the visits in each route. The customer selection is driven by a profit associated with each customer that makes such a customer more or less attractive.

To the best of our knowledge [11, 16, 17], the TOP framework is never applied to the modelling and the solution of health care logistics problems. We present two problems arising in the health care logistics. The former is a problem arising in the digital contact tracing system as a measure for the containment of the Covid-19 pandemic, that is the Daily Swab Test Collection Problem (DSTCP) [3]. The latter is a problem arising in post-disaster management, that is the Ambulance Routing Problem (ARP) to transport the injured to hospitals [4]. We present an integer linear programming formulation for the DSTCP while a graph formulation is presented for the ARP. We discuss the novelty of some of their features with respect to the current literature. Finally we present and discuss the mathematical formulation of a new variant of the TOP that includes such new features. Conclusions close the work.

2 The Daily Swab Test Collection Problem

Basically the DSTCP consists in organising the daily collection of swab tests reaching the house of the contact(s) of a positive case detected the day(s) before in accordance with the framework validated in [9] and depicted in Fig. 1. A set of teams are in charge of collecting the swab tests. We suppose that the number of required tests are larger than the daily capacity of all teams in terms of working time. We would remark that this hypothesis represents the situation in which the pandemic is rapidly spreading over a given geographic area. Each team travels around the city collecting the swab tests. The selection of which tests should be collected is driving by a priority associated to each person in accordance with her/his health status and social connections. The priority represents the need of testing some people before other since they could become a spreader of the virus and/or they belong to more risky class of people (e.g., elderly and/or frail people). The priority accounts also the time spent waiting for the swab for those people not selected the days before. Finally, time is crucial since we have to take into account both travel times and service times for collecting the swab(s) at home.

Let $P = \{1, \dots n\}$ be a set of places where a number b_p ($p \in P$) of swab tests should be collected. The collection of the swab tests follows an integer priority r_p: the greater the priority is, the greater the importance of collecting such a swab test is. After reaching a place $p \in P$, we suppose to have an estimate of the time t_p^+ and t_p^- respectively needed to dress and to undress the *personal protective equipment*, the time t_p^h to reach the house (which can not be negligible in the case of large buildings), and the time t_p^s to collect a single swab test. Therefore, the overall time t_p needed to perform all the operations needed to collect a swab test at the place $p \in P$ is equal to $t_p = t_p^+ + 2\,t_p^h + (b_p\, t_p^s) + t_p^-$.

Let $M = \{1, \dots k\}$ be a set of medical teams in charge of collecting swab tests during their work-shift whose maximum duration is equal to $t_{\max}$. The teams start their work-shift from a *depot* 0 ending at the laboratory $n + 1$. Depot and laboratory

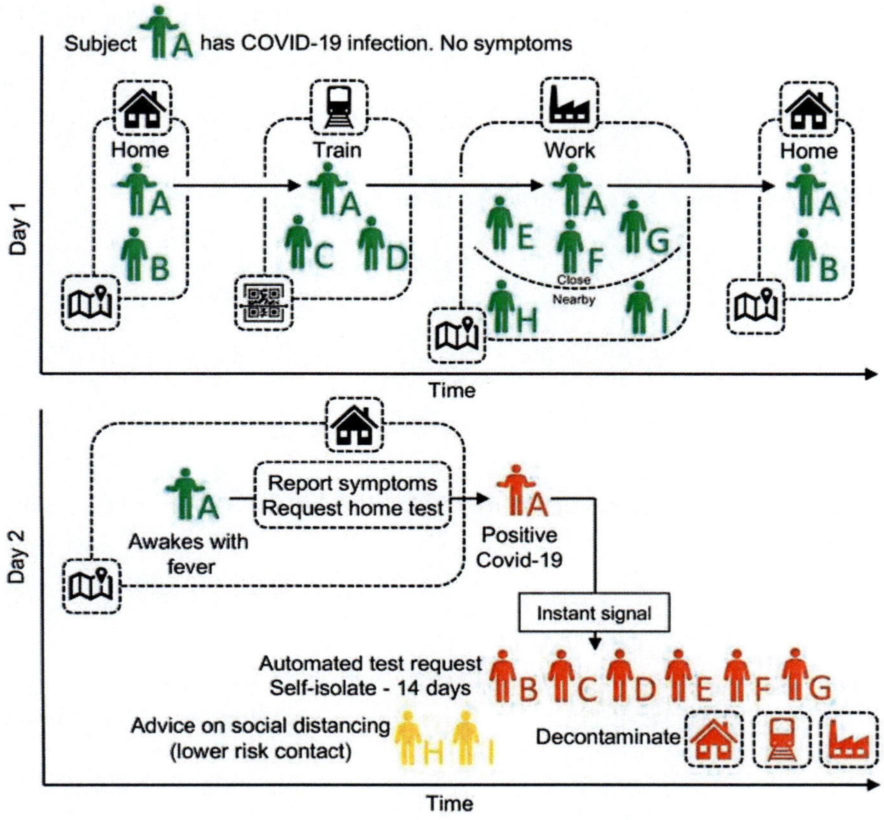

Fig. 1 DSTCP: description of digital contact tracing (source [9])

could be the same place. Considering $P^+ = P \cup \{0, n+1\}$, let t_{pq} be the travelling time to reach $q \in P^+$ from $p \in P^+$.

We are now ready to propose the integer linear programming model for the DSTCP. Let us introduce the following decision variables:

- $x_{pqm} = 1$ if the team $m \in M$ visits the place $q \in P^+$ after visiting the place $p \in P^+$, 0 otherwise;
- $y_{pm} = 1$ if the place $p \in P^+$ is visited by team $m \in M$, 0 otherwise;
- u_{pm} is an integer representing the position of the place $p \in P^+$ in the path of the team $m \in M$.

The objective function seeks to maximise the overall priority of the swab tests collected:

$$\max z = \sum_{m \in M} \sum_{p \in P} r_p \, b_p \, y_{pm}. \tag{1}$$

Constraint 2 ensures that each team starts their work-shift from the depot ending at the laboratory:

$$\sum_{m \in M} \sum_{q \in P} x_{0qm} = \sum_{m \in M} \sum_{p \in P} x_{p(n+1)m} = k. \tag{2}$$

Constraints 3 ensure that every place is visited at most once:

$$\sum_{m \in M} y_{pm} \leq 1, \qquad p \in P. \tag{3}$$

Constraints 4 guarantee the connectivity of the work-shift of each medical team:

$$\sum_{q \in P \cup \{0\}} x_{qpm} = \sum_{q \in P \cup \{n+1\}} x_{pqm} = y_{pm}, \qquad p \in P, m \in M. \tag{4}$$

Constraints 5 ensure that the duration of each work-shift is less than or equal to the maximum duration:

$$\sum_{p \in P \cup \{0\}} \sum_{q \in P \cup \{n+1\}} t_{pq} x_{pqm} + \sum_{p \in P} t_p y_{pm} \leq t_{\max}, \qquad m \in M. \tag{5}$$

Finally, the constraints 6 and 7 are necessary to prevent subtours in accordance with the Miller-Tucker-Zemlin formulation for the Travelling Salesman Problem (TSP) [13]:

$$2 \leq u_{pm} \leq n + 2, \qquad p \in P \cup \{n + 1\}, m \in M. \tag{6}$$

$$u_{pm} - u_{qm} + 1 \leq (n + 1)(1 - x_{pqm}), \qquad p, q \in P \cup \{n + 1\}, m \in M. \tag{7}$$

We remark that the above formulation is an adaptation of that presented in [17].

3 Ambulance Routing Problem

When a disaster occurs, initial data about the damages and injuries is collected as fast as possible. The dispatcher classifies patients' requests according to their severity and locations. The ambulances staffed by medical crew are dispatched to affected areas immediately to treat wounded people and transport patients to hospitals as needed. Two groups of patients based on a triage system can be considered in the affected area [15]. *Red* patients should be transported to hospitals because they suffer from serious injuries. *Green* patients are people who are slightly injured and need first aid directly in the field, which can be provided by the ambulances or by other rescue services.

Priority scores for green patients are introduced to define which of them is less or more urgent, that is a preference about who should be rescued in a short time. Therefore, each green patient has a score that can be interpreted as urgency level weights. Maximising the overall score for green patients can be considered as a goal to provide an efficient service. As travel time can provide the limitation for each ambulance, all the green patients cannot be visited and those with the higher scores should have the priority. To satisfy the needs of patients, an ambulance starts its tour from a depot (e.g., hospitals or medical centres) to visit patients in affected areas and returns to a hospital. The personnel of the ambulance treats a green patient in the field and visit another green patient in its tour while after visiting the red patient, the ambulance have to pick the patient and take her/him to a hospital.

The ARP based on the TOP can be formulated to find the optimal tour for a number of ambulances in the following graph to deliver the services to patients in a disaster.

Note that in the TOP each node can be visited once except for the source node and the destination node. Then, we can define a network on a complete graph $E = (N, A)$, where N is the set of nodes representing locations over the considered area and $A = (i, j) : i, j \in N$ is the set of arcs indicating the connections between each pair of such locations. Three main types of nodes are considered in this problem: (i) green patients (set G), (ii) red patients (set R), and (iii) actual hospitals (set O). To these node types, we add several dummy nodes in order to bring the problem in the TOP framework, that is: (iv) the dummy source and destination depot nodes 1 and n, and (v) several dummy hospitals (set D), which are replications of the actual hospital nodes (i.e. they have same coordinates of the original nodes of the set O) that allows us to visit two or more times the same actual hospital visiting two different nodes. The number of replications of each actual hospital node is computed in such a way to allow the visit of such nodes as starting hospital for the assigned ambulances and for the medical care of all the red patients (worst case). Furthermore, we indicate with $P = G \cup R$ the set of all patients and with $H = O \cup D$ the set of all (actual and dummy) hospitals. Therefore, $N = P \cup H \cup \{1, n\}$. We set $n = |N|$ and we enumerate all nodes from 1 to n. The priority score of green patients is denoted by s_i, while the set of available ambulances is indicated with K. An example of the introduced graph and the related solution is illustrated in Fig. 2.

Nodes of sets G, R and O are coloured in green, red and white, respectively. Labels on green nodes indicate the scores $s_i, i \in G$. In correspondence of each node of O, several replications are provided in order to suit our problem to the TOP framework; such nodes, in grey, are the elements of D. Black nodes represent the dummy depots $\{1, n\}$. Two ambulance tours are represented by the sequence of coloured arcs (orange and blue) from the source depot to the destination depot: continuous arcs indicate the actual moving of the ambulance between two physical places represented by the nodes, while dashed arcs are only used to connect the dummy depots to the actual hospital where the ambulance is located at the beginning and at the end. The connections of the node 1 are set in accordance with the instance of the problem: the first arc visited by the tour of an ambulance is that between 1 and one of the nodes related to the hospital in which it actually is located at the

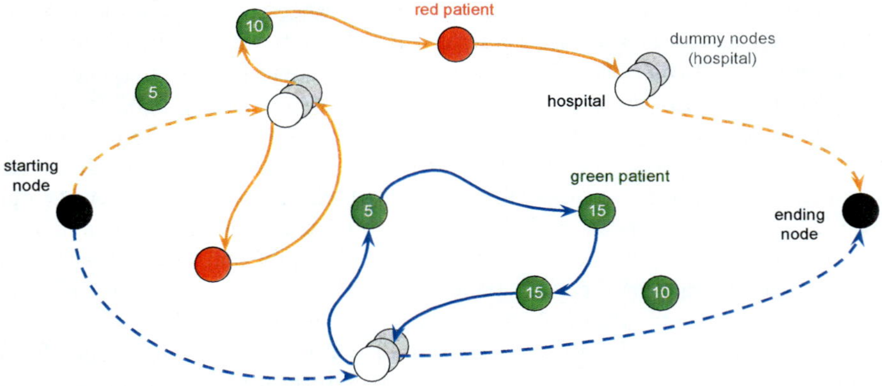

Fig. 2 ARP: efficient solution representing the ambulance tours in a disaster

beginning. Such an initial location is given by the parameter l_{hk}, which is equal to 1 if the ambulance $k \in K$ is initially parked at the hospital. The connections of the node n have the same meaning, but they are part of the solution because there is not any constraint about the final destination of the ambulances (except that it is on a hospital node). Furthermore, we observe that each intermediate node is visited at most once: if a hospital needs to be visited twice or more, then a different node belonging to H, and related to that hospital, is visited every time; for instance, the orange tour visits the actual hospital node firstly and a dummy node for the second visit. We remark that red nodes must be visited once, while green, white and grey nodes could be not visited.

Since the locations of patients and hospitals are identified, the travel time between a pair of nodes $i, j \in N$ is defined as t_{ij}, where:

- $t_{ij} = t_{ji}$ for each $i, j \in N$;
- $t_{ij} = 0$ if $i \in \{1, n\}, j \in H$;
- $t_{ij} = 0$ if $i, j \in H$ are replications of the same actual hospital.

We take into account also the time spent by the ambulance on the nodes, to which we will always refer as the service time f_i, $i \in N \setminus \{1, n\}$. Such time represents: (i) the on-place treatment duration when $i \in G$, (ii) the preparation time spent on the place when $i \in R$, and (iii) the time spent to release a red patient to the hospital when $i \in H$. We assume that hospitals can accept red patients regardless of the hospital capacity, and an ambulance can only carry one red patient at a time without giving other patients treatment in the meantime. We fix a maximum time T_{max} for each ambulance to provide treatments to patients and to complete its tour moving to a hospital.

4 A New Variant of the Team Orienteering Problem

Erdogan and Laporte [8] introduces the Orienteering Problem with Variable Profits (OPVP) in which a single vehicle can collect the whole profit at the customer after a discrete number of "passes" or spending a continuous amount of time. As in the classical orienteering problem, the objective is to determine a maximal profit tour for the vehicle, starting and ending at the depot, and not exceeding a travel time limit. An attempt to extend this approach to the TOP is reported in [18] in which the score of visiting an attraction is different depending on the time of visit.

Salazar-Aguilar et al. [1] introduce an extension of the TOP by considering the multi-district aspect, a set of mandatory and optional tasks located in several districts and some incompatible tasks which cannot be carried out during the same day. The problem is called the Multi-District TOP (MDTOP). It is required to perform all mandatory tasks over the planning horizon, while the optional tasks are only executed if time permits.

The DSTCP and the ARP have service times at nodes as in [8] while the ARP has mandatory and optional nodes as in [1]. Similarly to [1], the ARP has subset of incompatible nodes, that is a subset of nodes from which is not possible to reach a different subset of nodes. To the best of our knowledge, this is the first research attempt to consider all these aspects at the same time. This attempt can be view as a new variant of the TOP, that we call TOP with Service Time and Mandatory and Incompatible Nodes (TOP-ST-MIN).

Let $P = \{1, \ldots n\}$ be a set of nodes to which is associated a score r_p: the greater the score is, the greater the importance of serving such a node is. For each node $p \in P$, we suppose to have an estimate of the time t_p required to serve the corresponding node. Let $P^* \subset P$ be the set of mandatory nodes, that is those nodes that must be visited by a team. Finally, let $C = [c_{pq}]$ be the compatibility matrix, that is $c_{pq} = 0$ when the pair of nodes $p, q \in P$ are incompatible, $c_{pq} = 1$ otherwise.

Let $M = \{1, \ldots k\}$ be a set of teams in charge of serving the nodes during their tour whose maximum duration is equal to $t_{\max}$. The teams start their tour from the *starting depot* 0 ending at the *ending depot* $n + 1$. Starting and ending depots could be the same node. Considering $P^+ = P \cup \{0, n + 1\}$, let t_{pq} be the travelling time to reach $q \in P^+$ from $p \in P^+$.

We are now ready to propose the integer linear programming model for the TOP-ST-MP. Let us introduce the following decision variables (similarly to those for the DSTCP):

- $x_{pqm} = 1$ if the team $m \in M$ visits the node $q \in P^+$ after visiting the node $p \in P^+$, 0 otherwise;
- $y_{pm} = 1$ if the place $p \in P^+$ is visited by team $m \in M$, 0 otherwise;
- u_{pm} is an integer representing the position of the node $p \in P^+$ in the path of the team $m \in M$.

$$\max z = \sum_{m \in M} \sum_{p \in P} r_p y_{pm}, \tag{8a}$$

$$\sum_{m \in M} \sum_{p \in P} x_{0qm} = \sum_{m \in M} \sum_{p \in P} x_{p(n+1)m} = k, \tag{8b}$$

$$\sum_{m \in M} y_{pm} \leq 1, \quad p \in P, \tag{8c}$$

$$\sum_{m \in M} y_{pm} = 1, \quad p \in P^*, \tag{8d}$$

$$\sum_{q \in P \cup \{0\}} x_{qpm} = \sum_{q \in P \cup \{n+1\}} x_{pqm} = y_{pm}, \quad P \in P, \ m \in M, \tag{8e}$$

$$\sum_{q \in P \cup \{0\}} \sum_{q \in P \cup \{n+1\}} t_{pq} x_{pqm} + \sum_{q \in P} t_p y_{pm} \leq t_{\max}, \quad m \in M, \tag{8f}$$

$$x_{pqm} \leq c_{pq}, \quad p, q \in P, \ m \in M, \tag{8g}$$

$$2 \leq u_{pm} \leq n + 2; \quad p \in P \cup \{n+1\}, \ m \in M, \tag{8h}$$

$$u_{pm} - u_{qm} + 1 \leq (n-1)(1 - x_{pqm}) \quad p, q \in P \cup \{n+1\}, \ m \in M, \tag{8i}$$

The mathematical model is similar to that presented in Sect. 2: actually, the constraints (8b), (8c), (8e), (8f), (8h) and (8i) have the same meaning of the corresponding constraints for the DSTCP. Constraints (8d) ensure that the compulsory nodes are visited by the tour of one team while constraints (8g) models the incompatibility among subsets of nodes.

The above formulation makes use of a compatibility matrix to model the information regarding the incompatibility among subsets of nodes. We are aware that this can result in a not efficient mathematical formulation but this choice seems to be the most general and suitable to represent several types of compatibility arising in the analysis of real problems.

Although an example of incompatibility is depicted for the ARP in Sect. 3, a further example come out of an adaptation of the Black-and-White Travelling Salesman Problem (BWTSP) [5, 10]. The BWTSP is defined on an undirected graph where the set of nodes is partitioned into two sets of nodes, the set of black nodes B and the set of white nodes W. Each edge $e \in E$ is associated with a cost and a distance. The objective is to find a TSP tour with minimum total length in which the path between every two consecutive black nodes contains at most Q white nodes and has a distance of at most L.

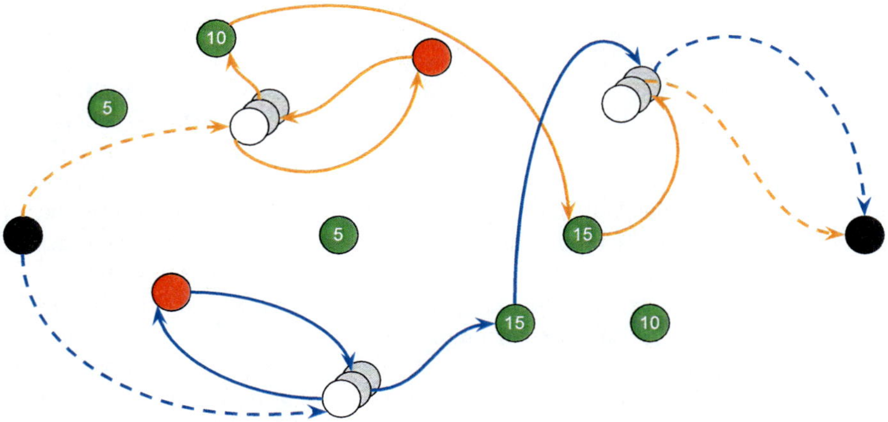

Fig. 3 ARP: fairness solution representing the ambulance tours in a disaster

4.1 Fairness

As discussed in Sect. 3, the objective function that maximises the overall green patient score is a goal to provide an efficient service. On the contrary, red patients are suffering from serious injuries and should be transferred as soon as possible to hospitals in such a way to receive the most appropriate treatment with an adequate medical equipment. From this perspective, minimising the maximum waiting time for red patients can be considered as a goal to provide fair services.

Considering the solution depicted in Fig. 2, we can observe that in the orange tour a red patient is visited after a green patient. Such a tour can not be an optimal solution using a fairness objective function: as a matter of fact, a change of the visit order of those patients can minimise the value of the maximum waiting time. Generally, we expect that setting a fairness objective function, red patients are all visited at the beginning of the tours, as depicted in Fig. 3.

As discussed in [12, 14], modelling fairness is in itself challenging since the concept of fairness can vary in accordance with the context of problem. This results in modelling approaches determining solutions with different quality. Examples of different concepts of fairness are (i) the Rawlisan approach in which usually the focus is on maximising the minimum outcome, (ii) the fairness recognised as deviation measure, and (iii) the deprivation approach, which is less used due to its difficulty.

Including fairness in the TOP and in the TOP-ST-MIN could be a research area which should be further emphasised considering also the *price of fairness*, a quality measure introduced by Nicosia et al. [14], which is a comparison between the system optimum solution and a fair solution. Further, our experience [4] showed that including fairness results in a more challenging optimisation problem.

5 Conclusions

In this paper we report two problems arising in the health care logistics. The former is a problem arising in the digital contact tracing system as a measure for the containment of the Covid-19 pandemic, that is the Daily Swab Test Collection Problem (DSTCP) [3]. The latter is a problem arising in post-disaster management, that is the Ambulance Routing Problem (ARP) to transport the injured to hospitals [4]. After discussing the novelty of some of their features with respect to the current literature, we introduced a new variant of the TOP including in the TOP framework three features, that is (i) service times at nodes, (ii) mandatory and optional nodes, and (iii) subset of incompatible nodes. We call such a variant TOP with Service Time and Mandatory and Incompatible Nodes (TOP-ST-MIN). We also highlighted the importance of including the concept of fairness in such a new variant of the TOP.

References

1. Angelica Salazar-Aguilar, M., Langevin, A., Laporte, G.: The multi-district team orienteering problem. Comput. Oper. Res. **41**, 76–82 (2014)
2. Archetti, C., Speranza, M.G., Vigo, D.: Vehicle Routing Problems with Profits, chap. 10, pp. 273–297 (2014)
3. Aringhieri, R., Bigharaz, S., Druetto, A., Duma, D., Grosso, A., Guastalla, A.: The daily swab test collection problem (2021). Submitted for publication
4. Aringhieri, R., Bigharaz, S., Duma, D., Guastalla, A.: Fairness in ambulance routing for post disaster management. Cent. Eur. J. Oper. 30, 189–211 (2022)
5. Bourgeois, M., Laporte, G., Semet, F.: Heuristics for the black and white traveling salesman problem. Comput. Oper. Res. **30**(1), 75–85 (2003)
6. Butt, S.E., Cavalier, T.M.: A heuristic for the multiple tour maximum collection problem. Comput. Oper. Res. **21**(1), 101–111 (1994)
7. Chao, I.M., Golden, B.L., Wasil, E.A.: The team orienteering problem. Eur. J. Oper. Res. **88**(3), 464–474 (1996)
8. Erdogan, G., Laporte, G.: The orienteering problem with variable profits. Networks **61**(2), 104–116 (2013)
9. Ferretti, L., Wymant, C., Kendall, M., Zhao, L., Nurtay, A., Abeler-Dörner, L., Parker, M., Bonsall, D., Fraser, C.: Quantifying sars-cov-2 transmission suggests epidemic control with digital contact tracing. Science **368**(6491), eabb6936 (2020)
10. Gouveia, L., Leitner, M., Ruthmair, M.: Extended formulations and branch-and-cut algorithms for the black-and-white traveling salesman problem. Eur. J. Oper. Res. **262**(3), 908–928 (2017)
11. Gunawan, A., Lau, H.C., Vansteenwegen, P.: Orienteering problem: A survey of recent variants, solution approaches and applications. Eur. J. Oper. Res. **255**(2), 315 – 332 (2016)
12. Karsu, O., Morton, A.: Inequity averse optimization in operational research. Eur. J. Oper. Res. **245**(2), 343–359 (2015)
13. Miller, C.E., Tucker, A.W., Zemlin, R.A.: Integer programming formulation of traveling salesman problems. J. ACM **7**(4), 326–329 (1960)
14. Nicosia, G., Pacifici, A., Pferschy, U.: Price of fairness for allocating a bounded resource. Eur. J. Oper. Res. **257**(3), 933–943 (2017)

15. Talarico, L., Meisel, F., Sorensen, K.: Ambulance routing for disaster response with patient groups. Comput. Oper. Res. **56**, 120–133 (2015)
16. Vansteenwegen, P., Gunawan, A.: Orienteering Problems: Models and Algorithms for Vehicle Routing Problems with Profits. Springer Nature Switzerland AG (2019)
17. Vansteenwegen, P., Souffriau, W., Oudheusden, D.V.: The orienteering problem: A survey. Eur. J. Oper. Res. **209**(1), 1–10 (2011)
18. Yu, V.F., Jewpanya, P., Lin, S.W., Redi, A.P.: Team orienteering problem with time windows and time-dependent scores. Comput. Ind. Eng. **127**, 213–224 (2019)

Optimizing a Dynamic Outpatient Facility System with Multiple Servers

Beatrice Bolsi, Arthur Kramer, Thiago Alves de Queiroz, and Manuel Iori

Abstract The management of queues is a complex problem, and it requires special attention in dynamic environments where information changes over time. This work focuses on an outpatient facility system where patients are attended by identical parallel servers offering different services. Each patient requires service and expects to receive it within a given target time, after which, a tardiness is created. The objective of the problem is to minimize the total tardiness while defining which services each server will offer during the working hours. The arrival of patients is dynamic, and the server's configurations of services can be updated from time to time. To solve the problem, we propose a local search-based heuristic that locally assigns a configuration to each server based on the improvement reached in terms of total tardiness. The heuristic is tested on realistic instances, considering different settings, showing its superiority over the solution currently implemented on the facility system.

Keywords Outpatient facility system · Dynamic environment · Multi-server · Local-search heuristic · FIFO policy

---

B. Bolsi (✉) · M. Iori
Department of Sciences and Methods for Engineering, University of Modena and Reggio Emilia, Reggio Emilia, Italy
e-mail: beatrice.bolsi@unimore.it; manuel.iori@unimore.it

A. Kramer
Department of Production Engineering, Federal University of Rio Grande do Norte, Natal, RN, Brazil
e-mail: arthur.kramer@ufrn.br

T. A. de Queiroz
Institute of Mathematics and Technology, Federal University of Catalão, Catalão, GO, Brazil
e-mail: taq@ufcat.edu.br

L. Amorosi et al. (eds.), *Optimization in Artificial Intelligence and Data Sciences*, AIRO Springer Series 8, https://doi.org/10.1007/978-3-030-95380-5_22

1 Introduction

Queue management is a widely known problem in many real-world situations, from everyday shopping to care services. This problem has been addressed by a variety of methods, from simple queuing rules to sophisticated algorithms of operations research [10, 11]. The optimization of queues is particularly challenging for healthcare systems that include multiple queues, dynamic arrival of patients, random service times, different priorities, multiple servers, among others [12].

In this work, we are interested in optimizing the queue of patients in an outpatient facility system, which differs, for example, from an emergency department in a hospital or fast food service systems. Our problem is motivated by a health care facility application, where patients arrive dynamically during the system operating hours. We have a queue with patients waiting for service. We know the service each patient demands (e.g., cardiology, radiology, blood analysis, or even a ticket payment) and the expected time to perform the service. The system comprises a single queue, and multiple identical servers working in parallel, each fulfilling a list of services. The servers are equipped with queue management software that provides information about services and patients. The work originates from joint research with an industrial partner and concerns data coming from a large hospital booking facility (*Centro Unico di Prenotazione—CUP*).

The configuration (i.e., list of services) associated with each server can change during the time horizon, so the server by which patients will be served may change during the time. This change may help at reducing the patients' waiting time in queue, balancing herding and congestion. In addition, each patient has a desired target time, depending on the requested service, within which the patient should be served. Patients requiring the same service are assisted according to a first-in, first-out (FIFO) policy. The objective is to decide which services each server will offer through the time horizon, so the list of services can change from time to time, so as to minimize the total tardiness based on the patients' target times.

The literature about problems where the flow of patients in the healthcare system have to be optimized is extensive. The survey by Abdalkareem et al. [1] discusses the recent contributions on healthcare scheduling, focusing on the scheduling of patients, nurses, operation rooms, surgeries, and others. The authors comment on problem characteristics, mathematical formulations, available data sets, solution methods, and open challenges as, e.g., considering the dynamic nature of the problems, using big data analytic, and solving problems in an integrated way. Worthington et al. [12] focused on queuing models assuming an infinite number of servers. The authors consolidated and simplified the existing theories, providing generic pseudocode and identifying future opportunities regarding infinite-server models.

At the same time, many contributions regarding multiple servers and single/-multiple queuing systems can be found in the literature. Kaboudan [3] proposed a dynamic-server algorithm to handle a multi-server system that operates a single queue with all servers always offering the same services. The algorithm considers a threshold parameter to control the number of people in the queue and then periodically adjusts the number of open servers. People are served following a FIFO policy. Lan et al. [5] investigated a system with a finite number of servers and a capacitated queue in a fast-food company. The authors developed a heuristic to allocate workers to the servers and then balance the queue size using a cumulative probability function. Kumar and Omar [4] worked on queuing model to deal with a stress testing system in a semiconductor assembling industry. The authors used the mean value analysis to determine the mean waiting time and the throughput rate for a five-stage queue system with a re-entrant line at the second stage.

Slegers et al. [7] modeled a cluster computing system with multi servers and multi queues. Jobs are separated by type and have specific queues entering the system according to an independent interrupted Poisson process. The authors described the system with a Markov decision process. They presented four heuristic policies, two of them requiring a priori knowledge of the system and the others requiring none. Su et al. [8] focused on improving the operational efficiency of the registration process in a hospital. This registration system has three counters and three respective queues. Through simulation experiments, the authors redesigned the system to have a single queue and multiple servers with a prepared queue, reducing the patients' waiting time and then improving the service quality. In Vass and Szabo [9] the objective is to reduce the patients' waiting time. The authors used queuing models to evaluate an emergency department in Romania with more than 50,000 registered patients, concluding that it is necessary to implement a dynamic allocation of the medical staff. Recently, Harper [2] investigated the influence of service times and workload to measure servers' behaviors. The author assumed the service times depending on the queue length and used switching thresholds to have different service speeds, the presence of workload fatigue, and service breakdown.

For the problem under investigation, we propose a local search-based heuristic that updates the servers' list of services (configurations). Differently from the literature (e.g., [3, 5], servers can hold different configurations (i.e., offer different services) over the time horizon. The heuristic simulates the impact on the facility's total tardiness for each configuration, from time to time, assigning the best configuration to each server. The FIFO policy is used to schedule patients with the same service, and the service time is estimated from previous historical data. The heuristic is applied to solve realistic instances, and its results are comparatively better than the facility solutions for all scenarios we have tested.

The remainder of this work is structured as follows. In Sect. 2, we formally describe the problem, introducing notation, constraints, and the objective function. In Sect. 3, we present the heuristic and comment on its main steps. In Sect. 4, we introduce the realistic instances and detail the comparison performed under different configurations. Finally, in Sect. 5, we give some concluding remarks and point out some directions for continuing this research.

2 Problem Description

The problem considers a time horizon T over which we need to determine the configuration (i.e., list of services) to be dynamically assigned to M available servers. Each server initially receives a configuration, and within Δ units of time, the configuration can be updated to reduce the expected total tardiness. There is a queue of patients, and each patient requires a single service. In other words, each required service imposes the patient to be served by one from a subset of admissible servers. There is a set S of services, and each server m offers a subset of services (configuration), including the possibility of having no service at a given period (i.e., the server is closed). Besides that, servers may offer the same services and can perform all services if necessary.

Each patient $i \in I$ has an arrival time r_i, which is when i enters the queue and starts waiting for a demanded service $s_i \in S$ with an expected service time p_i, which depends on the service, and a due (target) time d_i, depending on the arrival time and the demanded service. Even if patients require different services, they are all part of a single queue. When a server calls a patient, the server expects to service the patient within p_i units of time, a value obtained from previous historical data. The service to patient i can be performed in the time interval $[t; t + p_i)$ only if there is at least one server whose list of services contains the service s_i in this time interval. The objective is to serve all patients minimizing the facility's total tardiness.

Patients with the same service are assisted by a FIFO policy shared among the servers offering such a service and managed by the facility's management software. If we consider that: (1) a list of services defines each server; (2) at each period there are at most M servers opened simultaneously in parallel; and (3) for each server, there is a number of possible configurations of services it offers; then our goal is to define which of these configurations are assigned to the servers at every period of Δ units of time in the horizon T, so as to minimize the total tardiness. Suppose t_i is the time at which patient i is called by a server, then the total tardiness is $\sum_{i \in I} \max(0; t_i - d_i)$.

We provide a mathematical model to the static version of the problem (i.e., when all data are precisely known in advance). The model uses two sets of binary variables. The first set, x_{ijt}, is defined for each patient $i \in I$, server $j \in J = \{1, 2, \ldots, M\}$, and time $t \in \{r_i, \ldots, T - p_i\}$. This variable takes value one if patient i starts being served by server j at time t, and zero otherwise. The second set, y_{jkt}, is defined for each server $j \in J$, configuration $k \in C$, where C is the set of configurations, and time $t \in \{0, \Delta, 2\Delta, \ldots, \lfloor T/\Delta \rfloor \Delta\}$. This variable assumes value one if configuration k is assigned to server j from time t to $t + \Delta - 1$, and zero otherwise. Therefore, we can define the mathematical model as follows:

$$\min \sum_{i \in I} \sum_{t=r_i}^{T-p_i} c_{it} \left(\sum_{j \in J} x_{ijt} \right) \tag{1}$$

subject to:

$$\sum_{j\in J}\sum_{t=r_i}^{T-p_i} x_{ijt} = 1 \qquad i \in I, \tag{2}$$

$$\sum_{i\in I}\sum_{s=t}^{t+p_i-1} x_{ijs} \leq 1 \qquad j \in J, t \in \{0,\ldots,T\}, \tag{3}$$

$$\sum_{k\in C} y_{jkt} = 1 \qquad j \in J, t \in \{0, \Delta, 2\Delta, \ldots, \lfloor T/\Delta \rfloor \Delta\}, \tag{4}$$

$$x_{ijt} \leq \sum_{k\in C_i} y_{jk\lfloor t/\Delta \rfloor \Delta} \qquad i \in I, j \in J, t \in \{r_i, \ldots, T - p_i\}, \tag{5}$$

$$y_{jkt} \in \{0, 1\} \qquad j \in J, k \in C, t \in \{0, \Delta, 2\Delta, \ldots, \lfloor T/\Delta \rfloor \Delta\}, \tag{6}$$

$$x_{ijt} \in \{0, 1\} \qquad i \in I, j \in J, t \in \{0, 1, \ldots, T\}. \tag{7}$$

Objective function (1) seeks the minimization of the total tardiness, where $c_{it} = \max\{0, t-d_i\}$ represents the tardiness of patient i if he/she starts being served at time t. Constraints (2) state that all patients must be served. Constraints (3) guarantee that a server can serve at most a patient at a time. Constraints (4) force each server to have a configuration assigned to it at each time the configurations can change. In Constraints (5), we assure that a patient i is a server j at time t only if at this time this server is assigned with a configuration $k \in C_i$, where $C_i \subseteq C$ represents the set of configurations that can serve the demanded service of i. Finally, Constraints (6) and (7) define the domain of the variables.

It is important to notice that model (1)–(7) is general and does not impose the FIFO rule on being followed. Besides that, if there exists only one configuration in C providing all services, we have a particular version of the problem, i.e., the identical parallel machine scheduling problem, with release dates and aiming to minimize the total tardiness. This problem is referenced as $P|r_j|\sum_j T_j$ in the scheduling notation and is known to be $\mathcal{NP}$-hard (see Lenstra et al. [6]), justifying the proposal of heuristic methods.

3 Proposed Heuristic

We propose a heuristic that locally searches for the services to assign to each server to minimize the total tardiness. Servers can offer services from the set S, in a way that is limited by a set C of configurations, where each configuration is a subset of services ($C \subseteq 2^S$). The set C is defined by the healthcare facility. The configurations include the possibility of a server having no service (closed server). Patients with the same service are served by a FIFO policy based on their arrival time, implying that patients with a smaller arrival time but requiring different services being served later on. The heuristic starts by randomly assigning to each server an initial configuration. Then, it updates the servers configurations every Δ units of time. A pseudocode is provided in Algorithm 1.

Algorithm 1: Local search-based heuristic

```
1: current_assignment ← initial list of services based on historical data
2: best_assignment ← current_assignment
3: current_cost ← Cost(current_assignment, 0)
4: for t ← Δ, 2Δ, 3Δ, ..., T do
5:   for m ← 1, 2, ..., M do
6:     best_list ← current list of services of m
7:     for all lists of services c ∈ C do
8:       current_assignment ← update the list of services of m to be c
9:       cost_m ← Cost(current_assignment, t)
10:      if cost_m < current_cost then
11:        current_cost ← cost_m
12:        best_list ← c
13:    best_assignment ← current_assignment with m having best_list
14: return best_assignment
```

In Algorithm 1, we start with pre-defined lists of services, each list assigned to a server. The loop in lines 4–13 iterates over the time horizon, assuming that decisions are taken every Δ units of time. These decisions are based on a local search that identifies, at time t, the best configurations to assign to each server m. The cost of changing the list of services of a server is calculated by the function *Cost()* described in Algorithm 2. We notice that some servers may not have their list of services updated in the case this not incur in a reduction of the total tardiness.

Algorithm 2: Return the **Cost()** of solution s given the time t

```
1: cost ← tardiness of solution given the patients serviced until time t
2: queue ← all patients that have not received a service
3: for time ← t, t + 1, t + 2, ..., T do
4:   for m ← 1, 2, ..., M do
5:     if m is free at time then
6:       i ← patient in the queue with the smallest release date (r_i ≥ t) and that can be
   serviced by m; otherwise, go to line 4
7:       schedule patient i to the server m starting at time time;
8:       if time + p_i > d_i then cost ← cost + ((time + p_i) − d_i)
9:       Update the queue by removing the scheduled patients
10: return cost
```

At line 1 of Algorithm 2, variable *cost* receives the tardiness solution s has until the time t, that is, considering all patients already serviced (before t) and patients receiving service during time t. The loop at lines 3–9 is responsible to reschedule all patients whose release time is greater than or equal to t and that are not receiving service. The idea is to update the solution, considering the current queue, once we have changed the list of services of a server. At line 6, we select a patient with the FIFO policy if there is a patient that can be served by server m; otherwise, we consider the next server. This patient is assigned to m and starts receiving service at time t, with the expected completion time $t + p_i$. If this generates tardiness, i.e., if $t + p_i > d_i$ then we update the solution cost (line 8).

4 Computational Results

We implemented the proposed heuristic in C++ and performed numerical tests on realistic instances obtained from the outpatient facility system. These instances corresponded to 15 working days in December 2019. They consider $M = 13$ servers, operating from 7 AM to 4 PM. The facility provides $|S| = 9$ services, and each server can perform any configuration (list of services) from the set C. The computer used in the experiments was an Intel Core i7-7500U processor with 2.70 GHz, 12 GB of RAM, and Windows 10 operating system.

We evaluate the performance of the proposed heuristic considering the total tardiness (Z), but we also discuss the total number of tardy patients (N_T) and the average tardiness per tardy patient (Z/N_T), comparing the heuristic solutions with the existing ones provided by the facility system.

We present the results in Tables 1 and 2. These tables contain the following information: instance number, instance size (i.e., number of patients on that day), facility solutions, and solutions obtained with the proposed heuristic under different values of Δ. The heuristic runs 10 times with different seeds, and we show the average results for each instance. The last line of the table contains the average values over all 15 instances. We omit in tables the computing time of the heuristic because in the worst case, it is less than 5 seconds, and just 0.5 seconds on average.

In Table 1, we analyze the heuristic with four different values of Δ (15, 30, 60, and 120 minutes). To provide a fair comparison with the evaluation of the solutions at the facility, we increased all service times by 30%. This value takes into account the losses of productivity due to the necessary breaks in the activities of the workers operating the servers. We assume the set C is composed of the configurations already defined by the facility system. The size of the instances shows an intense flow of patients throughout the facility system, with more than 900 requests being served per day. Observing the total tardiness, the difference is large between the solutions. The facility solutions have an average tardiness value equal to 16,815.8, while the proposed heuristic achieves much better values, which ranges from 1179.0 when $\Delta = 15$ to 1388.0 when $\Delta = 120$. The improvement is indeed impressive, but it might be caused by many different causes. First of all, the facility starts with a set of configurations that is the same for every day, and very rarely performs dynamic changes during the execution of the activities. Secondly, if performed,

Table 1 Results for $\Delta = 15, 30, 60$, and 120 minutes

		Total tardiness (Z)					Number of tardy patients (N_T)					Z/N_T				
Inst.	Size	Fac.	15	30	60	120	Fac.	15	30	60	120	Fac.	15	30	60	120
1	1309	46,743.3	392.7	403.2	478.6	735.8	954.0	64.5	68.6	99.6	77.8	49.0	6.1	5.9	4.8	9.5
2	1276	26,592.7	560.9	575.3	577.1	704.4	960.0	181.4	185.0	185.6	174.6	27.7	3.1	3.1	3.1	4.0
3	1124	17,811.1	322.1	329.8	358.6	819.0	756.0	56.1	57.8	60.3	68.7	23.6	5.7	5.7	5.9	11.9
4	1193	21,490.5	31.1	36.0	54.1	307.6	829.0	15.9	17.9	20.1	22.7	25.9	2.0	2.0	2.7	13.6
5	1123	5395.3	144.9	145.7	145.7	143.9	615.0	35.4	35.4	35.4	35.2	8.8	4.1	4.1	4.1	4.1
6	1034	12,599.2	317.6	316.7	316.9	354.0	767.0	57.4	56.3	56.8	57.3	16.4	5.5	5.6	5.6	6.2
7	1160	6153.6	230.0	228.2	355.2	996.8	631.0	112.2	105.2	101.0	102.8	9.8	2.0	2.2	3.5	9.7
8	1052	5394.3	70.6	73.0	81.1	81.1	634.0	29.3	29.8	29.8	29.8	8.5	2.4	2.4	2.7	2.7
9	914	9457.3	262.0	267.4	266.4	387.9	479.0	44.0	45.1	44.9	47.7	19.7	6.0	5.9	5.9	8.1
10	1193	21,754.2	129.2	129.2	199.9	531.5	900.0	38.1	38.1	43.1	45.7	24.2	3.4	3.4	4.6	11.6
11	1105	17,975.6	81.9	82.8	121.6	739.9	861.0	20.7	21.2	26.4	35.2	20.9	4.0	3.9	4.6	21.0
12	959	2604.2	0.0	5.5	52.2	87.7	461.0	0.0	1.7	3.7	5.8	5.6	0.0	3.2	14.1	15.1
13	974	3592.8	51.0	51.0	51.0	104.4	485.0	18.0	18.0	18.0	19.6	7.4	2.8	2.8	2.8	5.3
14	1217	28,370.7	6071.5	6048.7	6026.6	6023.6	1037.0	553.2	551.5	552.0	551.8	27.4	11.0	11.0	10.9	10.9
15	1336	26,302.5	9019.6	9027.5	8994.3	8802.9	1043.0	847.1	847.7	849.5	821.8	25.2	10.6	10.6	10.6	10.7
Average		16,815.8	1179.0	1181.3	1205.3	1388.0	760.8	138.2	138.6	141.7	139.8	20.0	4.6	4.8	5.7	9.6

Table 2 Results for different sets C with the lists of services and $\Delta = 60$ minutes

Inst.	Size	Total tardiness (Z)					Number of tardy patients (N_T)					Z/N_T				
		C_1	C_2	C_3	C_{cur}	C_{all}	C_1	C_2	C_3	C_{cur}	C_{all}	C_1	C_2	C_3	C_{cur}	C_{all}
1	1309	5941.0	1746.9	922.8	203.0	179.0	396.0	264.7	169.6	49.0	47.0	15.0	6.6	5.4	4.1	3.8
2	1276	3238.0	1379.0	873.4	161.7	150.0	262.0	181.6	123.0	45.1	44.0	12.4	7.6	7.1	3.6	3.4
3	1124	5995.0	1780.7	862.2	164.1	148.0	348.0	237.8	148.7	46.3	44.0	17.2	7.5	5.8	3.5	3.4
4	1193	3147.0	726.2	405.1	12.7	6.0	260.0	117.1	64.6	5.0	4.0	12.1	6.2	6.3	2.5	1.5
5	1123	2591.0	885.9	626.7	64.8	63.0	273.0	161.6	125.9	24.6	25.0	9.5	5.5	5.0	2.6	2.5
6	1034	1824.0	1017.5	611.9	154.0	154.0	226.0	173.0	128.7	42.0	42.0	8.1	5.9	4.8	3.7	3.7
7	1160	7892.0	2516.1	439.2	113.2	16.0	459.0	235.2	78.2	14.9	9.0	17.2	10.7	5.6	7.6	1.8
8	1052	942.0	1079.8	376.9	22.0	22.0	92.0	128.9	77.4	13.0	13.0	10.2	8.4	4.9	1.7	1.7
9	914	1901.0	695.1	314.9	131.8	127.0	223.0	108.4	77.6	40.0	39.0	8.5	6.4	4.1	3.3	3.3
10	1193	1059.0	1687.5	662.8	58.6	53.0	145.0	208.6	103.5	23.1	22.0	7.3	8.1	6.4	2.5	2.4
11	1105	6265.0	1692.7	919.0	61.2	36.0	405.0	207.1	133.9	17.6	16.0	15.5	8.2	6.9	3.5	2.3
12	959	1198.0	504.0	252.5	38.4	0.0	150.0	76.9	52.3	2.7	0.0	8.0	0.0	4.8	14.2	0.0
13	974	3243.0	1461.5	913.2	19.0	19.0	320.0	189.7	127.5	10.0	10.0	10.1	7.7	7.2	1.9	1.9
14	1217	2890.0	1313.8	713.4	282.0	297.0	388.0	235.4	159.1	116.7	126.0	7.4	5.6	4.5	2.4	2.4
15	1336	9396.0	3356.3	2901.6	50.5	28.0	598.0	305.0	224.4	16.7	15.0	15.7	11.0	12.9	3.0	1.9
Average		3834.8	1456.2	786.4	102.5	86.5	303.0	188.7	119.6	31.1	30.4	11.6	7.0	6.1	4.0	2.4

these changes are due to manual interventions of the head of the system while visually looking at the current situation, and is thus not based on solid and complete information. Finally, we adopt mean values from the distribution of the service times that are derived from the real-world historical data; however, this distribution might be inaccurate in some cases, as it does not contemplate different speeds of the workers that might be used when, for example, the system is not crowded. The heuristic can return better values of total tardiness when the server configuration is updated every small time interval (i.e., $\Delta = 15$). Regarding the number of tardy patients, the facility solutions have an average of 760.8 against 138.2 patients of the heuristic with $\Delta = 15$. Although the heuristic achieves the best overall average of the number of tardy patients when the servers' list of services is updated every large interval of time (i.e., $\Delta = 120$), this happens for only 3 instances. In most of the remaining instances, the number of tardy patients is smaller when $\Delta = 15$. In terms of average tardiness per tardy patient, our experiments confirm that better results can be achieved for small values of Δ. Besides that, we notice that there is no direct relation between the number of requests per day and the total tardiness, meaning that even instances with fewer requests may lead to large tardiness. The latter is more dependable on the number of patients arriving close to each other or with the same arrival time.

The results of the second analysis we perform with the heuristic are shown in Table 2. We evaluate the impact of changing set C, considering different lists of services than those defined by the facility system. We denominate the current lists of services by C_{cur}, having an average of five services per list. On the other hand, we define the following configurations: C_1, where each configuration contains exactly one of the services in the set S; C_2, where each configuration contains exactly two of the services in S; C_3, where each configuration contains exactly three of the services in S; and, C_{all}, containing only one configuration with all services in S. We consider all possibilities of lists of services with one (C_1), two (C_2), and three services (C_3). For these tests, we consider 13 servers, the expected service time, and $\Delta = 60$ minutes. Although this value of Δ has not allowed the best results in Table 1, it is following the facility preferences to update the server configuration.

It is important to recall that the results of Table 2 do not assume the increase of 30% in the expected service times, as previously assumed in Table 1, since the aim of this new table is to compare different configurations among them, and not to contrast the results with those obtained in the real-world facility solution. Observing the results, the larger the number of services per list, the better are the results. In terms of total tardiness, we decrease from an overall value of 3834.8, with C_1, to 86.5, with C_{all}. The same behavior happens considering the number of tardy patients, going from 303.0 to 30.4, respectively. We also notice a large decrease in the tardiness per tardy patient when going from C_1 to C_{all}. The heuristic, using the current (lists of services) of the facility (C_{cur}), presents competitive solutions in terms of total tardiness when compared with the case where each server can offer all services (C_{all}).

5 Concluding Remarks

Concerned about reducing the total tardiness of a health care facility that services patients over a time horizon, we propose a local search-based heuristic to assign lists of services (configurations) to identical parallel servers. As patients are dynamically arriving over time, the servers can have their list of services updated accordingly. The proposed heuristic decides, for each server and from time to time, the list of services that may generate the smallest total tardiness. This choice is based on simulating the facility behavior with the patients in the queue, where their expected service time is deducted from historical data.

The results show that the heuristic can reduce the total tardiness of the facility system if decisions are updated between small intervals of time (e.g., every 15 minutes), including the number of tardy patients. On the other hand, using small intervals to update decisions is not convenient for the facility system, which prefers to update the lists of services every 60 minutes. Even in this case, the total tardiness is much better compared to the facility solutions, decreasing from an overall average of 16,815.8–1205.3 (about 93%). Another positive impact we observe in reducing the total tardiness is when each server can offer all available services. Both total tardiness and number of tardy patients reduced significantly from the case with a single service to the one with all possible services per list. It is important to mention that the results obtained with the facility current lists of services are comparable with the latter, being an interesting choice to not overload the servers.

Future works may consider the patients receiving multiple services. It would also be interesting to evaluate the impact of different policies (e.g., based on the earliest due date rule) compared to the FIFO one, besides the priority that some patients have over others. Attention should be given to other heuristics to handle the problems of assigning services to servers and of scheduling patients in an integrated way.

Acknowledgments Thiago Alves de Queiroz would like to thank the support given by the National Council for Scientific and Technological Development (CNPq grant number 311185/2020-7).

References

1. Abdalkareem, Z.A., Amir, A., Al-Betar, M.A., Ekhan, P., Hammouri, A.I.: Healthcare scheduling in optimization context: a review. Health Technol. **11**, 445–469 (2021)
2. Harper, P.R.: Server behaviours in healthcare queueing systems. J. Oper. Res. Soc. **71**(7), 1124–1136 (2020)
3. Kaboudan, M.A.: A dynamic-server queuing simulation. Comp. Oper. Res. **25**(6), 431–439 (1998)
4. Kumar, S., Omar, M.K.: Stochastic re-entrant line modeling for an environment stress testing in a semiconductor assembly industry. Appl. Math. Comput. **173**(1), 603–615 (2006)
5. Lan, C.H., Lan, T.S., Chen, M.S.: Optimal human resource allocation with finite servers and queuing capacity. Int. J. Comput. Appl. Technol. **24**(3), 156–160 (2005)

6. Lenstra, J., Kan, A.R., Brucker, P.: Complexity of machine scheduling problems. In: Hammer, P., Johnson, E., Korte, B., Nemhauser, G. (eds.) Studies in Integer Programming. Annals of Discrete Mathematics, vol. 1, pp. 343–362. Elsevier, Amsterdam (1977)
7. Slegers, J., Mitrani, I., Thomas, N.: Evaluating the optimal server allocation policy for clusters with on/off sources. Perform. e Eval. **66**, 453–467 (2009)
8. Su, Q., Yao, X., Su, P., Shi, J., Zhu, Y., Xue, L.: Hospital registration process reengineering using simulation method. J. Healthcare Eng. **1**(1), 67–82 (2010)
9. Vass, H., Szabo, Z.K.: Application of queuing model to patient flow in emergency department. case study. Procedia Econ. Finance **32**, 479–487 (2015)
10. Weiss, E.N., Tucker, C.: Queue management: elimination, expectation, and enhancement. Bus. Horiz. **61**(5), 671–678 (2018)
11. Worthington, D.: Reflections on queue modelling from the last 50 years. J. Oper. Res. Soc. **60**(sup1), S83–S92 (2009)
12. Worthington, D., Utley, M., Suen, D.: Infinite-server queueing models of demand in healthcare: a review of applications and ideas for further work. J. Oper. Res. Soc. **71**(8), 1145–1160 (2020)

Global Optimization of a Turbine Design via Neural Networks and an Evolutionary Algorithm

Pranath Kumar Gourishetty, Giovanni Pesare, Walter Lacarbonara, and Giuseppe Quaranta

Abstract This work discusses an effective approach to find the optimal solution for constrained engineering design problems. Specifically, the computational platform herein implemented exploits a neural network and a differential evolution algorithm, and it leverages on a parametric finite element modelling for the fully automation of the design process. The presented approach is applied to the design of the rear flange of a low-pressure turbine casing for an aircraft engine, whose shape is optimized in order to reduce the manufacturing cost while preserving the overall integrity through the fulfilment of stress-based constraints.

Keywords Global optimization · Differential evolution · Structural design · Parameter optimization

1 Introduction

This study is concerned with the search for the optimal solution in an engineering design problem subjected to constraints. Within this framework, the optimal design of the scallops in the rear flange of a low pressure turbine (LPT) of an aircraft engine is tackled. The design problem is aimed to reduce the manufacturing cost and it includes constraints related to the maximum stress experienced when a hoop pressure is applied radially in outward direction (Fig. 1).

The manufacturing cost can be minimized by shape optimization of the flange through reduction of the scallops number, redistribution of the scallops and modification of the scallop shape. A sub-model of the casing including the rear flange

P. K. Gourishetty (✉) · W. Lacarbonara · G. Quaranta
Sapienza University of Rome, Rome, Italy
e-mail: pranathkumar.gourishetty@uniroma1.it; walter.lacarbonara@uniroma1.it; giuseppe.quaranta@uniroma1.it

G. Pesare
Altran Italia, Rome, Italy
e-mail: giovanni.pesare@altran.it

L. Amorosi et al. (eds.), *Optimization in Artificial Intelligence and Data Sciences*, AIRO Springer Series 8, https://doi.org/10.1007/978-3-030-95380-5_23

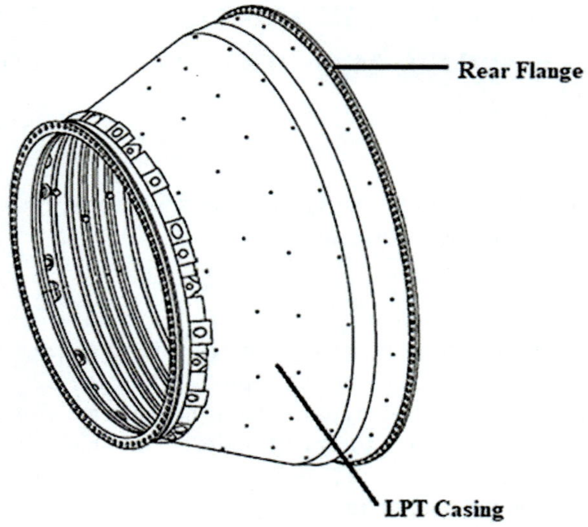

Fig. 1 Rear flange of a low pressure turbine of an aircraft engine

only is developed by using shell elements in order to reduce the computational time of the optimization process. Neural network and differential evolution algorithm [1] are employed to address the design problem. Most of the previous works require mathematical modelling of the design space to solve for the shape optimization. This hybrid neural network and differential evolution algorithmic approach can automate the mathematical modelling by parametric design and numerical simulation of the design space using commercially available CAE software. The input parameters and output data of the simulation models are used to train the neural network. Differential evolution algorithms use the trained neural network for global optimization of the parameters in the design space.

2 Finite Element Model

A 3D sub-model of the rear flange of the LPT has been first elaborated (Fig. 2). Next, a sub-model with shell elements is carried out by using Altair Hyperworks in order to reduce the elaboration time of the optimization process (Fig. 3). Suitable boundary conditions are enforced in the sub-model, and a hoop pressure of 5.66 MPa with total force equal to 100 kN is applied along the radially outward direction (Fig. 4). The rear flange is designed to have 72 scallops, out of which 13 are not included in the optimization domain in order to achieve optimal stress distribution. Therefore, the shape of 59 scallops (Fig. 5) can be optimized in order to reduce the manufacturing cost without compromising the structural integrity of the flange. A total of 5 possible levels is considered for each scallop (Fig. 6).

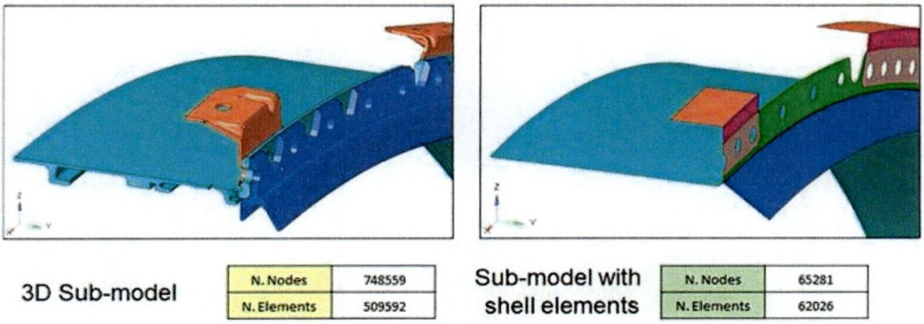

Fig. 2 3D sub-model and sub-model with shell elements

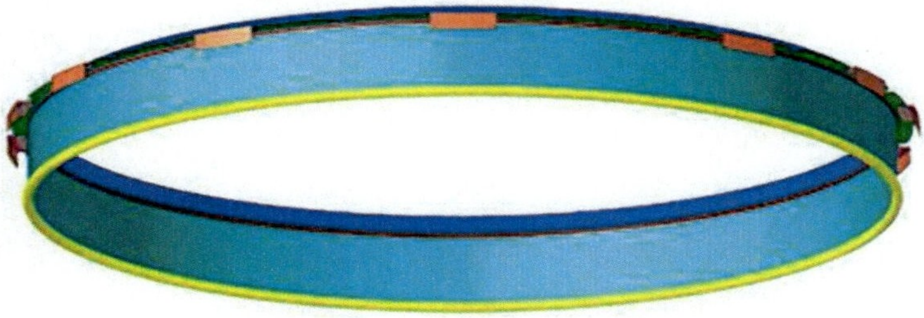

Fig. 3 Dimensions and mass of the sub-model. The diameter is 0.715 m while the height is 0.078 m. The mass is 4.943 kg

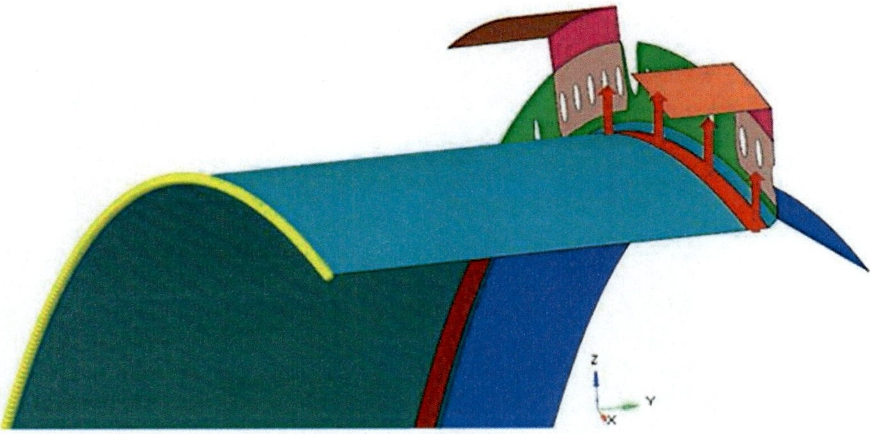

Fig. 4 Boundary and loading conditions. The yellow boundary is clamped. The radial hoop pressure is 5.66 MPa resulting in a force of 100 kN

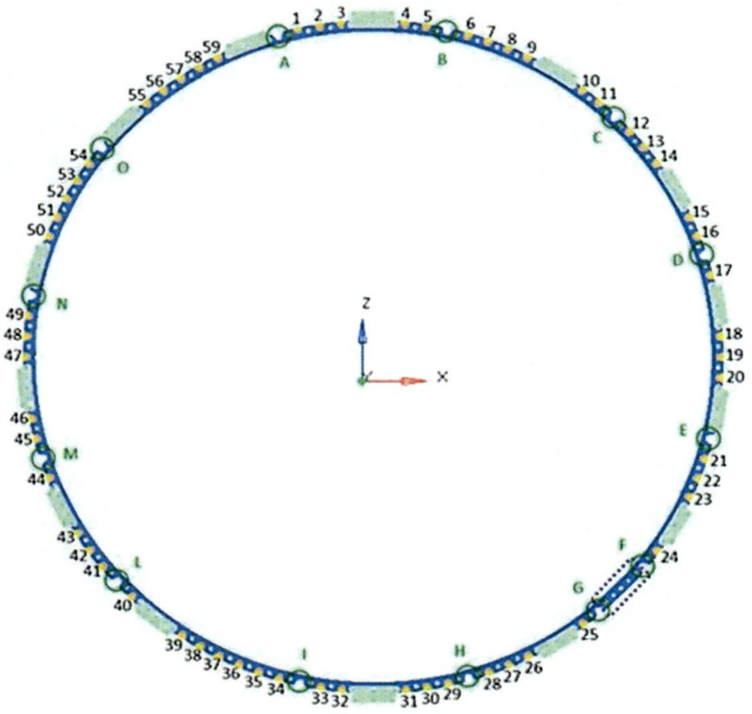

Fig. 5 Scallops distribution

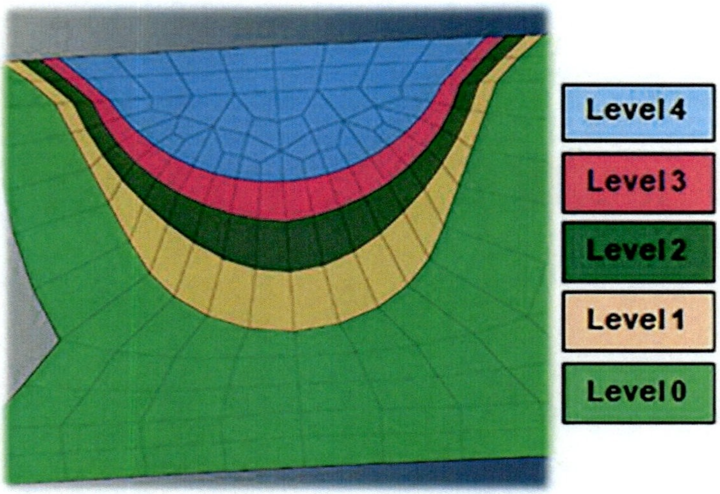

Fig. 6 Details about the five levels in the scallop

3 Elaboration of the Training Dataset

Element sets are created for each scallop level. These sets are passed as arguments to each scallop level. Models with random levels of scallops are generated using a Python sub-routine. These random models are simulated in ANSYS APDL by fixing the boundary conditions and the outward hoop pressure. At the end of each random model simulation, another Python sub-routine sorts out the element with maximum von Mises stress and returns the corresponding stress value as output. The models with the random levels in the scallops are saved as X matrix whose dimension is $\#\, random\, models \times \#\, scallops$. The corresponding maximum von Mises stress values are saved as Y vector whose dimension is $\#\, random\, models \times 1$. Saved data are converted into pickle files using Pandas framework and employed to train a neural network with the aim of predicting the maximum von Mises stress for a given scallop configuration. This, in turn, will support the constrained optimum design stage. A dataset including 5000 random scallop configurations is created.

4 Neural Network

The dataset is used to train a neural network consisting of 3 hidden layers by means of Keras and Scikit learning frameworks [2]. The final settings adopted for neural network training are listed in Table 1. These parameters are tuned using the Hyperopt framework available in Python. Figure 7 shows training and validation loss curves whereas Table 2 provides the details about the loss functions for training, validation and testing. The efficiency of the trained network is finally validated by comparing the results with ANSYS APDL.

Table 1 Settings for neural network training

Parameter	Choice
No. of hidden layers	3
Layer sizes	30, 15, 4
Layer activations	[‘relu’, ‘relu’, ‘relu’]
Dropouts	[0.1, 0.2, 0.3]
Kernel regularizers	[0, 0, 0]
Activity regularizers	[0, 0, 0]
Default learning rate (LRDEF)	1e-4
Optimizers	Adam(lr=LRDEF)
Loss	Mean squared error
No of epochs	3000
Batch size	16

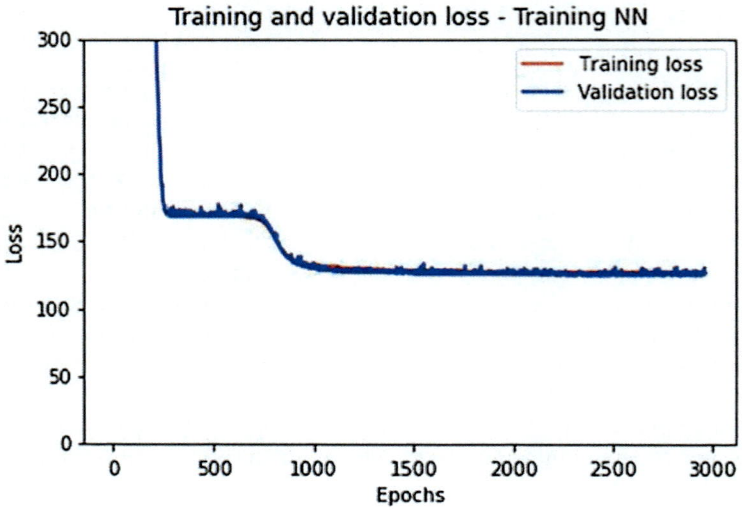

Fig. 7 Training and validation loss curves

Table 2 Outcomes of the neural network training

Results	
Time taken to train the network	2525.3 s
RMSE of the training	11.212226584
RMSE of the validation	11.1598499726
RMSE of the testing	11.6003906503
MAPE of the training	2.00932532841%
MAPE of the validation	1.9883635694%
MAPE of the testing	2.0658584809%

5 Global Optimization Results

The main goal of the design optimization is to reduce the number of scallops or decrease the scallops depth by increasing the levels in each scallop while fulfilling the imposed stress-based constraints. In fact, this allows to reduce the manufacturing cost of the rear flange of the LPT while ensuring its integrity. Hence, the design variables are the number of levels in all scallops whereas constraints are imposed for the maximum von Mises stress. The trained neural network is employed to predict the maximum von Mises stress value for a given scallop configuration. A differential evolution algorithm [3] is employed to solve this constrained optimization problem. The final results are listed in Table 3 for different threshold values on the maximum von Mises stress.

Table 3 Optimal results

	Stress=500 MPa	Stress=480 MPa	Stress=520 MPa
Constraint on Von Mises	Levels		
Scallop 1	3	0	4
Scallop 2	3	2	3
Scallop 3	2	2	4
Scallop 4	4	3	4
Scallop 5	3	3	4
Scallop 6	3	4	4
Scallop 7	4	4	4
Scallop 8	4	4	4
Scallop 9	4	3	4
Scallop 10	4	4	4
Scallop 11	4	4	4
Scallop 12	4	4	4
Scallop 13	3	4	4
Scallop 14	3	4	3
Scallop 15	3	4	4
Scallop 16	4	4	3
Scallop 17	4	2	4
Scallop 18	4	3	4
Scallop 19	4	4	4
Scallop 20	4	4	4
Scallop 21	4	4	3
Scallop 22	4	2	3
Scallop 23	4	3	4
Scallop 24	3	4	4
Scallop 25	4	4	4
Scallop 26	4	4	4
Scallop 27	4	3	4
Scallop 28	4	4	3
Scallop 29	4	4	3
Scallop 30	3	4	4
Scallop 31	4	3	4
Scallop 32	4	3	3
Scallop 33	3	3	4
Scallop 34	3	3	4
Scallop 35	3	3	4
Scallop 36	4	3	3
Scallop 37	3	4	4
Scallop 38	4	4	4
Scallop 39	3	3	4
Scallop 40	4	4	3

(continued)

Table 3 (continued)

	Stress=500 MPa	Stress=480 MPa	Stress=520 MPa
Constraint on Von Mises	Levels		
Scallop 41	4	4	4
Scallop 42	4	3	3
Scallop 43	4	3	4
Scallop 44	4	3	4
Scallop 45	3	4	3
Scallop 46	3	3	3
Scallop 47	3	4	4
Scallop 48	4	2	3
Scallop 49	3	3	4
Scallop 50	4	2	4
Scallop 51	4	4	3
Scallop 52	4	4	3
Scallop 53	4	3	3
Scallop 54	2	4	4
Scallop 55	4	3	4
Scallop 56	3	4	4
Scallop 57	3	3	4
Scallop 58	3	2	3
Scallop 59	3	3	4

6 Conclusions

This study proposed a computational intelligence-aided approach for the solution of a constrained engineering design problems. The case study tackled the optimum design of the rear flange of a low-pressure turbine casing for an aircraft engine. A neural network was employed to provide a surrogate model, and it was trained by means of results obtained, thanks to a parametric finite element sub-model. Next, a differential evolution algorithm was adopted to look for the scallops configuration that minimizes the manufacturing cost while fulfilling stress-based constraints. The results have demonstrated that the implemented platform is effective in solving complex constrained optimization design problems in an affordable way.

Acknowledgments The authors would like to thank Altran Italia for having provided the infrastructure employed in this study in terms of hardware and commercial software (Ansys APDL and Altair Hyperworks).

References

1. G. Quaranta, W. Lacarbonara, S.F. Masri, A review on computational intelligence for identification of nonlinear dynamical systems. Nonlinear Dynamics **99**(2), 1709–1761 (2020)
2. P. Virtanen, R. Gommers, T.E. Oliphant, M. Haberland, T. Reddy, D. Cournapeau, E. Burovski, P. Peterson, W. Weckesser, J. Bright, S.J. van der Walt, M. Brett, J. Wilson, K. Jarrod Millman, N. Mayorov, A.R.J. Nelson, E. Jones, R. Kern, E. Larson, C.J. Carey, İ. Polat, Y. Feng, E.W. Moore, J. VanderPlas, D. Laxalde, J. Perktold, R. Cimrman, I. Henriksen, E.A. Quintero, C.R. Harris, A.M. Archibald, A.H. Ribeiro, F. Pedregosa, P. van Mulbregt, SciPy 1.0 Contributors, SciPy 1.0: Fundamental algorithms for scientific computing in Python. Nature Methods **17**(3), 261–272 (2020)
3. R. Storn, K. Price, Differential evolution - A simple and efficient heuristic for global optimization over continuous spaces. J. Global Optim. **11**, 341–359 (1997)